Advanced Biological Treatment Processes for Industrial Wastewaters

Integrated Environmental Technology Series

The *Integrated Environmental Technology Series* addresses key themes and issues in the field of environmental technology from a multidisciplinary and integrated perspective.

An integrated approach is potentially the most viable solution to the major pollution issues that face the globe in the 21st century.

World experts are brought together to contribute to each volume, presenting a comprehensive blend of fundamental principles and applied technologies for each topic. Current practices and the state-of-the-art are reviewed, new developments in analytics, science and biotechnology are presented and, crucially, the theme of each volume is presented in relation to adjacent scientific, social and economic fields to provide solutions from a truly integrated perspective.

The *Integrated Environmental Technology Series* will form an invaluable and definitive resource in this rapidly evolving discipline.

Series Editor

Dr Ir Piet Lens, The University of Wageningen, P.O. Box 8129,
6700 EV Wageningen, The Netherlands. (piet.lens@ wur.nl)

Published titles

Biofilms in Medicine, Industry and Environmental Biotechnology: *Characteristics, analysis and control*
Biofuels for Fuel Cells: *Renewable energy from biomass fermentation*
Decentralised Sanitation and Reuse: *Concepts, systems and implementation*
Environmental Technologies to Treat Sulfur Pollution: *Principles and engineering*
Phosphorus in Environmental Technology: *Principles and applications*
Pond Treatment Technology
Soil and Sediment Remediation: *Mechanisms, technologies and applications*
Water Recycling and Resource Recovery in Industries: *Analysis, technologies and implementation*

Forthcoming

Waste Gas Treatment for Resource Recovery

www.iwapublishing.com

Advanced Biological Treatment Processes for Industrial Wastewaters

Principles and Applications

Edited by

Francisco J. Cervantes, Spyros G. Pavlostathis and Adrianus C. van Haandel

IWA Publishing

London • Seattle

Published by IWA Publishing, Alliance House, 12 Caxton Street, London SW1H 0QS, UK

Telephone: +44 (0) 20 7654 5500; Fax: +44 (0) 20 7654 5555; Email: publications@iwap.co.uk;
Web: www.iwapublishing.com

First published 2006

Typeset by Gray Publishing, Tunbridge Wells, UK
Index prepared by Indexing Specialists, Hove, UK

British Library Cataloguing in Publication Data
A CIP catalogue record for this book is available from the British Library

Library of Congress Cataloging-in-Publication Data
A catalog record for this book is available from the Library of Congress

ISBN-10: 1843391147
ISBN-13: 9781843391142

Contents

Contributors

Editors

Francisco J. Cervantes
Department of Water &
Environmental Sciences
Instituto Tecnológico de Sonora
Av. 5 de Febrero 818 Sur
Cd. Obregón, Sonora
85000 Mexico
fcervantes@itson.mx

Spyros G. Pavlostathis
School of Civil & Environmental
Engineering
Georgia Institute of Technology
311 Ferst Drive Atlanta
GA 30332-0512, USA
spyros.pavlostathis@ce.gatech.edu

Adrianus C. van Haandel
Department of Civil Engineering
Universidade Federal de Paraíba
R. Aprigio Veloso 882-58.109-970
Campina Grande, Pb Brazil
prosab@uol.com.br

Contributors

Felipe Alatriste-Mondragon
División de Ingeniería Ambiental y
Manejo de Recursos Naturales
Instituto Potosino de Investigación
Científica y Tecnológica (IPICyT)
San Luis Potosí, SLP, Mexico
falatriste@titan.ipicyt.edu.mx

Nicolas Bernet
Laboratoire de Biotechnologie de
l'Environnement
INRA, Avenue des Etangs
11100 Narbonne, France
bernet@ensam.inra.fr

Iemke A. E. Bisschops
Lettinga Associates Foundation
6700 AM Wageningen
The Netherlands
iemke.bisschops@wur.nl

Germán Buitrón
Institute of Engineering
National University of Mexico
UNAM
04510 Mexico, D.F., Mexico
gbm@pumas.ii.unam.mx

Berenice Celis
División de Ingeniería Ambiental y
Manejo de Recursos Naturales
Instituto Potosino de Investigación
Científica y Tecnológica (IPICyT)
San Luis Potosí, SLP
celis@ipicyt.edu.mx

Henk Dijkman
Paques BV
T.de Boerstraat 24
8560 AB Balk, The Netherlands
h.dijkman@paques.nl

Andre B. dos Santos
Department of Hydraulic &
Environmental Engineering
Universidade Federal do Ceará
Campus do Pici-Bloco 713
60455-760, Fortaleza
Ceará, Brazil

María Fernández-Polanco
Department of Chemical Engineering
and Environmental Technology
Universidad de Valladolid
47011 Valladolid Spain
maria@iq.uva.es

Pedro A. García Encina
Department of Chemical Engineering
and Environmental Technology
Universidad de Valladolid
47011 Valladolid Spain
pedro@iq.uva.es

Jacco Huisman
Paques BV
T.de Boerstraat 24
8560 AB Balk, The Netherlands
j.huiman@paques.nl

Piet Lens
Sub-Department of Environmental
Technology
Wageningen University
6700 EV Wageningen
The Netherlands
Piet.Lens@wur.nl

Hervé Macarie
IRD, Unité BioTrans, IMEP
Université Paul Cézanne
Faculté des Sciences de St. Jérôme
13397 Marseille cedex 20, France
herve.macarie@ird.fr

Mónica Meraz
Department of Biotechnology
Universidad Autónoma Metropolitana
Iztapalapa, D. F. Mexico
meraz@xanum.uam.mx

Oscar Monroy
Department of Biotechnology
Universidad Autónoma Metropolitana
Iztapalapa, D. F. Mexico
monroy@xanum.uam.mx

François Morier
Proserpol
8, rue Jean-Pierre Timbaud
78180 Montigny le Bretonneux
France
francois.morier@proserpol.fr

Etienne Paul
National Institute For Applied Sciences (INSA)
Département of Chemical and Environmental processes (GPE)
135 Avenue de Rangueil, 31077 Toulouse cedex 04
France

Elías Razo-Flores
División de Ingeniería Ambiental y Manejo de Recursos Naturales
Instituto Potosino de Investigación Científica y Tecnológica (IPICyT)
San Luis Potosí, SLP, Mexico
erazo@ipicyt.edu.mx

Elena Rustrian
Facultad de Ciencias Químicas
Universidad Veracruzana
94340 Orizaba, Veracruz Mexico
erustrian@uv.mx

Mathieu Sperandio
National Institute For Applied Sciences (INSA)
Département of Chemical and Environmental processes (GPE)
135 Avenue de Rangueil, 31077 Toulouse cedex 04
France

Adrie Veeken
Wageningen University
6700 EV Wageningen
The Netherlands
adrie.veeken@wur.nl

Jan Weijma
Paques BV
T. de Boerstraat 24
8560 AB Balk, The Netherlands
j.weijma@paques.nl

1

Strategies for industrial water pollution control

F.J. Cervantes and S.G. Pavlostathis

1.1 INTRODUCTION

An increasing demand for fuel, industrial chemicals, pesticides, fertilizers, pharmaceuticals, processed food, textiles and similar essential products, has been associated with society's desire to improve the quality of life. Unfortunately, to meet such demands, industrialization offering commodities and services to the public, has resulted in waste products, which are released into the environment through wastewaters, gaseous emissions, and solid residues, leading to environmental pollution and a deterioration of natural resources.

Contamination and the irrational use of natural resources have forced authorities to take drastic measures in several industrial sectors to achieve a sustainable development. A number of strategies for industrial water pollution control, described in this chapter, have been implemented during the past few decades in order to prevent environmental pollution and to achieve a sustainable utilization of natural resources. This chapter also provides a global perspective on water scarcity and water pollution.

1.2 GLOBAL PERSPECTIVE ON WATER SCARCITY AND WATER POLLUTION

1.2.1 Water scarcity

As the biosphere is now supporting life to over 6 billion people, water availability has become a major concern. From a historical perspective, the global population has tripled in the past 70 years causing water demand to increase six-fold because of industrial development, widespread irrigation and lack of conservation. Consequently, nearly 2.2 billion people in more than 62 countries (i.e. a third of the world's population), are lacking sufficient water supplies.

Even though around 70% of Earth's surface is covered by water, less than 3% of it is fresh, and most of it is in polar ice or too deep underground to be harvested. The amount of available fresh water in lakes, rivers and reservoirs is less than a quarter of 1% of the total fresh water quantity (Rosegrant *et al.* 2002).

This scenario represents a serious limitation for industrial development in several regions of the world. Certainly, these factors indicate a strong relationship between our very limited water resources and future industrial growth since 23% of water reservoirs are designated for industrial use worldwide. The remaining water is supplied for irrigation, 69%, and for domestic purposes, 8%. The distribution of water, however, varies considerably from one region to another. In Africa, 88% of fresh water is assigned for agriculture, 7% for domestic use, and 5% for the industry; in Asia, the distribution is 86%, 8% and 6%, respectively. In contrast, European industry requires 54% of total water available, whereas 33% and 13% is currently used for agriculture and domestic purposes, respectively (Falkenmark and Widstrand 1992).

Table 1.1 shows the total industrial water demand and industrial water use intensity under different circumstances. Industrial water use intensity under a business-as-usual scenario (BAU) refers to conventional water demand. The water crisis scenario (CRI) is projecting a worsening of the current situation for water and food policy, and the sustainable water use scenario (SUS), projecting a more positive future with greater environmental water reservation. Under SUS, industrial water demand declines compared with BAU, through technical improvements in water use and recycling, and increased water prices that induce reductions in demand. Under CRI, with weakened incentives and regulations and lower investment in technology, industrial water demand increases compared with BAU, as more water is needed to produce a unit of output. In 2025, total worldwide industrial water demand under CRI is projected to be 33% higher than under BAU, while it is 35% lower under SUS. Compared with BAU, global industrial water use intensity is $1.2\,m^3$ per thousand US dollars higher under CRI, and $1.3\,m^3$ per thousand US dollars lower under SUS.

Sustainable water use will play a crucial role on the development of most industrial areas worldwide. Several strategies will need to be implemented for allowing a sustainable use of water reservoirs linked to manufacturing. Section 1.3 presents several tactics, which are currently considered in various industrial sectors.

Table 1.1 Total industrial water demand and industrial water use intensity under BAU, CRI and SUS scenarios, 1995 and 2025 (Rosegrant *et al.* 2002).

Region/country	Industrial water demand (km^3)				Industrial water use intensity (m^3/US$1000)			
	1995 Baseline estimates	2025 Projections			1995 Baseline estimates	2025 Projections		
		BAU	CRI	SUS		BAU	CRI	SUS
Asia	48.9	92.6	148.5	55.1	16.2	6.7	11.3	3.8
China	13.2	32.1	74.8	18.5	16.0	6.2	14.5	3.6
India	7.3	16	23.1	9.8	19.6	7.9	11.5	4.9
Southeast Asia	11.5	21.3	23.2	11.6	20.4	8.9	9.7	4.9
South Asia[a]	1.9	4.7	5.7	2.6	18.3	11.7	14.0	6.5
Latin America	18	30.2	36.7	16.1	10.6	5.9	7.1	3.1
SS[b]-Africa	0.9	2.5	2.3	1.3	6.3	5.8	6.2	3.0
WANA[c]	4.6	8.8	9.7	4.5	8.4	5.1	5.7	2.6
Developed countries	96.6	115.7	133.2	85.6	4.3	2.5	2.8	1.8
Developing countries	62.9	123.8	186.4	69.1	13.2	6.4	9.6	3.6
World	159.5	239.5	319.6	154.6	5.9	3.6	4.8	2.3

[a] Excluding India; [b] Sub-Saharan; [c] West Asia/North Africa.

1.2.2 Water pollution

Until the middle of the last century, wastewater treatment objectives were mainly concerned with (1) the removal of suspended, colloidal and floatable material, (2) the destruction of biodegradable compounds, and (3) the elimination of pathogenic microorganisms. Since the early 1970s, great strides in analytical techniques have been made allowing for new and more sophisticated instrumentation. Detection methods have become more sensitive and broader ranges of compounds are now monitored in water bodies. Accordingly, there are a large number of scientific reports documenting the effects of many contaminants and microorganisms present in wastewaters on both human health and the environment. While most constituent concentrations are reported in milligrams per liter (mg/L), measurements in micrograms per liter (μg/L) and nanograms per liter (ng/L) are ordinary at present (Metcalf and Eddy 2003).

As new evidence demonstrates the impact of many different pollutants originating from a broad range of industrial activities on human health and on the environment, it becomes critical to know the following when an industrial wastewater is discharged (Crites and Tchobanoglous 1998):

(1) constituents of concern in the wastewater;
(2) impacts of these constituents when the wastewater is dispersed into the environment;
(3) the transformation and long-term fate of these constituents in treatment processes;

Table 1.2 Principal constituents of concern in wastewater treatment and the reason for their importance (Metcalf and Eddy 2003).

Constituent	Reason for importance
Suspended solids	Lead to the development of sludge deposits and anaerobic conditions when untreated wastewater is discharged in aquatic environments
Biodegradable organics	Generally expressed as biochemical oxygen demand (BOD) or chemical oxygen demand (COD). Their biological stabilization can originate depletion of natural oxygen resources and creation of septic conditions
Pathogens	Potential diseases can be disseminated by the pathogenic organisms that may be present in wastewater
Nutrients	Cause eutrophication of aquatic environments. When discharged in excess amounts on land, they can also cause pollution of groundwater
Priority pollutants	Organic and inorganic compounds selected on the basis of their known or suspected carcinogenicity, mutagenicity, teratogenicity, or high acute toxicity
Refractory organics	Compounds that tend to resist conventional wastewater treatments. Typical examples include pesticides, phenols, and surfactants
Heavy metals	Usually found in wastewater from many different origins. They are generally toxic for humans or have to be removed if water reuse is considered
Dissolved inorganics	Inorganic constituents, such as calcium, sodium, and magnesium are commonly found in different effluents and may have to be removed if the wastewater is to be reused

(4) methods that can be used to remove or modify the constituents found in the wastewater; and
(5) methods for beneficial use or disposal of solids generated by the treatment systems.

The principal constituents that are currently of concern in wastewater treatment, as well as the reason for their importance are summarized in Table 1.2. Priority pollutants, refractory organics and heavy metals deserve special attention for industrial effluents, since many of these constituents are discharged by several manufacturing sectors, and their impact on human health and the environment is well documented (US EPA 2000). Table 1.3 lists the industrial sources for the main aromatic priority pollutants of concern according to the US Environmental Protection Agency (EPA). In Chapters 8–10 several case studies are described in which wastewater treatment systems are applied for the removal of priority pollutants from different industrial sectors. Biodegradable organic matter, suspended solids, as well as nutrients are also common constituents of several industrial effluents, such as those coming from food-processing plants, distillers, breweries, fertilizers and agro-businesses.

The main strategies currently applied for the removal of these constituents are illustrated in different case studies in Chapters 5–7.

Table 1.3 Sources of toxic anthropogenic aromatic pollutants (Field *et al.* 1995).

Pollutants	Industrial origin
BTEX	Fossil fuels, solvents, industrial feedstock
Styrene	Plastics
PAH	Fossil fuels, wood preservation
Alkyl-phenols	Surfactants, detergents
Aromatic sulphonates	Surfactants, detergents, sulphite pulping, dyes
Aromatic amines	Pesticides, dyes, pigments, pharmaceuticals
Azo-aromatics	Dyes
Nitro-aromatics	Explosives, pharmaceuticals, pesticides, dyes
Chloro-phenols and dioxins	Wood preservation, pesticides, pulp bleaching effluents
Chloro-aromatics and PCB	Pesticides, solvents, dielectric and hydraulic fluids

BTEX: benzene, toluene, ethyl-benzene and xylene; PAH: polycyclic aromatic hydrocarbons; PCB: polychlorinated biphenyls.

1.3 POLLUTION PREVENTION APPROACHES

1.3.1 Introduction

The objectives of industry have generally been considered incompatible with the preservation and improvement of the environment for many years (Graedel and Allenby 1995). However, the focus on this issue has progressively changed. In fact, the pressure to provide a suitable quality of life for the Earth's population will not involve less, but more industrial activities and contribute to serious environmental problems. Thus, providing a sustainable world will require closer industry–environment interactions. The disregard for the protection of the environment cannot be justified by the economical benefits. Consequently, the question on "how to protect and improve the environmental quality" has risen several decades ago.

Since regulation on pollution was implemented, both industry and government regulatory agencies have focused their efforts on the reduction of toxic wastes by controlling discharges at the point where they enter the environment. Thus, end-of-pipe treatments were initially adopted for most industrial processes (Khan *et al.* 2001). However, the advent of strict environmental legislation in recent years, combined with the ineffectiveness and relatively high cost of several end-of-pipe treatment technologies, have, in many cases, resulted in making this concept inadequate to deal with the magnitude and complexity of environmental deterioration. Therefore, during the past few years, the concept of "industrial ecology", which mission is to design zero-emission industrial processes by focusing on cleaner production and waste minimization, was implemented in several industrial sectors (Ayres and Simonis 1994; Graedel and Allenby 1995; Ayres and Ayres 1996; Allenby 1999). These pollution prevention concepts have significantly contributed not only to reducing pollution, but also to improving environmental performance, raising profitability and enhancing competitiveness in several industries (My-Dieu 2003).

This section provides an overview of the major pollution prevention concepts and analyzes their strengths and weaknesses. In Chapters 5–10, several case studies illustrate the application of some of these approaches to achieve sustainable industrial development in different manufacturing areas.

1.3.2 End-of-pipe treatments

Linked to growing industrialization, which has resulted in the rise of pollution, there is an increasing concern about the quality of the biosphere. To comply with environmental quality standards, industries are forced to treat their wastes before discharge. Since the treatment is applied after generating the waste, this type of treatment is called "end-of-pipe" treatment. Many treatment technologies attempt to decrease the risk posed by pollutants using various strategies such as confining the contaminants in a defined area (e.g. landfilling), reducing toxic effects by dilution (e.g. smokestacks), transferring pollutants from one medium to another (e.g. air-stripping of contaminated water), or converting pollutants to inert materials (e.g. mineralization) (Huesemann 2001; My-Dieu 2003). A large number of treatment plants applying biological, physical–chemical or chemical processes to treat different kinds of industrial wastewaters are in use all over the world. Although the end-of-pipe treatment approach is being seriously questioned, it is still one of the pollution control methods used to handle unavoidable wastes and emission of contaminants from industrial processes.

Several reports emphasize different strengths of the end-of-pipe treatments. It is noteworthy that, even considering the best precautions during manufacturing of products, yet some contaminants will be generated, which will need to be treated properly before discharging into the environment. In those cases, end-of-pipe treatment is the only suitable environmental protection solution. Several reports indicate the application of cleaner production strategies for paper, textile, and electroplating industries, which allowed for the reduction of the volume of wastes generated (Bajracharya 1995; Chaan-Ming 1995; Chandak *et al.* 1995; Cheung 1995; Hwa 1995; Lee 1995; Nataagiin 1995; Oka 1995; Roestamsjah and Cahyaningsih 1995; Tapaneeyangkul 1995). Nevertheless, a wastewater treatment plant was required to complete the removal of contaminants in all cases. In some cases, it is recognized that end-of-pipe treatments demand less capital investment, less industrial development, and less disruption (Bahat 1996; Howes *et al.* 1997; Hilson 2000) than replacement of existing infrastructure with new equipment and structures (Duchin *et al.* 1995; Sarkis and Cordeiro 2001). Rapid and less costly approaches make end-of-pipe treatment methods more attractive for a company with a strict budget, under pressure to meet legislative compliance (My-Dieu 2003). This scenario can be illustrated by the case of power plants. Despite retrofitting power plants with scrubbers for removing sulphur from flue gas, which virtually always increases the cost of production due to additional requirements on equipment, energy, maintenance and other inputs, this is still considered widely a preferred solution in the short-term (Duchin *et al.* 1995, My-Dieu 2003).

Although end-of-pipe pollution control strategies have certainly contributed to reduce negative environmental consequences of industrial processes, they focus on the symptoms and not the origin of environmental problems (Khan *et al.* 2001).

Other important disadvantages of end-of-pipe pollution control methods are:

(1) they are not adequate to allow an efficient use of limited resources;
(2) they cause greater consumption of materials and energy, more capital expenditure and more work hours compared to measures taken at source and
(3) their use generally creates new environmental problems, such as the need for disposing wastes from treatment facilities.

Furthermore, there are several reasons that make end-of-pipe treatment methods highly costly to operate and maintain, and ineffective to revert environmental damage. For example, reducing toxic effects by emitting flue gas through sufficiently high smokestacks will certainly reduce the risk of localized acute toxicity, but is likely to increase the probability of more widespread chronic effects that are currently unknown or difficult to monitor. Likewise, it is uncertain whether transfer of a pollutant from one medium to another will actually reduce the overall risks (Huesemann 2001). Despite the fact that chemical and biological treatment technologies are effective in limiting or reversing contaminant dispersal, it is important to realize that many of these technologies also have undesirable side-effects. To some extent, even bioremediation, a low impact technology relying on native bacteria to metabolize pollutants, can cause the formation of intermediates that are more toxic than the original contaminants (e.g. carcinogenic vinyl chloride during dechlorination of polychlorinated solvents), or aromatic amines produced during azo dyes cleavage (Baker and Herson 1994, Field *et al.* 1995).

The implementation of end-of-pipe treatment approaches depends profoundly on the pressure applied by environmental authorities on the industry to control pollution at their facilities. For instance, about 90% of major water polluters in China preferred to violate discharge standards during the 1990s because pollution charge rates were generally very low and not effective to provide incentives for the industry to invest in end-of-pipe treatments (My-Dieu 2003). Moreover, strict enforcement to comply with discharge standards might lead to improper handling of end-of-pipe treatment systems. In fact, several industrial sectors in developing countries, in which legislation may be ambiguous, dilute contaminants by adding fresh water to wastewater in order to meet the legal concentration standards. In this way, industry avoids penalties from state regulatory agencies and reduces the costs of waste treatment, but this hampers the environment and increases the scarcity of natural resources. To avoid the old approach, "dilution is the solution to pollution", effluent standards in developed countries, such as USA, have been switched from concentration to mass load rate basis and the total maximum daily load (TMDL) approach (US EPA 2000). Considering all these limitations, end-of-pipe treatment methods are judged less sustainable than other environmental protection concepts.

1.3.3 Cleaner production

Cleaner production, also referred to as "waste minimisation", differs from end-of-pipe treatments in that it minimizes wastes and emissions by reducing them at their sources. Cleaner production can generally be defined as the continuous application of an integrated preventive environmental strategy to production processes in order to avoid wastes and emissions at the source, to preserve energy and raw materials, to eliminate the use of toxic materials and to improve working conditions. Cleaner production contributes to optimization of resources, therefore reflecting environmental improvement on financial and economic benefits, as well as on technological progress. The following measures have been successfully taken to achieve cleaner production:

(1) improved housekeeping around the existing processes;
(2) recycling, recovery, and reuse of materials/by-products/wastes;
(3) changing input materials;
(4) changing production process; and
(5) changing product.

Appropriate housekeeping implies that managers and employees of an industry are diligent in ensuring that they meet the terms for all environmental regulations, keeping to a minimum generated wastes and resources demand. Good housekeeping measures can often be implemented at little or no cost. Improved management of raw materials and products inventory, reduction in raw materials and product loss, and training of employees can be effective means to improve industrial organization. The following examples illustrate correct measures to achieve an efficient maintenance. In the case of beverages production, precise adjustment of bottle fillers or installation of a metal sheet under the fillers can minimize losses of product during the filling stage (EUROPEN 1997). Correct scheduling of the process in view of equipment cleaning can also reduce waste generation. For example, preparing light paints before dark ones, or arranging fabric requiring similar dyeing and finishing process in sequential order, will make cleaning of vats unnecessary before starting a new batch. In many industrial sectors, using the generated wastewater from bottle washing to wash the casks, or for other purposes, will reduce water supply demand.

Recycling, recovery, and reuse of materials are the next most preferable strategy of cleaner production as many waste materials generated can be reused either onsite, in the original process with or without treatment to remove impurities, or offsite, in other plants. For instance, in the finishing stage of clothing manufacture, internal recycling can reduce the amount of dyestuffs and printing pastes discharged with wastewater. It is possible to apply the recycled printing pastes in processes where a lower quality is acceptable. Also, less current colors can be mixed to darker or black colors (My-Dieu 2003). Organic solvents used in cleaning and pharmaceutical manufacturing processes are often collected, distilled, and reused in the original process (Chiu and Peters 1994). Further examples of recovery of materials/by-products/wastes include: milk powder recovery during its production, dye in textile industry, copper in electroplating, paint and water in car painting industry, cutting oil in machine workshop. Several wastewater treatment facilities also allow for the recovery of by-products, for which there is demand either internally or off-site. Recovery and utilization of methane from anaerobic digesters has resulted in making several wastewater treatment plants energetically self-sufficient or less demanding for electrical power (Kansal *et al.* 1998; Keller and Hartley 2003). Recovery of nutrients, such as nitrogen and phosphorus, from concentrated effluents like piggery wastewaters, in the form of struvite ($MgNH_4PO_4 \cdot 6H_2O$), is also a common practice in some industrialized countries in which the mineral is being applied as a fertilizer (Doyle and Parsons 2002). Heavy metals can also be recovered from wastewater treatment facilities by physical–chemical or biological methods (Karabelas *et al.* 2001; Lamba *et al.* 2002; Park *et al.* 2005), as well as elemental sulphur, which can be commercialized, using sulphide oxidizing bioreactors (Arena *et al.* 2000; Lens *et al.* 2002). Another example of offsite recycling and recovery is using biodegradable wastes, wherever possible as animal feedstock or soil conditioners and fertilizers. In a study of waste minimization in the palm oil industry, the following strategies were considered:

(1) use of palm oil waste as animal feed;
(2) recovery and use of fiber as fuel for production of steam in boilers;
(3) sludge as fertilizer; and
(4) recovery of methane from anaerobic digestion (Vigneswaran *et al.* 1999).

Thus, the exploitation of treatment by-products can certainly decrease the operational costs of wastewater treatment plants.

Changing input materials is another cleaner production strategy to reduce generation of pollutants from used raw material. Several examples are reported in the literature. In printing, water-based inks can be a substitute for chemical-solvent-based inks to minimize waste solvents and to prevent air pollution caused by their volatilization (Vigneswaran *et al.* 1999). Similarly, the water-based film developing system can be replaced by dry systems in electronic components and powder paints can be used instead of organic-solvent-based paints in painting the electrical light components. Replacement of current chemicals, including detergent, acetic acid, and Stabicol A by Hostapal EF-X, formic acid, and Stabilizer SG-X, respectively, in a bleaching and dyeing factory, reduced the effluent COD to 6% without adversely affecting the production quality (Chaan-Ming 1995). The application of less aggressive cleaning agents and biodegradable detergents, instead of recalcitrant compounds used for vats cleaning, can also contribute to decrease the costs of wastewater treatment at several industries.

Production procedure change plays an important role in reducing waste volume and strength. This includes equipment modification or modification of technology in order to decrease wastes and emissions during manufacturing. Examples of this practice consist of alteration of washing/cleaning procedure such as using counter current washing, replacement of single-pass processes by closed-loop processes, use of mechanical means to transport waste (e.g. from poultry farm) instead of using water, etc. Modifying the technology used in a factory is one of the most effective approaches to prevent pollution. However, as technology changes involve greater investments than procedural changes, and usually take longer time to be implemented, they are commonly explored after procedural changes have been applied (My-Dieu 2003).

Table 1.4 Benefits from practical applications of cleaner production strategies.

Project	Cleaner production measures	Achievement	Benefits
Tapioca starch industry	Reuse of root wash water after sedimentation Reuse tapioca wastewater for Torula yeast production	Reduction of 73% on COD, while produce 0.5 kg yeast/kg COD removed	Reduction of waste and treatment cost
Palm oil industry	Reuse turbine cooling water Use of floating valve in the press station and oil clarification room to stop overflow Modification at clarification process to recover oil		Prevent loss of water and increase production efficiency
Cheese making manufacture	Collection of whey and reuse in cattle food production		Reduction of waste and treatment cost
Electroplating factory	Recovery and recycling heavy metals (nickel and chrome) by equipment modification	Recovered 5.2 kg Ni/day (~90% Ni lost if not recovered) Recovered 2.1 kg Cr/day (~97% Cr lost if not recovered)	Reduction of raw material use Reduction of waste and treatment cost due to recovery
Dyeing and finishing factory	Use counter current washing processes to reduce water demand Install heat exchanger to recover waste heat Onsite recover and reuse of soda and chemicals Change hard metallic rollers by synthetic material rollers to eliminate heavy metals in generated sludge Computerized dyestuff handling devices to prevent wastage of dyestuffs	Operation profit of 10.7% compared to 2.3% in case of no cleaner production	Reduction of water use Prevent loss of energy Reduction of raw material use (dyestuffs and line) Reduction of waste and treatment cost
Siam paper industry	Installation of settling tank to reduce the extent of waste treatment and allow wastewater recycling	Use 100% waste paper as raw material and recycling 15,000 m^3/day of treated wastewater. Saving of 73% of annual operational cost	Reduction of raw material use Reduction of waste and treatment cost

Adapted from My-Dieu (2003).

Cleaner production is now recognized as a cost-effective complement for pollution control in meeting environmental regulations. Several case studies reported in the literature provide evidence on the benefits of the application of cleaner production in many emerging countries such as China, Singapore, Thailand,

Hong Kong, etc. Table 1.4 presents some of these case studies. The major benefits obtained from the implementation of cleaner production are:

(1) increase of production efficiency in terms of quantity of product per unit of raw material;
(2) improvement of the environmental quality;
(3) improvement of product quality;
(4) reduction in generation and discharge of pollutants; and
(5) enhanced positive public perception of the company.

Despite the fact that cleaner production offers major benefits for many industrial sectors, its implementation faces several barriers and constraints. In some cases, there is no available technology ready that can be adopted directly (Chiu and Peters 1994). In some cases, cleaner production strategies cannot completely substitute end-of-pipe solutions. Absence of strict environmental regulations is also one of the main legislative barriers that prevents the implementation of cleaner production practices.

Furthermore, according to Frijns (2001), the main constraint on the implementation of a cleaner production relies on financial matters. Major companies may have sufficient capital to upgrade inefficient processes, but small-and medium-size firms usually do not. To some extent, application of cleaner production strategies perhaps has higher short-term costs, not only due to investments in technology, but also due to the necessary revamping of organizational processes and the higher risk accompanying process modification.

1.3.4 Industrial ecology

Despite several principles considered in the concept of industrial ecology (IE) were already in practice a few decades ago, they have taken root in the past few years, particularly since the publication by Frosch and Gallopoulos (1989). The concept of IE includes the transformation of the traditional model of industrial activity into a more integrated model – an industrial ecosystem, in which wastes from one process can serve as raw material for the others. IE is an innovative strategy for sustainable industry involving design of industrial systems to minimize waste and maximize the cycling of materials and energy (Karamanos 1995). Several authors indicate three main key elements of the IE perspectives (Boons and Baas 1997; Krrishnamohan and Heart 2000; Erkman 2001):

(1) It is a systematic and integrated view of all components of the industrial economy and their relations with the biosphere.
(2) It emphasizes the bio-physical substratum of human activities (i.e. the complex patterns of material flows within and outside the industrial system), in contrast with the current approaches, which mostly consider the economy in terms of abstract monetary units, or alternatively energy flows.
(3) It considers technological dynamics, that is, the long-term evolution (technical trajectories) of clusters of key technologies as a crucial, although not exclusive, element for the transition from the actual unsustainable industrial system to a viable industrial ecosystem.

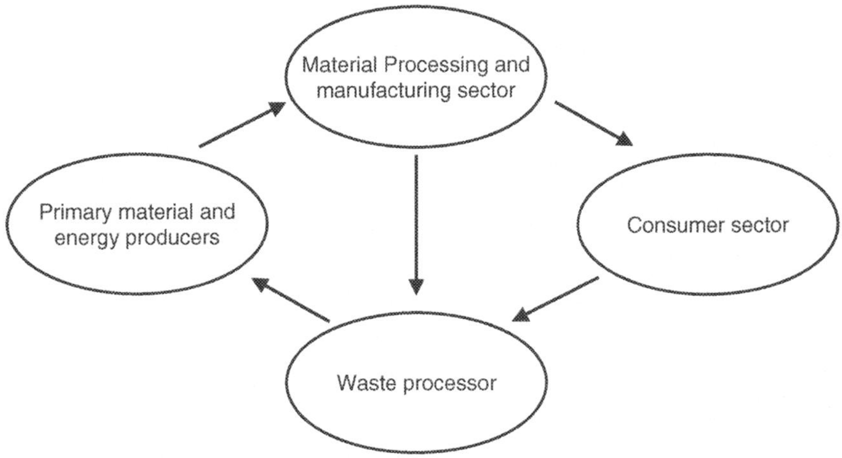

Figure 1.1 The major components of an industrial ecosystem (Manahan 1999; My-Dieu 2003).

Industrial ecosystems can be defined as including all types of production, processing and consumption, for instance agricultural production, as well as purely industrial operation (Manahan 1999). The four major components of an industrial ecosystem are illustrated in Figure 1.1. The primary material and energy producers may consist of one or several enterprises providing the basic materials that sustain the industrial ecosystem. Virgin materials are converted into finished products, and energy and wastes may be generated via the various manufacturing steps, such as extraction, concentration, separation, refining, processing, preparation, and finishing. The finished materials from the primary material producers are fabricated to make products in the material processing and manufacturing sector. This sector shows several opportunities for recycling such as process recycled streams (materials are recycled in the manufacturing operation itself) and external recycle streams (materials are recycled from other manufacturers or from post-consumer products). In the consumer sector, products are sold or leased to the consumers who use them. In all cases, the end of useful lifetime of the product is reached and it is either (1) discarded or (2) recycled. Finally, the waste-processing sector consists of enterprises that deal specifically with the collection, separation, and processing of recyclable materials and wastes.

Several advantages of implementing the IE concept can be underlined. Firstly, by creating linkages among firms or industrial sectors and moving them forward together, IE overcomes the shortcoming of end-of-pipe treatment, cleaner production and waste minimization approaches, which deal with the environmental problems in an individual perspective. Moreover, IE strives to optimize resource flows rather than just preventing pollution, and to promote sustainability rather than only reduce risks (Olderburg and Geiser 1997). Practical limitations of other pollution prevention concepts (e.g. end-of-pipe treatments and cleaner production) to achieve zero emissions from every production process lead to the need of broader loop-closing aspect of IE. IE offers a bridge between the specific innovations occurring in cleaner production and the attainment of an industrial system satisfying human needs within the constraint of global and local carrying capacity (Lowe and Evans 1995). Cleaner production and environmental management are identified as process-oriented, whereas IE is system-oriented and it covers a longer time frame and the whole arrangement of manufacturing (Brattebø 1996).

To accomplish a correct implementation of IE, the following challenges and risks should be taken into account (Lowe 1997; Oldenburg and Geiser 1997):

- Companies using other's residual products as inputs, face the risk of losing a critical supply or market if a plant closes down or changes its production protocol.
- Confidential information could become available to competitors.
- Uneven quality of by-products could cause damage to equipment or decrease the quality of products.
- Exchange of by-products could lock in continued reliance on toxic materials.
- Industrial recycling of materials does not always prevent by-products from being discharged as wastes.
- Possible innovations in regulation to enable an eco-industrial park development may not be allowed by regulatory agencies.

REFERENCES

Allenby, B.R. (1999) *Industrial Ecology: Policy Framework and Implementation*, Prentice Hall, New Jersey.

Arena, B.J., Sahlin, R.K., Janssen, A. and Buisman, C.J. (2000) The THIOPAQ™ process for sulfur removal and recovery from gas, light hydrocarbon and spent caustic streams, In *Proceedings of the 11th CHT Refinery Technology Conference*, Hyderabad, India, 9–11 February.

Ayres, R.U. and Ayres L.W. (1996) *IE: Towards Closing the Materials Cycle*, Edward Edgar.

Ayres, R.U. and Simonis, U.E. (1994) *Industrial Metabolism: Restructuring for Sustainable Development*. University Press, New York.

Bahat, V.N. (1996) *The Green Corporation: The Next Competitive Advantage*, Quorum Books, London.

Bajracharya, P. (1995) National reports Nepal on cleaner production for green productivity, In *Cleaner Production for Green Productivity*, Asian Productivity Organization.

Baker, K.H. and Herson, D.S. (1994) *Bioremediation*, McGraw-Hill, New York.

Boons, F.A.A. and Baas, L.W. (1997) Types of industrial ecology: the problems of co-ordination. *J. Cleaner Prod* **5**(1–2), 79–86.

Brattebø, H. (1996) Industrial ecology and sustainable product design, In *Proceedings of the Seminar and Workshop of Norwegian Academy of Technology Sciences*, Trondheim, 1–2 February.

Chaan-Ming, L. (1995) National reports of Hong Kong on cleaner production for green productivity. In *Cleaner Production for Green Productivity*, Asian Productivity Organization.

Chandak, S.P., Kuttiapan, N.K. and Gupta, P.K. (1995) National reports: India, In *Cleaner Production for Green Productivity*, Asian Productivity Organization.

Cheung, P.S. (1995) National reports of Republic of China on cleaner production for green productivity, In *Cleaner Production for Green Productivity*, Asian Productivity Organization.

Chiu, S.Y. and Peters, B. (1994) *Industrial Pollution Prevention*. Asian Development Bank.

Crites, R. and Tchobanoglous, G. (1998) *Small and Decentralised Wastewater Management Systems*, McGraw-Hill, New York.

Doyle, J.D. and Parsons, S.A. (2002) Struvite formation, control and recovery. *Wat. Res.* **36**, 3925–3940.

Duchin, F., Lange, G.M. and Kell, G. (1995) Technical change, trade and the environment. *Ecol. Econ.* **14**, 185–193.

Eckenfelder, W.W. (1989) *Industrial Water Pollution Control*, McGraw-Hill, Singapore.

Erkman, S. (2001) Industrial ecology: a new perspective on the future of the industrial system. *Off. J. Swiss Soci. Infect. Dis. Swiss Soci. Intern. Med. Swiss Soc. Pneumol.* **131**(37–38), 531–538.

EUROPEN (1997) Optimal waste management calls for flexible choice of recovery options. Retrieved January 2003 from http://www.europen.be/issues/optiwast.html

Falkenmark, M. and Widstrand, C. (1992) Population and water resources: a delicate balance. *Popul. Bull.* **47**(3), 1–36.

Field, J.A., Stams, A.J.M., Kato, M. and Schraa, G. (1995) Enhanced biodegradation of aromatic pollutants in cocultures of anaerobic and aerobic bacterial consortia. *Antoni. van Leeuwen.* **67**, 47–77.

Frijns, J. (2001) Pollution control of small-scale industry in Ho Chi Minh City: to relocate or to renovate? *J. Corp. Environ. Strat. Practi.* **34**, 55–73.

Frosch, R.A. and Gallopoulos, N.E. (1989) Strategies for manufacturing. *Managing Planet Earth* 97–108.

Graedel, T.E. and Allenby, B.R. (1995) *Industrial Ecology*, Prentice Hall, New Jersey.

Hilson, G. (2000) Pollution prevention and cleaner production in the mining industry: an analysis of current issues. *J. Cleaner Prod.* **8**, 199–226.

Howes, R., Skea, J. and Whelan, B. (1997) *Clean and Competitive*, Earthscan Publications, UK.

Huesemann, M.H. (2001) Can pollution problems be effectively solved by environmental science and technology? An analysis of critical limitations. *Ecol. Econ.* **37**, 271–287.

Hwa, T.J. (1995) National reports of Republic of Singapore on cleaner production for green productivity, In *Cleaner Production for Green Productivity*, Asian Productivity Organization.

Kansal, A., Rajeshwari, K.V., Balakrishnan, M., Lata, K. and Kishore, V.V.N. (1998) Anaerobic digestion technologies for energy recovery from industrial wastewater – a study in Indian context. *TERI Inform. Mon. Environ. Sci.* **3**(2), 67–75.

Karabelas, A.J., Yiantsios, S.G., Metaxiotou, Z., Andritsos, N., Akiskalos, A., Vlachopoulos, G. and Stavroulias, S. (2001) Water and material recovery from fertilizer industry acidic effluents by membrane processes. *Desalination* **138**, 93–102.

Karamanos, P. (1995) Industrial ecology: new opportunities for the private sector. *J. Indus. Environ.* **18**(4), 38–44.

Keller, J. and Hartley, K. (2003) Greenhouse gas production in wastewater treatment: process selection is the major factor. *Water Sci. Technol.* **47**(12), 43–48.

Khan, F.I., Natrajan, B.R. and Revathi, P. (2001) GreenPro: a new methodology for cleaner and greener process design. *J. Loss Prevent. Proc. Indus.* **14**, 307–328.

Krrishnamohan, K. and Heart, S. (2000) Industrial ecology and sustainable development – A viewpoint. *Int. J. Environ. Stud.* **57**(4), 387–400.

Lamba, N., Murthy, Z.V.P. and Kumar, R. (2002) Membrane processing of an aqueous waste stream from a catalyst manufacturing plant. *Separ. Sci. Technol.* **37**(1), 191–202.

Lee, C.H. (1995) National reports of Republic of Korea on cleaner production for green productivity, In *Cleaner Production for Green Productivity*, Asian Productivity Organization.

Lens, P., Vallero, M., Esposito, G. and Zandvoort, M. (2002) Perspectives of sulfate reducing bioreactors in environmental biotechnology. *Rev. Environ. Sci. Biotechnol.* **1**(4), 311–325.

Lowe, E.A. (1997) Creating by-products resource exchange: strategies for eco-industrial parks. *J. Cleaner Prod.* **5**(1–2), 57–65.

Lowe, E.A. and Evans, L.K. (1995) Industrial ecology and industrial ecosystem. *J. Cleaner Prod.* **3**(1–2), 47–53.

Manahan, S.E. (1999) *Industrial Ecology: Environmental Chemistry and Hazardous Waste*, Lewis Publishers, New York.

Metcalf and Eddy (2003) *Wastewater Engineering Treatment and Reuse*, McGraw-Hill, Mexico City.

My-Dieu, T.T. (2003) *Greening Food Processing Industry in Vietnam: Putting Industrial Ecology to Work.* PhD Thesis, Wageningen University, The Netherlands.

Nataagiin, O. (1995) National reports of Mongolia on cleaner production for green productivity, In *Cleaner Production for Green Productivity*, Asian Productivity Organization.

Oka, T. (1995) National reports of Republic of Japan on cleaner production for green productivity, In *Cleaner Production for Green Productivity*, Asian Productivity Organization.

Oldenburg, K.U. and Geiser, K. (1997) Pollution prevention and … or industrial ecology? *J. Cleaner Prod.* **5**(1–2), 103–108.

Park, D., Yun, Y.S., Park, J.M. (2005) Use of dead fungal biomass for the detoxification of hexavalent chromium: screening and kinetics. *Proc. Biochem.* **40**, 2559–2565.

Roestamsjah, E.M. and Cahyaningsih, S. (1995) National report: Indonesia, In *Cleaner Production for Green Productivity*, Asian Productivity Organization.

Rosegrant, M.W., Cai, X. and Cline, S.A. (2002) Implications for the future: meeting the challenge of water scarcity, In *World Water and Food to 2025 Dealing with Scarcity*, IFPRI and IWMI Publication.

Sarkis, J. and Cordeiro, J.J. (2001) An empirical evaluation of environmental efficiencies and firm performance: pollution prevention versus end-of-pipe practice. *Euro. J. Opera. Res.* **135**, 102–113.

Tapaneeyangkul, P. (1995) National reports of Thailand on cleaner production for green productivity, In *Cleaner Production for Green Productivity*, Asian Productivity Organization.

US EPA (2000) Total maximum daily load (TMDL), U.S. Environmental Protection Agency, EPA 841-F-00-009, Washington, D.C.

Vigneswaran, S., Jegatheesan, V. and Visvanathan, C. (1999) Industrial waste minimization initiatives in Thailand: concepts, examples and pilot scale trials. *J. Cleaner Prod.* **7**, 43–47.

2

Basic concepts of biological processes

S.G. Pavlostathis

2.1 INTRODUCTION

Biological processes have long played an important role in municipal wastewater treatment trains where these processes have been a common practice worldwide. Biological treatment processes, when compared to other competing physico-chemical processes, are simpler and less expensive because they do not require extremes in pH, temperature, or oxidation potential. Recognition of the ability of biological processes to remove a wide range of contaminants, both organic and inorganic, has led to the integration of biological processes into many industrial waste/wastewater treatment systems.

Specific information relative to the use of biological processes in various industries is given in subsequent chapters. The aim of this chapter is to familiarize the reader with basic concepts, common to all biological treatment processes regardless of the application: types of microorganisms encountered in such treatment systems, microbial metabolism and energetics, microbial growth, kinetics, as well as basic mathematical expressions for the quantitative description of biological processes typically used for both the design and operation of such systems.

2.2 MICROORGANISMS

All living matter is composed of one or more fundamental units called cells. Cells are dynamic, open systems continually taking materials from their environment and discarding waste products. Five cell characteristics set living cells apart from non-living systems (Brock *et al.* 1994): self-feeding, self-replication, differentiation, chemical signalling, and evolution. Each cell is isolated from its surroundings by a cell membrane (and a cell wall for some types of cells) and contains a variety of chemical substances and subcellular structures.

The genetic information needed for cell growth and function is stored in the deoxyribonucleic acid (DNA). Based on the arrangement of DNA within the cell, two types of cells have been recognized, prokaryote and eukaryote. In eukaryotes, several DNA molecules are contained in the nucleus, which is enclosed by a nuclear membrane. By contrast, in the prokaryotes, a single DNA molecule is found in the nuclear region, called nucleoid, which is not surrounded by a membrane. In addition to the difference in the DNA arrangement, prokaryotes and eukaryotes also distinctly differ in terms of size and the presence of membrane-enclosed internal structures (called organelles) found only in eukaryotes.

The microbial structural and functional diversity seen today is the result of a long, selective evolutionary process. Bacteria and Archaea are the only two prokaryotic, distinct evolutionary lineages, the third being the eukaryotes (Brock *et al.* 1994). The eukaryotes include microorganisms such as algae, fungi, and protozoa as well as multicellular life forms (plants and animals). Viruses are not cells, but rather obligate, intracellular parasites, unable to reproduce without a host cell. The following discussion is a general introduction to the various microbial groups related to water quality and wastewater treatment.

Bacteria These are single-celled prokaryotes with amazing morphological and functional diversity. Most bacteria can be grouped into five morphological categories: spheroid (called cocci, 1 to 3 μm in diameter); rod-shaped (called bacilli, 0.3 to 1.5 μm in width and 1 to 10 μm in length); curved, rod-shaped (called vibrios, 0.6 to 1 μm in width and 2 to 6 μm in length); spiral (called spirilla, up to 50 μm in length); and filamentous (100 μm and longer). Bacteria can use either light (phototrophic), organic compounds (organotrophic), or inorganic chemicals (lithotrophic) as energy source. A common molecular formula used to represent bacteria is $C_5H_7O_2N$. Most bacteria live in oxic as well as anoxic habitats. Bacteria include most pathogenic prokaryotes and most microorganisms commonly found in soil, water, animal digestive tracts, and other environments (Brock *et al.* 1994). Certain bacterial species produce endospores that are differentiated cells very resistant to heat, drying, radiation, acids, and chemical disinfectants. Endospore formation (or sporulation) is the response of the spore-forming bacteria to an environmental signal (e.g., carbon, nitrogen limitation, and desiccation). Endospores can remain dormant for extremely long periods of time and can convert back to vegetative cells fairly rapidly (Brock *et al.* 1994).

Archaea Most Archaea are anaerobes and many live in extreme environments such as hot springs (at temperatures above the boiling point of water), salty water bodies, highly acidic or alkaline soils and water (Brock *et al.* 1994). Distinctive differences exist in the chemical composition of both cell wall and cell membrane between bacteria and Archaea, which explains the existence of Archaea in such extreme environments.

Algae These range from unicellular to large aggregates of filamentous cells. They are commonly found in aquatic systems but because they are able to survive periods of desiccation they are also found in terrestrial systems (i.e., soil, tree bark, and rocks). Algae contain chlorophylls and other pigments, which serve as light-gathering molecules and allow photosynthesis to take place. They are primary producers since they require only inorganic chemicals for growth. A common molecular formula used to represent algal biomass is $C_{5.7}H_{9.8}O_{2.3}N$. In addition to photosynthesis, algae also use oxygen for respiratory metabolism, especially during periods of darkness, resulting in the removal of oxygen from the aquatic system. Excessive growth of algae (algal blooms) in eutrophic aquatic systems (i.e., systems rich in nutrients, especially N and P) is a nuisance and can lead to a severe depletion of oxygen during periods of darkness resulting in death of fish and other macroorganisms. Some algae produce toxins that are toxic to both fish and human beings. Several algal species produce organic compounds, which are associated with taste and odor in either surface water supplies or during raw water filtration. Algal growth can also cause variation in pH, hardness, color, and organic matter content of water.

Fungi These are aerobic, multicellular, non-photosynthetic, heterotrophic eukaryotes. Unicellular fungi, called *yeasts*, are facultative. Although there are many aquatic fungal species, most are terrestrial, living in the soil. Most fungi are saprophytic, that is, they use dead organic matter as the carbon and energy source. A common molecular formula used to represent fungal biomass is $C_{10}H_{17}O_6N$. Therefore, the nitrogen requirement is about one-half of that for bacteria (based on the molecular formula $C_5H_7O_2N$). Fungi can tolerate more severe stresses than other microorganisms: high osmotic pressure, low pH, and high water tension (i.e., desiccation).

Protozoa These are unicellular, non-photosynthetic, typically motile eukaryotic microorganisms without cell walls. Most protozoa are aerobic or facultative chemoheterotrophs, although some anaerobic protozoa have also been found. Protozoa are abundant in soil as well as in fresh and salt water systems. Protozoa feed on bacteria and other microscopic microorganisms, and aid in the clarification and purification of streams and secondary wastewater treatment effluents. Protozoa form cysts, some in response to exhaustion of food supply or desiccation, and others as the normal part of the reproductive cycle. Their optimum pH range is 6–8, although they can survive in environments with pH values as low as 2 and as high as 8.7.

Viruses These are obligate intracellular parasites that contain genetic material – DNA or RNA – necessary for their replication. However, they are unable to synthesize compounds, but instead they invade living cells (hosts) where they take over and redirect the cell activities to produce new viral particles at the expense of the host cell. Viruses, which infect bacteria, fungi, protozoa, plants and animals, have been found. Invasion of the host cell does not always lead to viral replication and lysis of the host cell (*virulent or lytic viruses*), but the viral genetic material can be incorporated into the host DNA and replicated, a process called *lysogeny* (*temperate or lysogenic viruses*). Therefore, viruses are considered as agents of either disease or heredity. Viruses cause a number of water-borne diseases, therefore, removal and control of viruses in public water supplies is a major concern.

In biological wastewater systems, the microorganisms that most commonly find application are bacteria and Archaea. Algae and protozoa are found in waste stabilization ponds and activated sludge systems, respectively. The application of fungi for wastewater treatment has not yet passed the development stage. Viruses are not applied with the objective of wastewater treatment; however, the efficiency of viruses' removal by biological treatment processes is of importance.

In wastewater treatment systems, the microorganisms mass is usually mixed with other solids (introduced with the influent or generated in the system) and form a slurry or sludge. This sludge is often considered as an equivalent bacterial suspension, although often the actual live microorganisms mass (viable biomass) is only a small fraction of the total sludge mass. Due to interactions of different types of microorganisms, mixed bacterial suspensions can have very different properties than pure bacterial cultures.

2.3 MICROBIAL METABOLISM

Microorganisms are dynamic units, continuously taking materials from their environment (i.e., nutrients), processing them, and ultimately incorporating some of them into biomass while excreting cellular materials as well as waste products back into their environment. Therefore, microorganisms are open systems, constantly undergoing change. All biochemical processes taking place within a cell are collectively called metabolism.

Anabolism (or biosynthesis) is the process by which the cell obtains simple nutrients from its environment and converts them to cellular constituents. A distinction is made between heterotrophic organisms that use organic material as the carbon source for biosynthesis and autotrophic organisms that build up their material from CO_2. In addition to carbon, the microbial mass is composed of several other elements. Although there are many nutrients required in trace amounts (called *micronutrients*, such as Co, Zn, Cu, Mo, Mn, Ni, Se, V, and W), major nutrients (required in relatively large amounts, called *macronutrients*) include the following elements: C, H, O, N, P, S, K, Mg, Ca, Na, Fe, and Cl. In addition to the above-mentioned inorganic nutrients, some microorganisms cannot synthesize certain organic compounds, known as *growth factors*, which need to be supplied, albeit in very small amounts. Such organic compounds belong to the following three classes: vitamins, amino acids, purines, and pyrimidines (Brock *et al.* 1994). However, the main chemical components of cells are polymers built up of monomers, which are

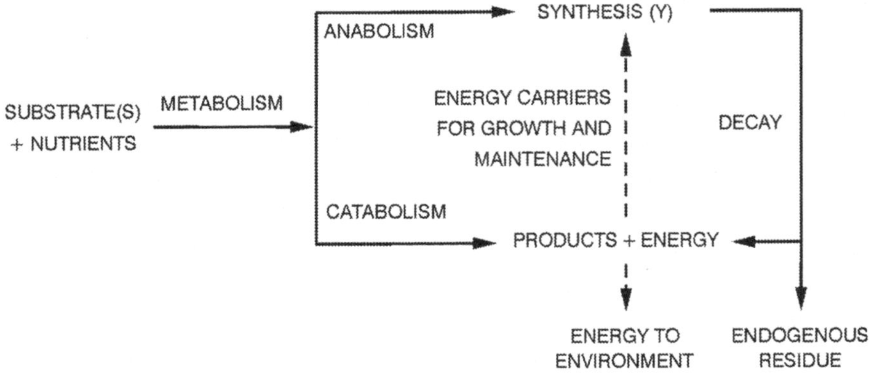

Figure 2.1 Relationship between catabolic and anabolic processes showing the key role of energy carriers.

considered the building blocks. The four main classes of monomers/polymers are: carbohydrates for polysaccharides, fatty acids for lipids, nucleotides for nucleic acids (DNA, RNA), and amino acids for proteins.

Biosynthesis is an energy-requiring process (i.e., endergonic process). This energy is obtained from the environment in three forms: light, inorganic chemicals, and organic chemicals. In addition to the energy requirements for biosynthesis, other non-growth-related cell functions (e.g., movement, transport of nutrients, osmotic pressure regulation, etc.) also require energy. The non-growth-related energy requirement is termed maintenance. The biochemical processes involved in the breakdown of organic and inorganic compounds resulting in energy release (i.e., exergonic processes) are collectively termed catabolism. Metabolism is therefore the sum of all anabolic and catabolic processes. The close relationship of the two metabolic processes is depicted in Figure 2.1. The energy released by the catabolic processes fuels the endergonic anabolic processes and also provides energy for cell maintenance. The energy flow from the catabolic to the anabolic reactions is mediated by the use of energy carriers – compounds which store energy in the form of high-energy phosphate bonds (e.g., adenosine triphosphate (ATP), adenosine diphosphate (ADP)).

The change in free energy (i.e., the energy which is available to do useful work, ΔG) of a reaction is the thermodynamic criterion of whether or not the reaction is feasible. Therefore, if $\Delta G < 0$, free energy will be released (exergonic reaction) and the reaction can proceed as written. On the other hand, if $\Delta G > 0$, free energy is required (endergonic reaction) for the reaction to proceed as written, or else the reverse reaction may occur spontaneously.

Utilization of chemical energy in living systems involves oxidation–reduction (or redox for short) reactions. Such reactions involve the intermolecular transfer of electrons from a relatively reduced substance (called reductant or electron donor) to a relatively oxidized substance (called oxidant or electron acceptor). The term redox reaction also implies that for any oxidation to occur, a reduction must also take place (i.e., production and consumption of electrons are balanced). Biochemical

Table 2.1 Reduction potentials of some redox pairs of interest[a,b].

Half reaction	$\Delta G^{o\prime}$ (kJ/eeq)[c]	$E_h^{o\prime}$(V)
(a) $1/4\ CO_2 + H^+ + e^- = 1/24\ C_6H_{12}O_6 + 1/4\ H_2O$	41.92	−0.43
(b) $H^+ + e^- = 1/2\ H_2$	40.46	−0.42
(c) $1/8\ SO_4^{-2} + 19/16\ H^+ + e^- = 1/16\ H_2S + 1/16\ HS^- + 1/2\ H_2O$	21.28	−0.22
(d) $1/8\ NO_3^- + 5/4\ H^+ + e^- = 1/8\ NH_4^+ + 3/8\ H_2O$	−39.50	0.36
(e) $1/2\ NO_3^- + H^+ + e^- = 1/2\ NO_2^- + 1/2\ H_2O$	−39.43	0.41
(f) $1/5\ NO_3^- + 6/5\ H^+ + e^- = 1/10\ N_2 + 3/5\ H_2O$	−71.66	0.74
(g) $1/4\ O_2 + H^+ + e^- = 1/2\ H_2O$	−74.14	0.81
(h) $1/3\ NO_2^- + 4/3\ H^+ + e^- = 1/6\ N_2 + 2/3\ H_2O$	−93.15	0.97

[a]Reactants and products at unit activity except $[H^+] = 10^{-7}$.
[b]McCarty (1972); Stumm and Morgan (1996).
[c]eeq: electron equivalent.

redox reactions involve not only the transfer of electrons, but also hydrogen atoms (H, i.e., $H^+ + e^-$). Thus, in these cases the terms oxidation and reduction are synonymous to the terms dehydrogenation and hydrogenation, respectively. The tendency of a substance to donate electrons and be oxidized or to accept electrons and be reduced is expressed by the reduction potential ($E_h^{o\prime}$, i.e., the reduction potential with reference to standard H_2 pressure (=1 bar) and pH (=7). Reduction potentials of some redox pairs of importance for biological treatment processes are given in Table 2.1. These potentials are arranged with the strongest reductants (or electron donors; $E_h^{o\prime} < 0$) at the top and the strongest oxidants (or electron acceptors; $E_h^{o\prime} > 0$) at the bottom. This arrangement is referred to as the *electron tower*, since electrons from the reductant can only be accepted by the oxidants below. The transfer of electrons in the redox reactions is mediated by intermediate substances called electron carriers (e.g., nicotinamide adenine dinucleotide, NAD^+; flavin mononucleotide, FMN; flavin-adenine dinucleotide, FAD). The change in reduction potential corresponds to a free energy change. Thus, the greater the difference in $E_h^{o\prime}$ values between the electron donor and electron acceptor, the more energy will be released. In catabolic reactions, the electron donor is also referred to as the energy source, although only the complete redox reaction involving both the electron donor and the electron acceptor leads to energy production.

Enzymes play an important catalytic role in all biochemical reactions by lowering the reaction activation energy and thus increasing the reaction rate by as much as 10^{20} times (Brock *et al.* 1994). The biocatalytic role of enzymes is appreciated if one considers that these reactions take place under normal temperature and pressure conditions. In addition to their energy-related, biocatalytic role, enzymes play an equally important role in the regulation of the extent of biochemical reactions, which occur simultaneously. The regulation of biochemical reactions is achieved by controlling the presence or absence of a particular enzyme (by the processes called *induction* and *repression*, respectively) and by controlling the activity of constitutive enzymes (i.e., enzymes which are always present).

Based on the effect of oxygen, as well as the energy and carbon source used, microorganisms are divided into several groups, as shown in Tables 2.2 and 2.3. Based on the electron and carbon flows, all metabolic processes can be divided into the following five types:

(a) aerobic respiration,
(b) anaerobic respiration,
(c) fermentation,
(d) chemolithotrophic metabolism,
(e) phototrophic metabolism.

The catabolic processes of the first three types of metabolism lead to the removal of organic material. A fundamental difference between aerobic or anaerobic respiration and fermentation is that in the former the organic material is completely oxidized (i.e., organic carbon is oxidized to CO_2), whereas during fermentation only partial oxidation of the organic material takes place. After fermentation, physical removal of gaseous products (e.g., methane) from the wastewater may take place due to their low solubility.

In the case of chemolithotrophic metabolism, there is an increase of organic material mass: relatively reduced inorganic chemicals serve as electron donors

Table 2.2 Microorganisms grouping according to oxygen relationships[a].

Group	O_2 effect
Aerobes	
Obligate	Required
Facultative	Not required but grow better with O_2
Microaerophilic	Required, but at levels lower than atmospheric
Anaerobes	
Aerotolerant	Not required; grow no better when O_2 present
Obligate (strict)	Harmful or lethal

[a]Adapted from Brock *et al.* 1994.

Table 2.3 Microorganisms grouping according to the energy and carbon source[a].

Group	Energy source	Carbon source
Autotrophic		
Chemolithotrophic	Inorganic redox	CO_2
Photolithotrophic	Radiant energy	CO_2
Heterotrophic (=Organotrophic)		
Chemoorganotrophic	Organic redox	Organic compounds
Photoorganotrophic	Radiant energy	Organic compounds
Mixotrophic	Inorganic redox	Organic compounds

[a]Adapted from Brock *et al.* (1994).

and CO_2 is used for the synthesis of biomass. Biomass is also produced in photolithotrophic metabolism where water serves as the electron donor for oxygen producing phototrophic systems and CO_2 is used for anabolism.

Both aerobic and anaerobic respirations require an external, terminal electron acceptor. In contrast, fermentation does not involve any external electron acceptor and redistribution of electrons takes place within the molecule(s) – intramolecular electron transfer. In other words, a relatively reduced part of the molecule serves as the electron donor and a relatively oxidized part of the molecule serves as the electron acceptor. For example, considering the acetoclastic methane fermentation:

$$CH_3COOH \rightarrow CH_4 + CO_2 \quad \Delta G^{o'} = -75.7\,kJ \tag{2.1}$$

The carbon oxidation state of the methyl group is −3 and that of the carboxyl group is +3 (resulting in a carbon mean oxidation state of zero). However, the carbon oxidation state of the CH_4 and CO_2 is −4 and +4, respectively. Therefore, the carboxyl-C serves as the electron donor and the methyl-C serves as the electron acceptor allowing this redox reaction to take place without any external terminal electron acceptor.

Since fermentation represents only a partial oxidation of the carbon source, generally a small amount of free energy is released, as compared to aerobic respiration, which leads to the complete oxidation of the fraction of the carbon source used for energy production (the remaining fraction serves as the carbon source and is assimilated). For example, the complete oxidation of acetic acid is given by the following reaction:

$$CH_3COOH + 2O_2 \rightarrow 2CO_2 + 2H_2O \quad \Delta G^{o'} = -893.7\,kJ \tag{2.2}$$

Therefore, methanogenesis from acetic acid (Equation (2.1)) releases only 8.5% of the potentially available free energy of acetic acid if it was completely oxidized. Note that these calculations are based on the catabolic reactions while microbial synthesis (i.e., anabolism) has been ignored. The difference in the free energy release between reactions (2.1) and (2.2) is the free energy of methane that will be released upon its oxidation:

$$CH_4 + 2O_2 \rightarrow CO_2 + 2H_2O \quad \Delta G^{o'} = -818.0\,kJ \tag{2.3}$$

Note that the demand of oxidant is the same before (Equation (2.2)) and after (Equation (2.3)) the fermentation, because during fermentation there is no intermolecular electron transfer or destruction of the organic material. Since Equation (2.3) shows that the chemical oxygen demand (COD) of methane is 64 g of oxygen per 16 g of methane or 4 g COD/gCH_4, when methane is generated from any type of organic material, there will always be destruction of 4 g COD to produce 1 g of CH_4.

Among the several types of respiratory metabolism, aerobic respiration is the process that releases most free energy (with the exception of nitrite reduction to N_2, see Table 2.1 above). For the same carbon source, the free energy release diminishes as the terminal electron acceptor becomes more reduced (i.e., moving upwards on the electron tower, see Table 2.1). The free-energy yields among the various types of respiratory metabolism positively correlate with the growth rate and cell yield of the microorganisms that mediate these processes. Table 2.4 shows the

Table 2.4 Metabolic processes of interest in biological treatment processes.

Process	Electron donor[a]	Electron acceptor	Carbon source
Oxidation of organics	Organics	O_2	Organics
Nitrification	NH_4^+, NO_2^-	O_2	CO_2
Sulfide/sulfite oxidation	S^{-2}/S^{+4}	O_2	CO_2/organics
Denitrification	Organics, H_2	NO_3^-/NO_2^-	Organics
Sulfate reduction	Organics, H_2	SO_4^{-2}	Organics
Acetoclastic methanogenesis	Acetate	None	Acetate
Hydrogenotrophic methanogenesis	H_2	CO_2	CO_2 (Acetate)

[a]Serve also as the energy source.

electron donor, electron acceptor, carbon and energy source for metabolic processes of interest in biological treatment processes.

2.4 MICROBIAL GROWTH

Microbial growth is defined as the increase in quantity of cellular constituents and structures, and is accompanied by an increase in size or number of individual cells, or both. If cells are grown in a batch culture, the growth pattern shown in Figure 2.2 is typically observed with four, more or less distinct phases.

Lag phase After the addition of a small quantity (compared to the substrate mass) of inoculum to the culture medium, the organisms must adjust to the new environment before they begin to divide. The lag phase may be required for the production of inducible enzymes, as well as the replacement and/or generation of growth factors not previously required for microbial growth.

Exponential phase Once the inoculum has adjusted to its new environment and all growth conditions have been set to their optimum values, the organisms will start multiplying as rapidly as possible, according to the following series:

$$N \rightarrow 2N \rightarrow 2(2N) \rightarrow \cdots \rightarrow 2^n N \tag{2.4}$$

where n = number of generations or replications. By defining t_d = minimum generation or doubling time, then $n = t/t_d$, where t = time. For $N = N_o$ at $t = 0$,

$$N = N_o 2^{t/t_d} \tag{2.5}$$

or

$$\ln N = \ln N_o + \frac{t}{t_d}\ln 2 \tag{2.6}$$

By defining $\hat{\mu} = (\ln 2)/t_d$ and substituting into Equation (2.6),

$$N = N_o e^{\hat{\mu} t} \tag{2.7}$$

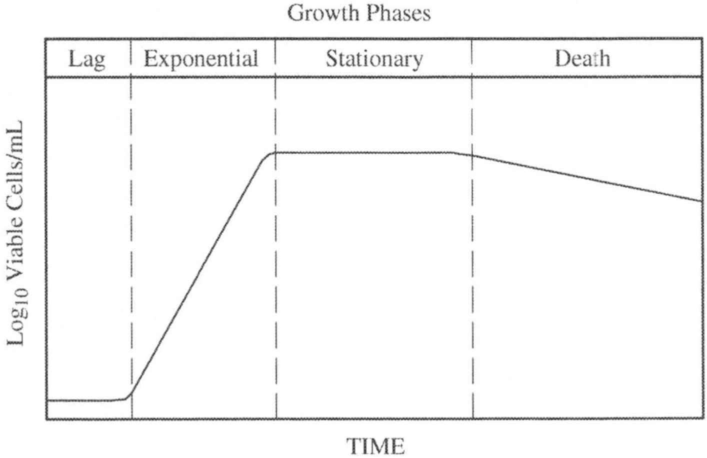

Figure 2.2 Typical bacterial growth curve.

By differentiation of Equation (2.7) with respect to maximum growth conditions, the following expression is obtained:

$$\left[\frac{1}{N}\frac{\mathrm{d}N}{\mathrm{d}t}\right]_{\max} = \hat{\mu} \tag{2.8}$$

which is the definition of $\hat{\mu}$, that is, maximum specific growth rate (T^{-1}).

Stationary phase In a batch system with time the substrate concentration will decrease and eventually become limiting. However, growth is still taking place along with cell death (see below) during the stationary phase, leading to a negligible net growth rate (defined as growth rate minus death rate). Accumulation of toxic end product(s) could also lead to the development of a stationary phase even though the substrate may not be growth limiting.

Death phase During this phase, the number of viable cells decreases over time, that is, the net growth rate becomes negative (see Figure 2.2). When the cell exhausts all extracellular and intracellular substrate(s), vital metabolic functions are disrupted and eventually the cell dies. However, as a result of cell lysis, secondary substrates and nutrients become available for other viable cells to grow. Therefore, even during the death phase, some cells are still replicating (cryptic growth). The rate of decrease of the number of viable cells during the death phase is typically assumed to be proportional to the number of viable cells present, as follows:

$$\frac{\mathrm{d}N}{\mathrm{d}t} = -k_{\mathrm{d}}\, N \tag{2.9}$$

where k_{d} is the specific microorganism death rate constant (T^{-1}), independent of the substrate concentration. Biomass decrease during the death phase may also be due to a decrease in the mass per viable cell through endogenous respiration (i.e., respiration supported by intracellular constituents).

2.5 KINETICS OF MICROBIAL GROWTH

The microbial growth rate in the case where all growth requirements are met is assumed to be first-order with respect to the concentration of viable microorganisms:

$$\frac{dX_a}{dt} = \mu X_a \tag{2.10}$$

where, X_a = active (i.e., viable) microorganisms concentration ($M\,L^{-3}$); t = time (T); $\mu = (1/X_a)(dX_a/dt)$ = specific growth rate (T^{-1}).

The microbial growth rate is proportional to the substrate utilization rate:

$$\frac{dX_a}{dt} = Y\left(\frac{-dS}{dt}\right) \tag{2.11}$$

where, Y = maximum microbial yield coefficient (mass of cells formed/mass of substrate utilized); S = growth-limiting substrate concentration ($M\,L^{-3}$). Combining Equations (2.10) and (2.11) yields:

$$\frac{-dS}{dt} = \frac{\mu}{Y} X_a \tag{2.12}$$

Based on Equation (2.12), the specific rate of substrate utilization (U, mass of substrate/mass of cells − time) can be expressed as follows:

$$U \equiv \frac{\mu}{Y} = \frac{1}{X_a}\left(\frac{-dS}{dt}\right) \tag{2.13}$$

2.5.1 Effect of substrate concentration

For pure, batch cultures fed with a soluble single substrate, Monod (1942) proposed the following empirical relationship to describe the specific microbial growth rate as a function of the growth-limiting substrate concentration:

$$\mu = \frac{\hat{\mu}\, S}{K_s + S} \tag{2.14}$$

where, $\hat{\mu}$ = maximum specific microbial growth rate (T^{-1}); K_s = half-saturation constant ($M\,L^{-3}$).

Equation (2.14) was developed concurrently with the Michaelis–Menten enzyme kinetic equations, although the Monod equation merely represents a good curve fit to experimental data. Later, Monod (1949) successfully applied this equation to continuous-flow, pure cultures. Two extreme cases with respect to the substrate concentration can be distinguished: (a) if $S >> K_s$, then $\mu \approx \hat{\mu}S^o = \hat{\mu}$, that is, the specific growth rate is zero-order with respect to substrate concentration (substrate saturation); (b) if $S \ll K_s$, then $\mu \approx (\hat{\mu}/K_s)S$, that is, the specific growth rate becomes first-order. Equation (2.14) can also be presented in another form, which uses dimensionless specific growth rate (i.e. $\mu/\hat{\mu}$) and substrate concentration (i.e., S/K_S):

$$\frac{\mu}{\hat{\mu}} = \frac{\dfrac{S}{K_s}}{1+\dfrac{S}{K_s}} \tag{2.15}$$

By substituting $S/K_S = 1$ into Equation (2.15), $\mu/\hat{\mu}$ becomes equal to 0.5 which defines K_S as the substrate concentration which leads to a specific growth rate at half of its maximum value. Note that the higher the value of K_s is for a particular substrate/microbe system, the lower the substrate affinity is.

By substituting Equation (2.14) into (2.12), the following equation is obtained:

$$\frac{-\mathrm{d}S}{\mathrm{d}t} = \frac{1}{Y}\frac{\hat{\mu}X_a S}{K_s + S} \tag{2.16}$$

By defining $k = \hat{\mu}/Y$ = maximum specific substrate utilization rate (mass of substrate/mass of cells − time), the following expression describing the substrate utilization rate based on the Monod equation is obtained:

$$\frac{-\mathrm{d}S}{\mathrm{d}t} = \frac{kX_a S}{K_s + S} \tag{2.17}$$

2.5.2 Net microbial growth

The above-presented Monod equation does not account for microbial decay. In the case of mixed microbial cultures, microbial decay includes processes that lead to a decrease of biomass and include endogenous respiration, predation, natural cell death and lysis. A portion of the energy released as a result of substrate utilization is used to support basal metabolic processes (such as motility, repair, ion transport, regulation of osmotic pressure, etc.), which collectively is called *basal metabolism* or *cell maintenance*. Therefore, the substrate utilization rate needs to be corrected for the biomass fraction either produced and lost (e.g., predation and endogenous respiration) or never produced (e.g., basal metabolism) which collectively will be called microbial decay.

Assuming that the microbial decay is first-order with respect to the active microorganisms concentration, the net microbial growth is as follows:

$$[\text{Net growth}] = [\text{Growth}] - [\text{Decay}] \tag{2.18}$$

$$\frac{\mathrm{d}X_a}{\mathrm{d}t} = Y\left(\frac{-\mathrm{d}S}{\mathrm{d}t}\right) - bX_a \tag{2.19}$$

where, b = specific microorganism decay rate constant (T^{-1}). By substituting Equation (2.17) into (2.19), the following equation is obtained:

$$\frac{\mathrm{d}X_a}{\mathrm{d}t} = \frac{YkX_a S}{K_s + S} - bX_a \tag{2.20}$$

Dividing Equation (2.20) by X_a, the following expression is obtained:

$$\mu_{net} = \frac{YkS}{K_s + S} - b \tag{2.21}$$

or by substituting $Yk = \hat{\mu}$, an equivalent equation is obtained:

$$\mu_{net} = \frac{\hat{\mu}S}{K_s + S} - b \tag{2.22}$$

Note that when $S >> K_s$, then $U \approx k$, thus:

$$\mu_{net} = YU - b \tag{2.23}$$

or

$$\left(\mu_{net}\right)_{max} = Yk - b \tag{2.24}$$

Figure 2.3 depicts the relationship between μ and μ_{net} as a function of substrate concentration. When microbial decay is taken into account (i.e., $b > 0$), as the substrate concentration decreases, a value of $S > 0$ is required to balance growth and decay, that is, where $\mu_{net} = 0$. This substrate concentration is termed S_{min}. For values of S less than S_{min}, biomass loss due to microbial decay surpasses microbial growth and the biomass will either never grow or will gradually disappear.

By definition, $S = S_{min}$ when $\mu_{net} = 0$, that is, $dX_a/dt = 0$. By setting Equation (2.20) equal to zero and letting $S = S_{min}$, the following expression is obtained:

$$S_{min} = \frac{K_s b}{Yk - b} \tag{2.25}$$

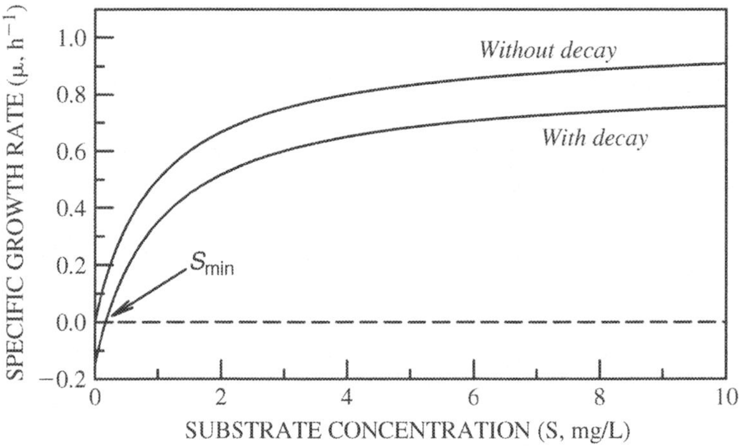

Figure 2.3 Specific growth rate and net growth rate as a function of the growth-limiting substrate concentration ($\hat{\mu} = 1\,h^{-1}$; $K_s = 1\,mg\,L^{-1}$; $b = 0.15\,h^{-1}$).

Therefore, S_{min} is a kinetic parameter, independent of reactor configuration. When $b = 0$ (i.e., microbial decay and maintenance are ignored), $S_{min} = 0$. In other words, the existence of an S_{min} value for a particular system is a result of microbial decay and maintenance. Therefore, the lowest limit of substrate that can be achieved in a particular system increases with increasing values of b.

2.5.3 Microbial yield coefficient

The yield coefficient (Y) is a function of the energy content of the electron donor relative to the electron acceptor and that of the carbon source relative to the microbial cell. Values of Y can be found in the literature, evaluated experimentally or estimated from thermodynamics (see Section 2.6). Another, convenient way to calculate the microbial yield coefficient for a particular system is by looking at the net biomass production and substrate utilization during a specified time interval and then calculating the observed yield coefficient (Y_{obs}):

$$Y_{obs} \equiv \frac{(dX_a/dt)_{net}}{-(dS/dt)} = \frac{(dX_a/dt)_{net}(1/X_a)}{-(dS/dt)(1/X_a)} = \frac{\mu_{net}}{U} \tag{2.26}$$

Based on Equation (2.23),

$$\mu_{net} = YU - b; \quad \text{hence } U = \frac{\mu_{net} + b}{Y} \tag{2.27}$$

By substituting into Equation (2.26), the following expression is obtained:

$$Y_{obs} = \frac{Y}{1 + \dfrac{b}{\mu_{net}}} \tag{2.28}$$

By taking the limits of Equation (2.28), as $\mu_{net} \rightarrow 0$, $Y_{obs}/Y \rightarrow 0$ and as $\mu_{net} \rightarrow \infty$, $Y_{obs}/Y \rightarrow 1$. Therefore, in general, $Y_{obs} \leqslant Y$. If $\mu_{net} \rightarrow \infty$ or $b = 0$, which also implies that $\mu_{net} = 0$, $Y_{obs} = Y$. In other words, as long as there is growth, microbial decay takes place, which in turn lowers the value of the maximum yield coefficient to that of the observed yield coefficient. As it is shown in Section 2.7, below, in continuous-flow systems, $1/\mu_{net} = \theta_c$, which is the solids retention time (T). Thus, Equation (2.28) becomes:

$$(Y_{obs})_a = \frac{Y}{1 + b\theta_c} \tag{2.29}$$

It should be emphasized that Equation (2.29) is based on the active (i.e., viable) biomass concentration (X_a). When the total biomass, X_v (=active + inactive) concentration is taken into account, it can be shown that:

$$(Y_{obs})_v = \frac{Y}{1 + b\theta_c}[1 + (1 - f_d)b\theta_c] \tag{2.30}$$

where, f_d = net biodegradable fraction of active biomass. Based on Equations (2.29) and (2.30), the *viability* (v) of a system, defined as the ratio of the active to the total biomass concentration, is obtained from the following equation:

$$v = \frac{1}{1 + (1 - f_d) b\, \theta_c} \tag{2.31}$$

2.5.4 Estimation of biokinetic coefficients

The above-developed relationships based on the Monod model require the values of four biokinetic coefficients: Y, k, K_s, and b. The values of these coefficients (assumed to be constant for a particular biological system), can be found as follows: from literature; evaluated experimentally for a batch or a continuous-flow system (Metcalf & Eddy Inc. 1991); or estimated from thermodynamics (see Section 2.6, values of Y and k only).

Based on experimental data of S and X_a versus t for a batch reactor, the system of Equations (2.17) and (2.20) can be used to perform a non-linear regression to simultaneously estimate the values of the four -biokinetic coefficients. In cases where the variation of biomass concentration during a batch incubation period is considered to be negligible, then Equation (2.17) can be rewritten as follows:

$$\frac{-dS}{dt} = \frac{k' S}{K_s + S} \tag{2.32}$$

where $k' = k X_a \approx$ constant, is the maximum substrate utilization rate ($M L^{-3} T^{-1}$) for a constant biomass concentration and other imposed experimental conditions. Integration of Equation (2.32) yields:

$$S = S_o - K_s \ln\left(\frac{S}{S_o}\right) - k't \tag{2.33}$$

In case microbial decay is insignificant (i.e., $b \approx 0$), $\hat{\mu}$ and K_s can be estimated by use of the Monod equation (Equation (2.14)) employing a non-linear regression technique, or by linear regression using one of the three, alternative equations (alternative linearizations of Equation (2.14)):

Lineweaver-Burk:

$$\frac{1}{\mu} = \frac{1}{\hat{\mu}} + \frac{K_s}{\hat{\mu}}\left(\frac{1}{S}\right) \tag{2.34}$$

Eadie-Hofstee:

$$\mu = \hat{\mu} - K_s \frac{\mu}{S} \tag{2.35}$$

Hanes:

$$\frac{S}{\mu} = \frac{K_s}{\hat{\mu}} + \left(\frac{1}{\hat{\mu}}\right) S \tag{2.36}$$

Although linear regression is a relatively simple technique, it does not necessarily lead to the best estimates of the biokinetic coefficients. For the same experimental data, the three types of linearization presented above will yield different estimates of the coefficients. Although $\hat{\mu}$ and K_s are presumed to be uncorrelated, results of data linearization have indicated relatively strong correlations between these two coefficients (Characklis 1990). Non-linear regression techniques which use the non-linear form of the model (Equation (2.14)) provide better estimates of the biokinetic coefficients and should be preferred over the above-presented linearized equations.

The values of Y and k can be estimated based on bioenergetics (see Section 2.6). However, there are no relationships available based on fundamentals able to predict the values of K_s and b. Mass-transfer effects are usually accounted for by using an "apparent" K_s value, which is larger than the intrinsic K_s value (i.e., in the absence of any mass-transfer limitations) (see Contois model, below, as well as Section 2.8). The value of b depends on the microbial species type and is positively correlated with the value of $\hat{\mu}$ (typically taken as equal to $0.1\,\hat{\mu}$ in the absence of a better estimate). Typical b values range from 0.05 to 0.3 per day for aerobic, heterotrophic species and from 0.01 to 0.05 per day for slowly growing species (e.g., anaerobic or lithotrophic species).

2.5.5 Kinetics of multiple substrates

It is possible that two or more reaction components may limit microbial growth. For example, both ammonia and oxygen can be limiting in the nitrification process. The following expression can be used to represent multiple substrate kinetics:

$$\mu = \hat{\mu}\left(\frac{S_1}{K_{s1} + S_1}\right)\left(\frac{S_2}{K_{s2} + S_2}\right) \cdots \left(\frac{S_n}{K_{sn} + S_n}\right) - b \tag{2.37}$$

where $S_1, S_2, \cdots, S_n$ = substrate concentrations ($\mathrm{M\,L^{-3}}$); $K_{s1}, K_{s2}, \cdots, K_{sn}$ = half-saturation coefficients for the respective substrates ($\mathrm{M\,L^{-3}}$). Typically, for relatively low K_s values and high substrate concentrations, several of the terms in Equation (2.37) approach a value of one (i.e., saturation).

2.5.6 Alternative substrate utilization rate models

Although the Monod model (Equations (2.14) and (2.17)) is widely used, several other kinetic models have been used and applied for special circumstances, as follows:

First-order model:

$$\frac{-\mathrm{d}S}{\mathrm{d}t} = k\,S \tag{2.38}$$

Grau model:

$$\frac{-\mathrm{d}S}{\mathrm{d}t} = \frac{kX_a S}{S_o} \tag{2.39}$$

where S_o = initial substrate concentration (or influent substrate concentration for a continuous-flow system ($\mathrm{M/L^3}$)).

Moser model:

$$\frac{-\mathrm{d}S}{\mathrm{d}t} = \frac{kX_a S}{K_s + S^{-\gamma}} \tag{2.40}$$

where γ = constant (unitless). When $-\gamma = 1$, the Moser model reverts to the Monod model.

Tessier model:
$$\frac{-\mathrm{d}S}{\mathrm{d}t} = k(1 - \mathrm{e}^{S/K_s})X_a \tag{2.41}$$

Contois model:
$$\frac{-\mathrm{d}S}{\mathrm{d}t} = \frac{kX_aS}{BX_a + S} \tag{2.42}$$

where B = constant ($M_{\text{substrate}}/M_{\text{biomass}}$). Note that as $X_a \rightarrow \infty$, the substrate utilization rate becomes first-order with respect to S and zero-order with respect to X_a. Typical values of the ratio k/B in the case of hydrolysis of particulate organic matter range from 1 to 3 per day. For $BX_a = K_s$, the Contois model reverts to the Monod model, except that K_s is a function of active biomass concentration.

Chen and Hashimoto model:
$$\frac{-\mathrm{d}S}{\mathrm{d}t} = \frac{\hat{\mu}X_aS}{KX_a + YS} \tag{2.43}$$

where K = constant (unitless).

2.5.7 Effect of temperature

Biological processes are affected by temperature. Generally speaking, the higher the temperature, the higher the microbial activity until an optimum temperature is reached. Further increase of the temperature beyond its optimum value results in a precipitous decrease of microbial activity. Although theoretically all four biokinetic coefficients may be affected by temperature, the effect of temperature on the kinetics of biological processes is usually depicted as affecting the maximum specific substrate utilization rate constant (k). The most widely used equation for k-temperature correlations is the Arrhenius equation:

$$k = A\exp\left(\frac{-E_a}{RT}\right) \tag{2.44}$$

where A = frequency factor (same units as k); E_a = apparent activation energy ($\mathrm{kJ\,mol^{-1}}$); R = gas constant (=0.00829 $\mathrm{kJ\,mol^{-1}K^{-1}}$); and T = absolute temperature (K). A commonly used parameter for the quantification of the temperature effect on the microbial activity and thus the biological process rate is the temperature coefficient (Q_{10}) for a 10°C temperature increase defined as the ratio of a process rate at two temperatures, T_2 and T_1, where $T_2 = T_1 + 10$. By use of Equation (2.44) and the above definition, the following equation is obtained:

$$Q_{10} = \exp\left[\frac{E_a(T_2 - T_1)}{R\,T_1\,T_2}\right] \tag{2.45}$$

The effect of substrate concentration on the value of Q_{10} can be shown by examining the specific substrate utilization rate (U) based on the Monod model evaluated at two different temperatures:

$$U = \frac{kS}{K_s + S} \tag{2.46}$$

$$Q_{10} = \frac{k_2(K_{s1} + S)}{k_1(K_{s2} + S)} \tag{2.47}$$

Empirical k vs. temperature relationships also exist in the following form:

$$k_T = k_{Tr} f_T^{(T-T_r)} \tag{2.48}$$

where k_T and k_{Tr} is the value of k at temperature T and at a reference temperature T_r (°C), respectively; f_T is the temperature coefficient (unitless). For a value of $f_T = 1.07$, $k_T/k_{Tr} = 2$, that is, the substrate utilization rate doubles for a 10°C increase in temperature.

The above k-temperature relationships can only describe the temperature positive effect up to the physiological optimum temperature for a particular process, but not the negative effect for temperature values above the optimum. Hinshelwood (1946) proposed a dual Arrhenius model by recognizing that there are two processes: a *synthetic* and a *degradative* process. Based on the Hinshelwood model, the following equation can be used to describe both the positive and negative effect of temperature on the maximum specific substrate utilization rate constant:

$$k = A_1 \exp\left(\frac{-E_{a1}}{RT}\right) - A_2 \exp\left(\frac{-E_{a2}}{RT}\right) \tag{2.49}$$

where $E_{a2} \gg E_{a1}$ (subscripts 1 and 2 refer to *synthetic* and *degradative* processes, respectively). This dual temperature effect can be explained by enzymatic activation (positive) and inactivation (negative) due to a temperature increase. In addition to the denaturation of proteins at relatively high temperatures, cell lysis has been observed to increase sharply with increasing temperature, especially when the substrate is exhausted (Allen 1950). Similarly to Equation (2.48), other empirical k-temperature relationships exist which describe both the positive and negative temperature effect (Pavlostathis and Giraldo-Gomez 1991).

2.5.8 Effect of pH

Regardless of the extracellular pH value in various biological processes, the intracellular pH is circumneutral. As with temperature, pH affects microbial activity, and thus the process rate, both positively and negatively. The optimum pH range for various processes varies and is related to the requisite microbial species. By analogy to an enzyme deactivation model (Bailey and Ollis 1986), the following expression can be used to depict the effect of pH (i.e., H^+ concentration) on the specific microbial growth rate:

$$\mu_{pH} = \frac{\mu_{pH=7}}{1 + \frac{[H^+]}{K_1} + \frac{K_2}{[H^+]}} \tag{2.50}$$

where μ_{pH} and $\mu_{pH=7}$ are the values of μ at a given pH value and at pH = 7, respectively (T^{-1}); $[H^+]$ is the proton concentration ($mol\,L^{-1}$); K_1 and K_2 are equilibrium constants for inactivation of requisite enzyme(s) by protonation and deprotonation, respectively ($mol\,L^{-1}$).

Several empirical pH-activity functions also exist. For example, in the case of nitrification, the pH factor (f_{pH}) is equal to 1 for $7.2 \leqslant pH \leqslant 9.0$ and for $pH \leqslant 7.2$, $f_{pH} = 1 - [0.833(7.2 - pH)]$.

2.5.9 Inhibition kinetics

Inhibition models used in biological processes can be divided into three categories (Pavlostathis and Giraldo-Gomez 1991): empirical, Monod-type with adjustable biokinetic constants, and inhibition coefficient models. The inhibition coefficient models are based on the Monod equation with the incorporation of an inhibition correction factor. In the case of reversible inhibition, three types of inhibition models have been proposed: competitive, uncompetitive, and non-competitive. The specific substrate utilization rate (U) can then be presented as follows:

Competitive:

$$U = \frac{kS}{K_s\left(1 + \frac{I}{K_i}\right) + S} \tag{2.51}$$

Uncompetitive:

$$U = \frac{kS}{K_s + S\left(1 + \frac{I}{K_i}\right)} \tag{2.52}$$

Non-competitive:

$$U = \frac{kS}{(K_s + S)\left(1 + \frac{I}{K_i}\right)} \tag{2.53}$$

where I = concentration of inhibitor; and K_i = inhibition coefficient (same concentration units as I). Competitive inhibition affects the value of K_s, that is, $(K_s)_{apparent} = K_s(1 + I/K_i)$. Uncompetitive inhibition affects both k and K_s, that is, $(k)_{apparent} = k/(1 + I/K_i)$ and $(K_s)_{apparent} = K_s/(1 + I/K_i)$. Non-competitive inhibition affects only k, that is, $(k)_{apparent} = k/(1 + I/K_i)$.

A special case of uncompetitive inhibition is the substrate inhibition where the inhibitor and the substrate are the same substance. In this case, Equation (2.52) for $I = S$ leads to the Haldane equation (Haldane 1930):

$$U = \frac{k}{1 + \frac{K_s}{S} + \frac{S}{K_i}} \tag{2.54}$$

As the substrate concentration increases, the value of U increases, but at the critical substrate concentration (S_c), U decreases with further increase of S. At $S = S_c$, $dU/dS = 0$, which based on Equation (2.54) leads to:

$$S = \sqrt{K_s K_i} \tag{2.55}$$

An extension of the Haldane inhibition model is the so-called generalized Haldane model for substrate uncompetitive inhibition:

$$U = \frac{k}{1 + \frac{K_s}{S} + \left(\frac{S}{K_i}\right)^n} \tag{2.56}$$

where n = constant (determines the order of inhibition). When $n = 1$, Equation (2.56) reduces to the Haldane equation and when $n = 0$, Equation (2.56) reduces to the Monod model (Equation (2.46)). Other special inhibition functions that have been used for the simulation of anaerobic biological processes can be found elsewhere (Pavlostathis and Giraldo-Gomez 1991; Batstone *et al.* 2002).

2.6 STOICHIOMETRY AND BIOENERGETICS

Stoichiometry relates quantities of consumed reactants to those of formed products. In the case of biochemical reactions, stoichiometry provides information about the amount of cell material formed, terminal electron acceptor used, and nutrients necessary for microbial growth. In other words, stoichiometry provides a mass balance for the system. Accurate stoichiometric equations describing microbial processes allow the calculation of all reactant and product concentrations based on the experimentally measured concentration of only one reactant or product. Nutrient limitations (e.g., nitrogen and phosphorus) in a system can easily be estimated based on stoichiometric equations. This is especially useful in the case of industrial wastewaters where nutrient imbalances are common. Nutrient oversupply or limitation may be the key to either undesirable biomass accumulation or lack of biomass growth. In addition, production or consumption of alkalinity and the resulting effect on the system pH can also be predicted based on stoichiometric calculations.

Along with the material balance, stoichiometry provides information about the energy changes and transformations, which occur during a reaction. Therefore, stoichiometry provides both a material and an energy balance for a system. Microorganisms obtain energy for growth and maintenance from the oxidation of organic and inorganic substances. Heterotrophic organisms use a portion of the organic material for energy production whereas the remaining portion is incorporated into biomass. On the other hand, autotrophic organisms oxidize inorganic substances and the energy released (i.e., ATP) is used to convert carbon dioxide to cellular organic constituents whereas another portion of the inorganic electron donor is used to generate the reducing power [i.e., NAD(P)H] required for the reduction of carbon dioxide. Thus, in both heterotrophic and autotrophic systems, a portion of the electron donor is used for energy production while the other is used for microbial synthesis.

McCarty (1972) presented a procedure based on thermodynamic and bioenergetic principles which allows the derivation of stoichiometric equations as well as the estimation of microbial yield coefficients and specific substrate utilization rates. A balanced and complete biological reaction (which includes microbial synthesis) can be presented as the sum of three half reactions (McCarty 1975; Rittmann and McCarty 2001):

$$R = f_e R_a + f_s R_c - R_d \tag{2.57}$$

where R_d, R_a and R_c represent reduction half reactions for one electron equivalent of electron donor, electron acceptor, and bacterial cells, respectively. Half reactions for selected, microbially mediated reactions can be found elsewhere (Rittmann and McCarty 2001). The fractions f_e and f_s represent the portion of the electron donor which is coupled with the electron acceptor (i.e., used for energy) and the portion of the electron donor which is coupled with cell formation (i.e., used for synthesis), respectively. By definition:

$$f_e + f_s = 1 \tag{2.58}$$

The values of f_s and f_e are a function of cell yield coefficient, microbial decay and solids retention time. For suspended-growth, continuous-flow systems, the following expression applies (McCarty 1975; Rittmann and McCarty 2001):

$$f_s = f_s^o \left(1 - \frac{f_d b\theta_c}{1 + b\theta_c}\right) \tag{2.59}$$

where $f_s^o = f_s$ without decay (i.e., $b = 0$) is the microbial growth yield coefficient (eeq cells formed/eeq electron donor used) (eeq: electron equivalent, and is the mass of a substance which releases 1 mol of electrons when completely oxidized); f_d = biodegradable fraction of an active microorganism (typically taken as equal to 0.80); b = specific microorganism decay coefficient (T^{-1}); and θ_c = solids retention time (T). Equation (2.59) can also be used for batch systems by replacing θ_c by actual time, as long as complete exhaustion of the electron donor (and therefore endogenous respiration) has not taken place. Based on Equation (2.59), as $\theta_c \rightarrow 0$, $f_s \rightarrow f_s^o$ and as $\theta_c \rightarrow \infty$, $f_s \rightarrow (1 - f_d)f_s^o$. The values of f_s^o can be estimated from previously reported experimental values of the microbial yield coefficient, from experimental measurements made for a particular system, or from thermodynamic considerations (Rittmann and McCarty 2001). By considering that 1 eeq of bacterial cells is equal to 1/20(113) = 5.65 g Volatile Suspended Solids (VSS) (the molecular weight of biomass is equal to 113 based on the empirical formula $C_5H_7O_2N$), the following relationship between the true microbial yield coefficient (Y) and f_s^o is obtained:

$$f_s^o = \frac{8Y}{5.65} \tag{2.60}$$

where Y = true microbial yield coefficient (g VSS/g COD used) (the coefficient 8 is the conversion of any electron donor to COD units, i.e., 8 g COD = 1 eeq of any electron donor). Therefore, using reported values of Y, the values of f_s^o can be calculated. In addition, experimentally determined microbial yield coefficients can be used to arrive at the f_s^o values. If values of the observed yield coefficients (Y_{obs}, g VSS/g COD used) are used, then the values of f_s can be calculated without the need for values of f_s^o, as follows:

$$f_s = 1.416\ Y_{obs} \tag{2.61}$$

The coefficient 1.416 is the conversion of bacterial mass from VSS to COD units based on the empirical formula $C_5H_7O_2N$.

The above-described procedure was used and complete stoichiometric equations as well as estimates of the true yield coefficient (Y) and the maximum substrate utilization rate (k) were obtained for a number of substrates typically encountered in anaerobic treatment processes (Pavlostathis and Giraldo-Gomez 1991), as well as for the autotrophic oxidation of thiosulfate and thiocyanate (Schreiber and Pavlostathis 1998; Hung and Pavlostathis 1999).

2.7 CONTINUOUS-FLOW SYSTEMS

Microbial growth and substrate utilization expressions can be incorporated into mass balances to yield equations that can be used to predict effluent microorganism and substrate concentrations, and thus process efficiency. Continuous-flow systems are grouped into two broad categories, suspended-growth and attached-growth processes, depending on whether the process microorganisms are maintained in suspension, or are attached to an inert medium (e.g., rocks, sand, granular activated carbon, or plastic materials). Attached-growth processes are also called fixed-film processes or biofilm processes.

2.7.1 Suspended-growth processes

Several reactor configurations have been used in suspended-growth biological treatment processes (Lawrence and McCarty 1970; Metcalf & Eddy 1991; Grady *et al.* 1999; Rittmann and McCarty 2001). The completely mixed stirred tank reactor (CSTR), without biomass recycle, will first be presented here as an illustrative example (Figure 2.4A). The influent comes into the reactor at a constant flow rate and is instantaneously and homogeneously mixed with the reactor contents. Withdrawal of reactor contents occurs at a rate equal to the influent rate, thus maintaining the reactor at a constant liquid volume. Due to the typically very low concentrations of active microorganisms in the influent stream, as compared to their concentration in the reactor, the influent biomass contribution in the microorganism mass balance will be ignored.

The mass balance for the net rate of microbial mass change in the system is as follows:

[Rate of biomass change] = [Net growth rate] − [Mass rate out]

$$V\frac{dX_a}{dt} = \left[Y\left(\frac{-dS}{dt}\right) - bX_a\right]V - QX_a \tag{2.62}$$

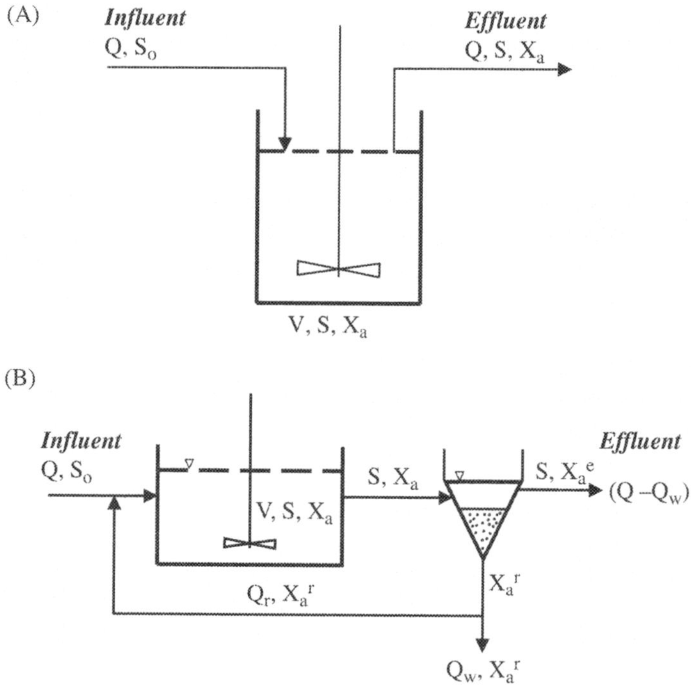

Figure 2.4 Schematic of a CSTR without recycle (A) and with recycle (B).

where, dX_a/dt = rate of change of microorganisms concentration in the reactor (M cells $L^{-3}T^{-1}$); V = reactor volume (L^3); Q = flow rate ($L^{-3}T^{-1}$); X_a = active (i.e., viable) microorganisms concentration in the reactor and effluent (M cells L^{-3}); Y = true microbial yield coefficient ($M_{cells}/M_{substrate}$); S = effluent (and reactor) substrate concentration (ML^{-3}); b = specific microorganisms decay coefficient (T^{-1}). At steady state, the reactor microorganisms concentration will reach a constant value (*i.e.*, $dX_a/dt = 0$). By applying the condition of steady state to Equation (2.62), we obtain:

$$\frac{Q}{V} = Y\frac{\left(-\frac{dS}{dt}\right)}{X_a} - b \tag{2.63}$$

But $V/Q = \theta$ = hydraulic retention time (T) and by using the definition of U (see Equation (2.16)), Equation (2.63) becomes:

$$\frac{1}{\theta} = YU - b \tag{2.64}$$

However, for this system, $\theta = \theta_c$ = solids retention time (or mean cell retention time, defined as the mass of microorganisms in the reactor divided by the mass of microorganisms removed per day) (T):

$$\frac{1}{\theta_c} = YU - b \tag{2.65}$$

A comparison of Equations (2.23) and (2.65) yields:

$$\frac{1}{\theta_c} = \mu_{net} \tag{2.66}$$

Equation (2.66) is a universal, inverse relationship between θ_c and μ_{net} for any biological system regardless of its configuration. This relationship is also the cornerstone of the design and operation of biological reactors. By choosing a value of θ_c, it fixes the net rate of microbial growth, that is, the net microbial growth for a particular system is controlled by the value of the mean cell retention time, within the physiological limits of μ_{net}.

By substituting Equation (2.21) (decay-corrected Monod model) into Equation (2.66) and solving for S the following expression is obtained:

$$S = \frac{K_s(1 + b\theta_c)}{\theta_c(Yk - b) - 1} \tag{2.67}$$

It should be noted that the effluent quality (i.e., S) is controlled by the value of θ_c (assuming the values of Y, k, K_s, and b are constants for a particular system). In addition, the value of S is independent of the influent substrate concentration (S_o) (based on the Monod model).

By use of the definition of U for the system:

$$U = \frac{S_o - S}{\theta_c X_a} \tag{2.68}$$

and substituting into Equation (2.65) the following expression is obtained:

$$X_a = \frac{Y(S_o - S)}{1 + b\,\theta_c} \tag{2.69}$$

It is noteworthy that X_a is a function of θ_c (assuming the values of Y, k, K_s, and b are constants for a particular system and the value of S is also fixed by θ_c). Therefore, both S and X_a are functions of θ_c alone, that is, the process efficiency (defined as the fraction of the influent substrate concentration removed) is controlled by θ_c alone.

Because of the positive effect of the substrate concentration on the specific microbial growth, the fastest microbial growth rate at which microorganisms can grow is when $S = S_o$, that is:

$$(\mu_{net})_{max} = \frac{YkS_o}{K_s + S_o} - b \tag{2.70}$$

However, based on the fundamental relationship between θ_c and μ_{net}, an upper limit on μ_{net} implies that there is a lower limit on θ_c:

$$\theta_c^{min} = \left[\frac{YkS_o}{K_s + S_o} - b \right]^{-1} \tag{2.71}$$

When $S_o \gg K_s$, the value of θ_c^{min} can be approximated by:

$$\theta_c^{min} = \frac{1}{Yk - b} = \frac{1}{\hat{\mu} - b} = \frac{1}{(\mu_{net})_{max}} \tag{2.72}$$

The above relationship implies that if the imposed θ_c value for a particular system is reduced below the θ_c^{min} value, the microorganisms will be removed from the system at a rate greater than they can possibly grow. If this condition is imposed, eventually the microorganisms will be washed out of the system and the process efficiency will drop to zero (Figure 2.5). The realization of a unique θ_c^{min} value for a particular microbial species/substrate system is of paramount importance for two main reasons. First, from the reactor design point of view, there is a cost and operational incentive to build the smallest possible reactor, which for a given influent flow rate means choosing the smallest possible θ_c value. However, there are several physical limitations which dictate a lower limit on the reactor volume (e.g., volumetric oxygen supply limit, settleability of biomass, etc.). Second, when the process microorganisms belong to several species or even microbial classes (i.e., mixed microbial populations), which is typically the case of biological treatment processes, a range of θ_c^{min} values exists. When the reactor θ_c value is gradually lowered, a gradual wash out of the slower-growing species will take place. Therefore, the choice of the reactor θ_c value will largely depend on the microbial species (or class), which has the largest θ_c^{min} value.

The above-presented relationships also apply to a CSTR with biomass recycle (see Figure 2.4B), except that the solids retention time expression becomes:

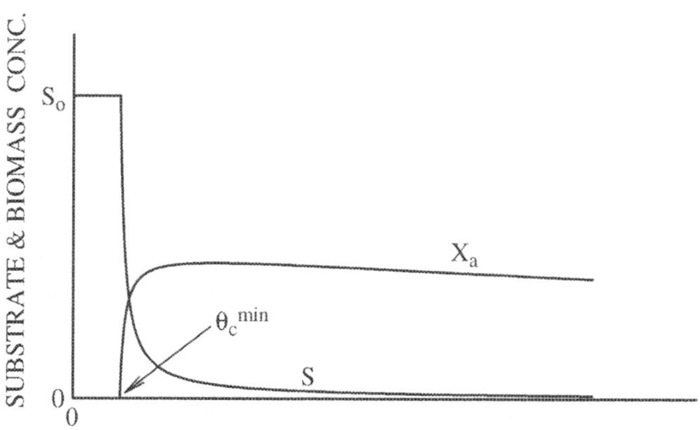

Figure 2.5 Effluent substrate (S) and active biomass (X_a) concentration as a function of solids retention time for a CSTR without recycle.

$$\theta_c = \frac{X_a V}{Q_w X_a^r + (Q - Q_w) X_a^e} \tag{2.73}$$

where Q_w = wastage rate ($L^{-3}T^{-1}$); X_a, X_a^r, and X_a^e = reactor, recycle, and effluent active biomass concentration ($M L^{-3}$).

Further details for CSTR configurations as well as plug-flow reactors can be found in several standard textbooks dealing extensively with the subject of biological treatment processes (e.g., Metcalf & Eddy 1991; 2003; Grady *et al.* 1999; Rittmann and McCarty 2001).

All of the above-presented relationships imply that the substrate is soluble or in the case of colloidal matter, the necessary hydrolysis is faster than the intrinsic substrate utilization rate. In the case of multistep processes (e.g., anaerobic digestion), the rate of hydrolysis determines, for a given retention time, the potential maximum substrate concentration possible, which in turn determines the maximum possible specific growth rate. For example, in anaerobic treatment processes, even in cases where acidogenesis or methanogenesis are considered to be the limiting steps, hydrolysis may affect the overall process kinetics, a point too often overlooked. More on the effect of the rate of hydrolysis on the minimum retention time of a biological process where the influent substrate is particulate, can be found elsewhere (Pavlostathis and Giraldo-Gomez 1991).

2.7.2 Attached-growth processes

In the case of dilute wastewater streams, high biomass concentrations cannot be achieved in suspended-growth, continuous-flow systems. Attachment and accumulation of biomass on support material and the development of biofilm leads to relatively high biomass concentrations and minimizes the potential for washout of the slowly growing microorganisms. Attached-growth processes have successfully been applied for the treatment of relatively high-strength wastewaters offering the advantage of biomass concentrations much higher than achieved in suspended-growth systems. In addition to the relatively long solids retention times achieved in biofilm systems, as compared to very short hydraulic retention times, biofilm systems offer the opportunity for spatial distribution of mixed microbial populations, thus further enhancing the efficiency of such systems and increasing their resilience against toxics.

Figure 2.6 shows a conceptual model typically used to represent a biofilm. S_b, S_s, S_f, and S_w are the substrate concentrations in the bulk liquid, at the outer biofilm surface, within the biofilm, and at the surface of attachment of the biofilm onto the support material, respectively. The biofilm has a thickness L_f and a microbial density X_f. The substrate passes through a liquid diffusion layer of thickness L before it reaches the outer biofilm surface and then further diffuses into the biofilm.

There are three processes which affect the substrate concentration within the biofilm:

(a) molecular diffusion from the bulk liquid through the diffusion layer described by Fick's first law (Equation (2.74));
(b) molecular diffusion through the biofilm layer described by Fick's second law (Equation (2.75));

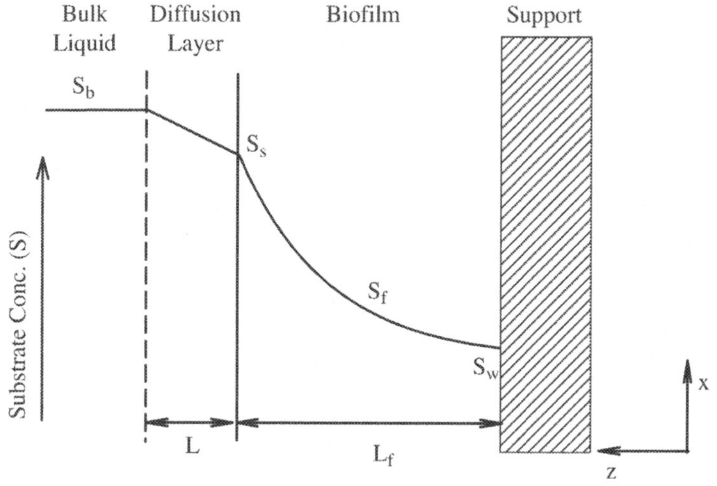

Figure 2.6 Schematic of a conceptual biofilm model showing the substrate concentration profile (*Source*: Adapted from Heath *et al.* 1990).

(c) substrate utilization within the biofilm described by a Monod relationship (Equation (2.76)).

$$J = -D\frac{\partial S}{\partial z} = \frac{-D}{L}(S_s - S_b) \tag{2.74}$$

$$\left(\frac{\partial S_f}{\partial t}\right)_{diff} = D_f \frac{\partial^2 S_f}{\partial z^2} \tag{2.75}$$

$$\left(\frac{\partial S_f}{\partial t}\right)_{rxn} = -\frac{kS_f X_f}{K_s + S_f} \tag{2.76}$$

where J = the substrate flux ($M\,T^{-1}$); D = the molecular diffusivity of the substrate in the liquid ($L^2\,T^{-1}$); L = diffusion layer (L); S_b, S_s, and S_f = substrate concentrations as previously described ($M\,L^{-3}$); D_f = molecular diffusivity within the biofilm ($L^2\,T^{-1}$); k = maximum specific substrate utilization rate (T^{-1}); X_f = biofilm mass density ($M\,L^{-3}$); and K_s = half-velocity coefficient ($M\,L^{-3}$).

Based on different assumptions and initial conditions made, several mathematical models have been presented in the literature (Rittmann and Huck 1989; Characklis *et al.* 1990 and citations within these references). Two distinct conditions exist, steady state and non-steady state. In the case of steady-state biofilms, growth of biofilm mass is balanced by biomass losses due to both maintenance-decay and detachment. Thus, the biofilm mass and thickness remain constant over time. During start-up as well as during long-term operation, the impact of shear losses from hydraulic loading on the system performance should be considered. Biomass detachment losses have a direct effect on the S_{min} value (Rittmann 1989):

$$S_{min} = \frac{K_s(b + b_{det})}{Yk - (b + b_{det})} \tag{2.77}$$

where b_{det} is the first-order specific detachment coefficient (T^{-1}). Therefore, increasing b_{det} makes S_{min} larger, that is, the lower limit to which a substrate concentration can be driven increases. The specific detachment coefficient can be expressed as a function of the shear stress and several correlations have been proposed in the literature (Rittmann 1989). Values of b_{det} exceeding those of the specific decay coefficients can be reached under very high shear stress conditions.

Three dependent parameters achieve unique values for a given biofilm (i.e., given a microbial population, a rate-limiting substrate and a set of reactor conditions) under steady-state conditions: substrate concentration (S), substrate flux (J), and biofilm thickness (L_f). Pseudo-analytical solutions for estimating J and L_f as a function of S have been developed (Rittmann and McCarty 1980; Rittmann 1982; Sáez and Rittmann 1988). In order for a biofilm to be sustained under steady-state conditions, the substrate concentration should be equal or greater than the value of S_{min}, as defined by Equation (2.77). Conditions, which do not meet the definition of a steady-state biofilm, lead to non-steady-state biofilms.

Examples of non-steady-state biofilms are cases where the biofilm is undergoing transient growth or loss during start-up, loading change or other perturbations, as well as when the substance being monitored is not the growth-rate-limiting substrate (e.g., secondary utilization and degradation of trace organic compounds in excess of a primary substrate). In the case of non-steady-state biofilms, L_f is an independent parameter and is not coupled to S and J. For a known L_f value, J can be predicted from S by the use of pseudo-analytical solutions (Rittmann 1982; Rittmann and McCarty 1981).

By combining Equations (2.75) and (2.76), and assuming steady state (i.e., $\partial S_f/\partial t = 0$) the following equation is obtained:

$$D_f \frac{\partial^2 S_f}{\partial z^2} = \frac{kS_f X_f}{K_s + S_f} \tag{2.78}$$

The boundary conditions for the above equation are:

$$\frac{\partial S_f}{\partial z} = 0 \ @\ z = 0 \tag{2.79}$$

$$S_f = S_s \ @\ z = L_f \tag{2.80}$$

In addition, under steady state, the rate of new cell growth per unit surface area (YJ) is balanced by the rate of biomass loss from the biofilm, which leads to the definition of a steady-state biofilm thickness (Heath *et al.* 1990):

$$L_f = \frac{JY}{b' X_f} \tag{2.81}$$

where Y = true yield coefficient ($\mathrm{M\,M^{-1}}$); and b' = total specific biofilm mass loss rate ($\mathrm{T^{-1}}$). The system of Equations (2.74), and (2.78)–(2.81) can be solved using numerical techniques. However, such solutions are not useful for design applications. Several pseudo-analytical solutions have been developed where a set of algebraic equations are used to approximate the numerical solutions (Rittmann and McCarty 1980; Rittmann 1982; Suidan and Wang 1985; Sáez and Rittmann 1988; Heath *et al.* 1990). Examples of using such pseudo-analytical solutions for the design and analysis of biofilm processes have been presented (Rittmann 1990; Rittmann and McCarty 2001). Values of S, S_{min}, and J for typical operating conditions of biofilm processes can be estimated and used to interpret loading conditions, process design and performance (Rittmann and McCarty 2001).

2.8 MASS-TRANSFER EFFECTS AND MIXING

An extensive coverage of the effect of mass transfer on the kinetics of substrate utilization, especially in anaerobic processes, has previously been presented (Pavlostathis and Giraldo-Gomez 1991). The problem of mass-transfer effects on the observed kinetics of substrate utilization has been considered in detail by several authors and the interested reader is referred to the literature (e.g., Horvath and Engasser 1974; Ngian *et al.* 1977; LaMotta and Shieh 1979). A brief discussion of the effects of mass transfer on the observed substrate utilization rates is presented below.

Based on the definition of an effectiveness factor (η), which is the ratio of the observed substrate utilization rate (U_o) to the intrinsic substrate utilization in the absence of any mass-transfer limitations (U_i), the observed, mass-transfer-limited substrate utilization rate based on the Monod model can be expressed as follows:

$$U_o = \eta \mathrm{U}_i = \eta \frac{kS_l}{K_s + S_l} \tag{2.82}$$

where S_l = substrate concentration in the bulk, liquid phase ($\mathrm{M\,L^{-3}}$). All other symbols are as previously defined. Equation (2.82) is linearized and usually used to obtain a double reciprocal plot according to the following equation:

$$\frac{k}{U_o} = \frac{K_s}{\eta}\left(\frac{1}{S_l}\right) + \frac{1}{\eta} \tag{2.83}$$

An apparent half-saturation constant can then be defined, which relates to the intrinsic one as follows:

$$(K_s)_{\mathrm{app.}} = \frac{K_s}{\eta} \tag{2.84}$$

Thus, the value of the apparent half-saturation constant would increase as the mass-transfer limitations become more severe. Contois (1959) observed that the growth rate of continuous cultures was a function of the population density and arrived at the dependence of the half-saturation constant on population density

(Equation (2.42)). One might now suspect the existence of mass-transfer limitations during these types of experiments. From a practical perspective, Contois kinetics provide a simple alternative means to model the effect of mass-transfer in microbial aggregates. It is noteworthy that, for a continuous-flow system, the Contois model predicts a dependence of the effluent substrate concentration on the influent substrate concentration (Pavlostathis and Giraldo-Gomez 1991), a result typical of mass-transfer-limited reactors.

Another approximation of the kinetics of transfer-limited biological reactions is that of the half-order kinetics (Harremoes 1978). It can be theoretically shown that zero-order reactions turn into half-order reactions when mass-transfer limitations become significant. When the Thiele modulus for the biofilm is greater than 4 and the dimensionless substrate concentration (S/K_s) is greater than 5, the half-order approximation for biofilm kinetics is considered satisfactory (Pavlostathis and Giraldo-Gomez 1991).

Mixing is of paramount importance in order to improve mass transfer between the substrate(s) and microorganisms, especially in case of microbial aggregation and/or biofilm formation. In addition to the mass-transfer considerations, mixing is also important in order to achieve uniform conditions throughout the reactor volume and avoid localized conditions. In many aerobic, suspended-growth bioreactors, mixing is achieved via aeration, whereas in expanded/fluidized reactors, mixing is achieved by a relatively high recirculation rate. Mixing is particularly important for high-rate anaerobic digesters, where in addition to overcoming mass-transfer limitations, sufficient mixing prevents scum formation and solids settling. Failure to prevent such undesirable conditions leads to localized conditions of high concentration of volatile fatty acids and ammonia, lower pH, and poor heating, which can all have a detrimental effect on the digester's performance.

REFERENCES

Allen, M.B. (1950) The dynamic nature of thermophily. *J. Gen. Physiol.* **33**, 205–214.

Bailey, B.J. and Ollis, D.F. (1986) *Biochemical Engineering Fundamentals*, 2nd edn, McGraw-Hill, New York, NY, USA.

Batstone, D.J., Keller, J., Angelidaki, R.I., Kalyuzhnyi, S.V., Pavlostathis, S.G., Rozzi, A., Sanders, W.T.M., Siegrist, H. and Vavilin, V.A. (2002) *Anaerobic Digestion Model No. 1*. 2002. IWA Task Group for Mathematical Modelling of Anaerobic Wastewater Processes, Scientific and Technical Report No. 13, IWA Publishing, London, UK.

Brock, T.D., Madigan, M.T., Martinko, J.M. and Parker, J. (1994) *Biology of Microorganisms*, 7th ed., Prentice Hall, Englewood Cliffs, NJ, USA.

Characklis, W.G. (1990) Kinetics of microbial transformations. In: *Biofilms* (eds. W.G. Characklis and K.C. Marshall). John Wiley & Sons, New York, NY, USA.

Characklis, W.G., Turakhia, M.H. and Zelver, N. (1990) Transport and interfacial transfer phenomena. In *Biofilms* (eds. W.G. Characklis and K.C. Marshall). John Wiley & Sons, New York, NY, USA.

Contois, D.E. (1959) Kinetics of bacterial growth – relationship between population density and specific growth rate of continuous cultures. *J. Gen. Microbiol.* **21**, 40–50.

Grady, C.P.L., Daigger, G. and Lim, H.C. (1999). *Biological Wastewater Treatment*, 2nd edn, Marcel Dekker, New York, NY.

Haldane, J.B.S. (1930) *Enzymes*, Longmans, London, UK.

Harremoes, P. (1978) Biofilm kinetics. In: *Water Pollution Control Microbiology* (ed. R. Mitchell). Wiley-Interscience, New York, NY, USA.

Heath, M.S., Wirtel, S.A. and Rittmann, B.E. (1990) Simplified design of biofilm processes using normalized loading curves. *Res. J. WPCF* **62**, 185–192.

Hinshelwood, C.N. (1946) *The Chemical Kinetics of the Bacterial Cell*, Oxford University Press, London, UK.

Horvath, C. and Engasser, J.M. (1974) External and internal diffusion in heterogeneous enzyme systems. *Biotechnol. Bioeng.* **16**, 909–923.

Hung, C-H. and Pavlostathis, S.G. (1999) Kinetics and modeling of autotrophic thiocyanate biodegradation. *Biotechnol. Bioeng.* **62**, 1–11.

LaMotta, E. and Shieh, W.K. (1979) Diffusion and reaction in biological nitrification. *J. Environ. Eng. Div. ASCE* **105**, 655–673.

Lawrence, A.W. and McCarty, P.L. (1970) Unified basis for biological treatment design and operation. *J. Sanit. Engin. ASCE* **96**, 757–778.

McCarty, P.L. (1969) Energetics and bacterial growth. *Proceedings of Fifth Rudolf Research Conference*, Rutgers, The State University, New Brunswick, NJ, USA.

McCarty, P.L. (1972) Energetics of organic matter degradation. In: (*Water Pollution Microbiology* (ed. R. Mitchell). Wiley-Interscience, New York, NY, USA.

McCarty, P.L. (1975) Stoichiometry of biological reactions. *Progr. Water Technol.* **7**, 157–172.

Metcalf & Eddy, Inc. (1991) *Wastewater Engineering – Treatment, Disposal, Reuse*, 3rd edn, McGraw-Hill, New York, NY, USA.

Metcalf & Eddy, Inc. (2003) *Wastewater Engineering – Treatment and Reuse*, 4th edn, McGraw-Hill, New York, NY, USA.

Monod, J. (1942) *Recherches sur la Croissance des Cultures Bacteriennes*, Herman et Cie., Paris, France.

Monod, J. (1949) The growth of bacterial cultures. *Ann. Rev. Microbiol.* **3**, 371–394.

Ngian, K.F., Lin, S.H. and Martin, W.R.B. (1977) Effect of mass transfer resistance on the Lineweaver-Burk plots for flocculating microorganisms. *Biotechnol. Bioeng.* **19**, 1773–1784.

Pavlostathis, S.G. and Giraldo-Gomez, E. (1991) Kinetics of anaerobic treatment: a critical review. *CRC Crit. Rev. Environ. Control* **21**, 411–490.

Rittmann, B.E. (1982) Comparative performance of biofilm reactor types. *Biotechnol. Bioeng.* **24**, 1341–1370.

Rittmann, B.E. (1989) Detachment from biofilms. In: *Structure and Function of Biofilms* (eds. W.G. Characklis and P.A. Wilderer). John Wiley & Sons, Inc., New York, NY, USA.

Rittmann, B.E. (1990) Analyzing biofilm processes used in biological filtration. *J. AWWA* **82**, 62–66.

Rittmann, B.E., and Huck, P.M. (1989) Biological treatment of public water supplies. *Crit. Rev. Environ. Control* **19**, 119–184.

Rittmann, B.E. and McCarty, P.L. (1980) Model of steady-state biofilm kinetics. *Biotechnol. Bioeng.* **22**, 2343–2357.

Rittmann, B.E. and McCarty, P.L. (1981) Substrate flux into biofilms of any thickness. *J. Environ. Eng. ASCE* **107**, 831–849.

Rittmann, B.E. and McCarty, P.L. (2001). *Environmental Biotechnology: Principles and Applications*. McGraw-Hill, New York, NY, USA.

Sáez, P.B. and Rittmann, B.E. (1988) Improved pseudo-analytical solution for steady-state biofilm kinetics. *Biotechnol. Bioeng.* **32**, 379–385.

Schreiber, D. and Pavlostathis, S.G. (1998). Biological oxidation of thiosulfate. *Water Res.* **32**, 1363–1372.

Stumm, W. and Morgan, J.J. (1996) *Aquatic Chemistry – Chemical Equilibria and Rates in Natural Waters*, 3rd edn, John Wiley & Sons, New York, NY, USA.

Suidan, M.T. and Wang, Y.T. (1985) Unified analysis of biofilm kinetics. *J. Environ. Eng.*, **111**, 634–646.

3

Principles of process design for industrial wastewater treatment systems

A.C. van Haandel, O. Monroy, B. Celis, E. Rustrian and F.J. Cervantes

3.1 INTRODUCTION

The first attempt to improve the declining surface water quality due to discharge of wastewaters was to separate the settleable solids. However, this primary treatment soon proved to be inadequate to cause a significant improvement and additional treatment was introduced to remove also the non-settleable organic material by applying biological methods. Initially both anaerobic and aerobic biological treatment methods were used, but gradually aerobic systems prevailed over anaerobic facilities.

By the second half of last century, it became clear that biological treatment (secondary treatment) to remove organic material by its own was insufficient to restore the quality of surface waters that had been deteriorated by excessive discharge of wastewaters. The poor quality that was observed in surface water bodies, even when effluents with a very low concentration of biodegradable organic material were

 Advanced Biological Treatment Processes for Industrial Wastewaters edited by F.J. Cervantes, S.G. Pavlostathis and A.C. van Haandel. ISBN: 1843391147. Published by IWA Publishing, London, UK.

discharged, was attributed to eutrophication: the excessive growth of aquatic life due to nutrients present in the effluents. Thus, a third treatment step was added to wastewater treatment systems, aimed at removing also these nutrients.

Initially these tertiary treatment systems were specifically designed to remove nitrogenous material by the sequential biological processes of nitrification and denitrification, which led to the conversion of ammonium nitrogen (predominant in most wastewaters) via nitrate to nitrogen gas. However, it became clear that in many cases the growth limiting factor of the aquatic life was phosphate, so that biological treatment systems were developed to remove also this nutrient. The method for biological phosphate removal is to create favourable conditions for the development of bacteria that have a very high percentage of phosphorus in their cellular mass and therefore can absorb high quantities of phosphate from the wastewater.

More recently another important development in biological wastewater treatment systems was the removal of sulphate by sequential processes of reduction to sulphide and partial oxidation to elemental sulphur. This biological treatment can be applied to improve the effluent quality of treated wastewaters for reuse, but its applicability can be extended to eliminate sulphur components from contaminated gases and soils as well as the elimination of the element from raw materials such as oil and minerals.

In this chapter basic considerations are discussed for the process design of biological wastewater treatment plants to remove not only organic material by anaerobic or aerobic methods, but also the nutrients nitrogen, phosphorus and sulphur.

3.2 AEROBIC CARBON-OXIDIZING PROCESSES

3.2.1 Introduction

Carbon-oxidizing processes can be divided basically between suspended growth and fixed-film processes. In practice, the former are much more widely applied than the latter and for that reason will be considered in more detail. The reason that suspended growth systems are preferred is that it is easy to control the sludge age in the system which is the fundamental operational parameter that controls the behaviour of biological systems. Suspended growth systems comprise the activated sludge process and its variants such as the oxidation ditch and the aerobic lagoon. These systems may be operated under continuous flow conditions or in sequential batch mode.

Suspended growth aerobic systems have been investigated most intensely and the developed models allow engineers to make a rational design of all aspects of aerobic treatment plants and the mass flows that pass through it. The basis for a rational description of the activated sludge process was given by Marais and Ekama (1976), which formed the basis for Activated Sludge Models 1 issued by the International Water Association (Henze *et al.* 1986).

The Marais–Ekama model insofar as the removal of organic material is represented schematically in Figure 3.1. It represents the simplified model, in which it is assumed that the removal of organic material is substantially complete. This represents the ideal condition of treatment, which can be closely approached in

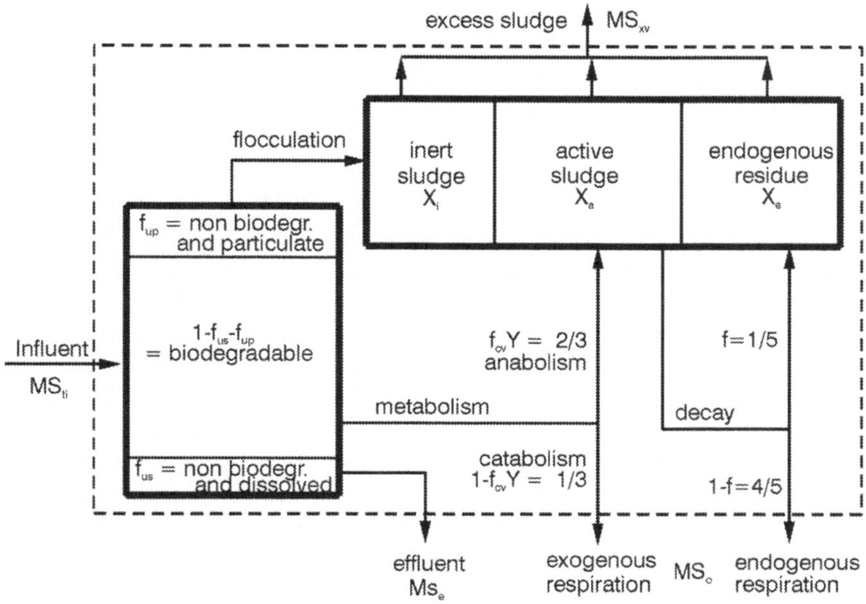

Figure 3.1 Schematic representation of the simplified model of aerobic suspended growth systems.

practice depending on the operational conditions, especially the sludge age. The simplified model that can be used to develop expressions for the activated sludge system is discussed below. The general model, in which the kinetics of organic material utilization is considered, is discussed later in this chapter.

In Figure 3.1, the processes that form the basis of the simplified model for the activated sludge process are represented. When a wastewater containing organic material is placed in contact with an activated sludge mass, the following processes will occur in an aerobic environment:

(a) *Metabolism*: The biodegradable organic material in the influent is removed from the liquid phase and metabolized by the sludge. It was seen in Section 2.2.2 that this process leads to both sludge growth (anabolism) and oxygen consumption (catabolism).
(b) *Decay*: It is postulated that sludge decay is independent of metabolic processes and that part of the decayed active sludge is oxidized to inorganic compounds, whereas the remainder accumulates in the reactor as endogenous residue, until it is discharged as excess sludge. The oxygen consumption due to oxidation of decayed biomass is called endogenous respiration, to distinguish it from the oxidation of influent organic material, which is called exogenous respiration.
(c) *Bioflocculation*: The particulate and non-biodegradable material in the influent is not affected by the metabolic activity of the sludge but it is removed physically from the liquid phase by flocculation. The flocculated material constitutes the inert sludge fraction.

In the model of Figure 3.1, the biodegradable fractions and the non-biodegradable and particulate fraction are removed from the liquid phase, but the fourth fraction, with non-biodegradable and dissolved material is not affected in any way by the activated sludge process and is discharged without modifications into the effluent.

3.2.2 The simplified activated sludge model

The simplified or ideal activated sludge model is characterized by three basic assumptions: (1) the system operates under constant flow and load conditions in a single completely mixed reactor with constant excess sludge discharge, (2) the removal of the biodegradable organic material is substantially complete and (3) the sludge accumulation in the settler is negligible in comparison with the sludge mass in the biological reactor.

In the previous section, it was shown that the organic influent material in activated sludge systems is divided into three fractions (1) discharged in the effluent, (2) discharged as excess sludge and (3) oxidized by oxygen. By equating the flux of influent organic material to the fluxes of the three fractions it is possible to perform a mass balance of organic material. The mass balance does not depend on any theory; it is simply a form of mass continuity.

While the fact that a closing mass balance for organic material is important to assess the reliability of the experimental data, it can only be carried out after the system has been constructed and put into operation. For design engineers it is much more important to be able to predict the values of these fractions before the construction of the plant, because the fractions are indicative for the most important aspects of activated sludge behaviour: (1) the flux of organic material in the effluent determines the effluent quality, (2) the flux of organic material in the excess sludge determines the sludge production and hence the size of the installations for sludge processing and (3) the flux of oxidized organic material is per definition equal to the flux of oxygen required for the oxidation and hence determines the aeration capacity that must be installed in the system.

The essence of the simplified activated sludge model is exactly to develop expression that can be used to calculate the three fractions of organic material. In order to develop the simplified model some assumptions must be made that are described below.

3.2.2.1 Composition of the influent organic material

(1) Organic influent material (expressed as chemical oxygen demand, COD) is considered to be composed of biodegradable and non-biodegradable material, each of which can be subdivided into a dissolved and a particulate fraction.
(2) The non-biodegradable and dissolved COD fraction of the influent is not affected by physical or biological action in the activated sludge process and leaves the activated sludge system together with the effluent.
(3) The non-biodegradable and particulate COD fraction is not affected by the biological action of the sludge, but it is flocculated and becomes part of the sludge, forming the inert sludge fraction.

(4) The biodegradable organic material can be metabolized and if the operational conditions are adequate, the effluent COD is composed essentially of non-biodegradable and dissolved COD.

Since the division of the influent organic material in the four main fractions will be used often in this text, it is convenient that each be indicated by a different symbol. Using S (substrate) as a generic symbol for organic material concentration (expressed as COD), the following parameters can be defined:

S_{ti} = (total) influent COD concentration
S_{te} = biodegradable influent COD concentration
S_{ui} = non-biodegradable influent COD concentration
S_{bi} = biodegradable and particulate influent COD concentration
S_{bsi} = biodegradable and dissolved influent COD concentration
S_{bpi} = biodegradable and particulate influent COD concentration
S_{upi} = non-biodegradable and particulate influent COD concentration
S_{usi} = non-biodegradable and dissolved influent COD concentration.

From the above definitions:

$$S_{ti} = S_{bi} + S_{ui} = S_{bsi} + S_{bpi} + S_{upi} + S_{usi} \tag{3.1}$$

It will prove to be convenient to introduce the following fractions:

$f_{us} = S_{usi}/S_{ti}$ = non-biodegradable and dissolved influent COD fraction
$f_{up} = S_{upi}/S_{ti}$ = non-biodegradable and particulate influent COD fraction
$f_{bs} = S_{bsi}/S_{bi}$ = soluble part of the influent biodegradable COD fraction.

Figure 3.2 is a graphical representation of Equation (3.1). The methods to determine the different fractions experimentally will be discussed in this chapter. The numerical values of the fractions vary significantly for different wastewaters, especially in the case of industrial wastes. Table 3.1 shows some examples. It goes without saying that differences in the composition of the organic material will lead to differences in the behaviour of the activated sludge system.

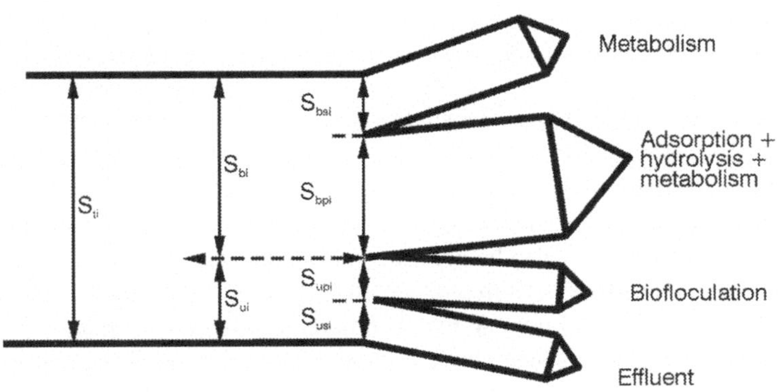

Figure 3.2 Representation of the four fraction of organic material in the influent of wastewaters.

Table 3.1 Values of the fractions of organic material for different wastewaters.

Wastewater	f_{us}	f_{up}	f_{bs}	Reference
Raw municipal sewage				
Campina Grande (Brazil)	0.12	0.06	0.25	Coura Dias *et al.* (1983)
Cape town (RSA)	0.10	0.12	0.25	Marais and Ekama (1976)
Burlington (Canada)	0.12	0.25	–	Sutton *et al.* (1979)
Industrial wastes				
Vinasse	0.02	0.02	0.80	Van Haandel (2004)
Black liquor (paper mill)	0.40	0.10	0.35	Macedo (1990)
Coal washing (steel mill)	0.10	0.01	0.47	Dombroski (2003)

The division of the influent organic material into four fractions is a simplification of a more complex reality, but it will be adopted, since a more refined model does not lead to better simulations and thus becomes an unnecessary sophistication for the purpose of developing a general description of the activated sludge behaviour.

3.2.2.2 Sludge composition

The suspended solids or sludge in the activated sludge process is composed of organic (volatile) and inorganic (fixed) material. The concentrations of total (TSS), fixed (FSS) and volatile suspended solids (VSS) are determined gravimetrically. The minimum fraction of FSS in biological sludge is about 80%, but due to flocculation of influent inorganic solids (silt, clay) it may be higher.

In order to describe the activated sludge behaviour, Marais and Ekama (1976) suggested a subdivision of the VSS into two basic fractions: (1) active sludge, composed of the live microorganisms that act in the metabolism of the influent organic material and (2) inactive sludge composed of organic material that does not exhibit metabolic activity. This division is theoretical and there is no test to determine directly the active or inactive sludge concentration: only the sum of the two can be determined experimentally. The division is justified by the fact that it leads to a rational model of the activated sludge processes, capable of predicting its measurable parameters under strongly varying operational conditions.

(a) The active sludge: The active sludge is generated from synthesis of influent organic material. The microorganisms in the activated sludge process are composed of a very large number of species of bacteria, fungi and protozoa. Depending on the operational conditions, more complex organisms like ciliates and rotifers may also be present. The composition of the live organisms may differ considerably from one system to the other, depending on the nature of the influent wastewater and the operational conditions. In spite of the complex nature of live organism mass, for the purpose of modelling, in this text it will be considered as an equivalent bacterial suspension. The validity of this assumption is tested by comparing the predictions generated by the model with the experimentally observed results. It must be stressed that although bacteria are predominant in the active sludge, its actual behaviour may be very different from a pure culture of bacteria.

(b) The inactive sludge: The inactive sludge is composed of non-biodegradable organic material and can be subdivided into two fractions in accordance with its origin: (a) the inert sludge and (b) the endogenous residue. The inert sludge fraction is generated from the non-biodegradable particulate influent material. This material is flocculated and becomes part of the solid phase, forming the inert fraction. The endogenous residue has its origin in the decay of live cells, which occurs in the activated sludge process.

Having defined the different sludge fractions, it is convenient to introduce symbols for each. Using X to indicate generically sludge concentration:

X_a = active sludge concentration (mg VSS/L)
X_e = endogenous sludge concentration (mg VSS/L)
X_i = inert sludge concentration (mg VSS/L)
X_v = organic or volatile sludge concentration (mg VSS/L)
X_m = mineral, fixed or inorganic sludge concentration (mg FSS/L)
X_t = total sludge concentration (mg TSS/L).

From the definitions it follows that:

$$X_t = X_v + X_m = X_v/f_v \tag{3.2}$$

and

$$X_v = X_a + X_e + X_i \tag{3.3}$$

The numeric value of the volatile sludge fraction, f_v, depends on the origin of the wastewater and is typically in the range of 0.65–0.80.

Along with the three organic sludge fractions mentioned above, another two may exist, depending on the operational conditions. If the sludge age is very short, the sludge wastage rate may be so high that there is not enough time for metabolism of all the influent biodegradable material. In that case flocculation of influent biodegradable and particulate material will take place and this material will be adsorbed (stored) on the active sludge mass. Thus it is possible that part of the discharged organic sludge is actually flocculated influent organic material. The stored material fraction depends on the metabolism rate, the sludge age and, of course, on the composition of the influent organic material. Removal of organic material from wastewaters by flocculation may be significant, especially in regions with a cold climate.

If nitrification takes place in the activated sludge process, a population of nitrifying bacteria (*Nitrobacter* and *Nitrosomonas*). In most cases, the mass of nitrifying bacteria is insignificant compared to the total organic sludge mass due to the relatively low ammonium influent concentration and its low yield coefficient (see Section 3.4.2).

Having defined the composition of the influent organic material and the sludge fractions, the simplified model can now be developed. The objective of the activated sludge system is to remove the influent organic material as completely as possible from the liquid phase at minimum costs. If the removal of the biodegradable material is complete, the proportion between the organic material in excess sludge and oxidized organic material depends on the extent of the decay of active sludge in the reactor, which in turn depends on the sludge age: the rate at which the sludge is discharged from the system. At a short sludge age there is little decay, so that the endogenous respiration does not consume much oxygen, but the wasted sludge represents a high fraction of the influent organic material and has a high fraction of active (putrescible) sludge. At long sludge ages, there is less excess sludge but now endogenous respiration requires more oxygen. Also the low rate of wastage leads to a large sludge mass, so that a large reactor is required.

Insofar as the COD concentration in the liquid phase of the reactor and the effluent is concerned, it can be equated to the influent COD fraction that is dissolved and non-biodegradable, that is:

$$S_{te} = f_{us} S_{ti} \tag{3.4}$$

This means that the fraction of influent organic material that is discharged together with the effluent is given by:

$$\mathrm{mS_{te}} = S_{te}/S_{ti} = f_{us} \tag{3.5}$$

The inert fraction of the volatile sludge can be calculated by considering that the non-biodegradable and particulate influent COD fraction will be flocculated and discharged (wasted) as inert sludge, so that:

$$Q_i f_{up} S_{ti} = \mathrm{f_{cv}} q X_i$$

or $$X_i = (f_{up}/f_{cv})(Q_i/q)S_{ti} = (f_{up}/f_{cv})(\theta/\theta_c)S_{ti} \tag{3.6}$$

where:

Q_i = influent flow rate
q = flow rate of wasted sludge (from the reactor)
θ = hydraulic or liquid retention time
θ_c = sludge age or solids retention time
X_i = inert sludge concentration in the reactor
f_{cv} = COD/VSS ratio of volatile sludge.

The active sludge concentration is calculated by considering that its growth rate is equal to the sum of the decay rate and the rate of discharge of excess sludge:

$$\mathrm{d}X_a/\mathrm{d}t = 0 = (\mathrm{d}X_a/\mathrm{d}t)_c + (\mathrm{d}X_a/\mathrm{d}t)_d + (\mathrm{d}X_a/\mathrm{d}t)_e \tag{3.7}$$

indices c, d and e refer to growth, decay and wastage, respectively.

In Section 2.5.3, it was shown that the growth rate of active sludge is proportional to the rate of biodegradable substrate utilization, the proportionality constant being Y, the yield coefficient.

$$r_c = (\mathrm{d}X_a/\mathrm{d}t)_c = Yr_u = Y(1 - f_{us} - f_{up})S_{ti}Q_i/V_r = Y(1 - f_{us} - f_{up})S_{ti}/\theta \tag{3.8}$$

In Section 2.5.2 the decay rate was given as:

$$r_d = (\mathrm{d}X_a/\mathrm{d}t)_d = -b_h X_a \tag{3.9}$$

where:
r_d = rate of change of the active sludge concentration due to decay
b_h = decay constant for active sludge (heterotrophs).

The rate of change of the active sludge concentration due to discharge of excess sludge can be expressed by the definition of the sludge age:

$$\theta_c = \text{(active sludge mass)/(rate of active wastage)} = V_r X_a/[V_r(-dX_a/dt)_e]$$

hence:

$$r_e = (dX_a/dt)_e = -X_a/\theta_c \qquad (3.10)$$

By substituting Equations (3.8, 3.9 and 3.10) in Equation (3.7) the following expression is obtained for the active sludge concentration:

$$YS_{bi}/R_h - b_h X_a - X_a/\theta_c = 0$$

or

$$X_a = (1 - f_{us} - f_{us})[Y\theta_c/(1 + b_h\,\theta_c)]S_{ti}/\theta = (1 - f_{us} - f_{up})C_r S_{ti}/\theta \qquad (3.11)$$

where:

$$C_r = Y\theta_c/(1 + b_h\theta_c) \qquad (3.12)$$

C_r is the inverse of the specific utilization rate U of organic material by the active sludge described in Section 2.7.1.

The concentration of the endogenous residue is calculated by considering that its production rate is equal to the rate of wastage.

$$(dX_e/dt) = 0 = (dX_e/dt)_d + (dX_e/dt)_e \qquad (3.13)$$

When active sludge decays a fraction "f" remains as endogenous residue, so that the production rate is given by:

$$(dX_e/dt)_d = -f(dX_a/dt)_d = fb_h X_a \qquad (3.14)$$

where:
f = fraction of non-biodegradable solids remaining after active sludge decay.

Since the rate of sludge wastage can be expressed as $(dX_e/dt)_e = -X_e/\theta_c$ (see above for active sludge), the endogenous sludge concentration is given as:

$$X_e = fb_h\theta_c X_a \qquad (3.15)$$

Now the volatile sludge concentration can be expressed as the sum of its three components:

$$X_v = X_a + X_e + X_i = [(1 - f_{us} - f_{up})\,C_r(1 + fb_h\theta_c) + f_{up}\,\theta_c/f_{cv}]S_{ti}/\theta \qquad (3.16)$$

The expression for the volatile sludge concentration is particularly important because its concentration can also be determined experimentally, so that it offers an opportunity to verify the validity of the model. With the aid of the volatile sludge concentration the mass of wasted volatile sludge can be calculated:

$$ME_v = V_r \cdot X_v/\theta_c = [(1 - f_{us} - f_{up})(1 + fb_h\,\theta_c)C_r/\theta_c + f_{up}/f_{cv}]MS_{ti} \quad (3.17)$$

where:
ME_v = daily produced volatile sludge mass (g VSS/day)
$MS_{ti} = Q_iS_{ti}$ = daily applied COD mass (g COD/day).

Knowing that there is a proportionality constant f_{cv} between the COD and mass of volatile sludge, the fraction of influent COD that is transformed into sludge and wasted as such is given as:

$$\begin{aligned} mS_{xv} = ME_v/MS_{ti} &= f_{cv} \cdot (V_rX_v/\theta_c)/(Q_a \cdot S_{ti}) \\ &= f_{cv}(1 - f_{us} - f_{up})(1 + fb_h\theta_c)C_r/\theta_c + f_{up} \end{aligned} \quad (3.18)$$

In order to calculate the oxidized fraction of influent organic material, expressions for the oxygen uptake rate (OUR) for exogenous and endogenous respiration must be derived. As for the exogenous OUR, it has been observed experimentally that a constant fraction $(1 - f_{cv}Y) \approx 1/3$ of the biodegradable organic material is oxidized during metabolism, independent of the operational conditions. Knowing that per definition 1 kg of oxygen is required to oxidize 1 kg of COD one has:

$$OUR_{ex} = (1 - f_{cv}Y)r_u = (1 - f_{cv}Y)\,(1 - f_{us} - f_{up})\,S_{ti}/\theta \quad (3.19)$$

where:
OUR_{ex} = oxygen uptake rate for exogenous respiration.

The OUR for endogenous respiration is calculated as follows: the oxidation rate of decayed active sludge is the difference between the rate of active sludge decay and the rate of endogenous residue production:

$$r_o = (dX_a/dt)_d - (dX_e/dt)_d = b_hX_a - fb_hX_a = (1 - f)b_hX_a \quad (3.20)$$

where:
r_o = oxidation rate of decayed active sludge.

Now, knowing that there is a proportionality constant "f_{cv}" between the COD and the volatile sludge mass, one has:

$$OUR_{en} = f_{cv}r_o = f_{cv}(1 - f)b_hX_a \quad (3.21)$$

where:
OUR_{en} = oxygen uptake rate for endogenous respiration.

The total OUR for oxidation of organic material is the sum of the two components, endogenous and exogenous respiration:

$$OUR_c = OUR_{ex} = (1 - f_{cv}Y)(1 - f_{us} - f_{up})/\theta_h + f_{cv}(1 - f)b_hX_a$$

By substituting for the active sludge concentration the expression becomes:

$$OUR_c = (1 - f_{us} - f_{up})(1 - f_{cv}Y + f_{cv}(1 - f)b_hC_r)S_{ti}/\theta_h \quad (3.22)$$

Now the fraction of the influent COD that will be oxidized in the activated sludge system can be expressed as:

$$mS_o = MO_c/MS_{ti} = (V_r OUR_c)/(Q_a S_{ti}) = (1 - f_{us} - f_{up})[(1 - f_{cv}Y) + f_{cv}(1 - f)b_h C_r] \quad (3.23)$$

Equation (3.23) concludes the development of the simplified activated sludge model: the three basic COD fractions are now expressed in simple equations:

(1) Influent COD fraction that remains in the liquid phase (Equation (3.5)):

$$mS_e = f_{us}$$

(2) Influent COD fraction transformed in sludge and wasted as a solid (Equation (3.18)):

$$mS_{xv} = f_{cv}(1 - f_{us} - f_{up})(1 + f b_h \theta_c)C_r/\theta_c + f_{up}$$

(3) Oxidized influent COD fraction (Equation (3.23)):

$$mS_o = (1 - f_{us} - f_{up})[(1 - f_{cv}Y) + f_{cv}(1 - f)b_h C_r]$$

It can be noted that the sum of the three fractions is always one:

$$mS_e + mS_{xv} + mS_o \equiv 1 \quad (3.24)$$

Table 3.2. shows the parameters that influence the values of the three basic fractions. Several of these parameters are constant or can readily be determined: the sludge mass parameters (Y, f and f_{cv}) are considered constant independent on the origin and nature of the wastewater. The wastewater temperature influences the kinetic decay constant and for that reason must be known. The decay constant is considered to be influenced only by temperature. The influent COD composition is important: the values of the non-biodegradable COD fractions influence the value of mS_e, mS_{xv} and

Table 3.2 Factors that influence the simplified activated sludge model and its numerical values when possible.

No	Factor	Symbol	Value	Unit
	Influent COD composition			
1	Non-biodegradable dissolved fraction	f_{us}	variable	–
2	Non-biodegradable particulate fraction	f_{up}	variable	–
	Sludge mass parameters			
3	Yield coefficient	Y	0.45	g VSS/g COD
4	Endogenous residue fraction	f	0.2	–
5	COD/VSS ratio	f_{cv}	1,5	g COD/g VSS
	Kinetic constants			
6	Decay constant	b	0.24* $1.04^{(t-20)}$	per day
	Wastewater characteristics			
7	Wastewater temperature	T	variable	°C
	Operational parameter			
8	Sludge age	θ_c	variable	day

mS_o. The f_{us} and f_{up} values tend to vary considerably in industrial wastewaters as can be seen from the data in Table 3.1. Pre-settling of wastewater may alter the non-biodegradable fractions considerably: f_{up} will tend to decrease (particulate matter is retained in the settler) and f_{us} will tend to increase (the total influent COD concentration to the activated sludge plant decreases, but the soluble biodegradable fraction does not).

Since the f_{us} and f_{up} values are not known, ideally they must be determined experimentally. Unfortunately this requires quite some effort because a (lab scale) activated sludge plant must be operated under steady state conditions and preferably at several sludge ages to determine mS_e, mS_{xv} and mS_o. Then, those f_{us} and f_{up} values are adopted that give the best fit between the experimental and theoretical mS_e, mS_{xv} and mS_o values.

The only operational variable that influences the values of mS_e, mS_{xv} and mS_o is the sludge age, so that the rational selection of this parameter is the key to satisfactory operation of activated sludge plants with a low effluent concentration and minimal costs for aeration and sludge disposal. There is a lower limit to this variable: when the sludge age is too short, the basic assumption of the simplified model (that the removal of biodegradable organic material is substantially complete) is no longer valid, because part of the biodegradable material will not be metabolized. Incomplete biodegradable material removal will occur first with the particulate fraction since the dissolved fraction can be metabolized at high rate. The incomplete removal of biodegradable material will not be noticed by an increase of the effluent fraction: the biodegradable material that is not metabolized will be flocculated (as is the non-biodegradable particulate material) and lead to an increase of the sludge production and a reduction of the oxygen consumption. At very short sludge ages not only the removal of the particulate biodegradable fraction is incomplete: the removal of the dissolved biodegradable will also be incomplete, leading to an increase of the effluent COD.

The value of the minimum sludge age to have a near complete removal of the organic material depends on the kinetics constants (particularly the specific substrate utilization rate constant (K_m) and the temperature, because the value of K_m depends on the temperature. In the case of sewage treatment the value of K_m has been established as $K = 2 \times (1.1)^{t-20}$ (Dold *et al.* 1980). For this value the sludge age for near complete biodegradable organic material removal is short: about 1 day at 20°C and 3 days at 14°C. In practice, the sludge age at which activated sludge plants are operated usually is longer than 1–3 days, so that the assumption that biodegradable organic material removal is substantially complete is justified.

3.2.3 Applications of the simplified model

3.2.3.1 Basic equations of aerobic treatment systems

The equations derived for the sludge concentrations and the OUR in the previous section all contain the hydraulic retention time θ, thus creating the false impression that this parameter is of fundamental importance. Therefore it is more convenient to use the basic equations, where the different COD fractions, the sludge concentration and the oxygen uptake are expressed as fractions of the daily applied COD

Table 3.3 Basic equations of the simplified model of the activated sludge system.

Equations
$mS_{bi} = S_{bi}/MS_{ti} = (Q_iS_{bi})/(Q_aS_{ti}) = (1 - f_{us} - f_{up})$
$mS_e = MS_{te}/MS_{ti} = (Q_aS_{usi})/(Q_iS_{ti}) = f_{us}$
$mX_i = (MX_i)/(MS_{ti}) = (V_rX_i)/(Q_aS_{ti}) = f_{up}\theta_c/f_{cv}$
$mX_a = (MX_a)/(MS_{ti}) = (V_rX_a)/(Q_aS_{ti}) = (1 - f_{us} - f_{up})C_r$
$mX_e = (MX_e)/(MS_{ti}) = (V_rX_e)/(Q_aS_{ti}) = (1 - f_{us} - f_{up})C_rfb_h\theta_c$
$mX_v = mX_i + mX_a + mX_e = (1 - f_{us} - f_{up})(1 + fb_h\theta_c)C_r + f_{up}\theta_c/f_{cv}$
$mX_t = (MX_t)/(MS_{ti}) = mX_v/f_v$
$mS_o = (1 - f_{us} - f_{up})(1 - f_{cv}Y + f_{cv}(1 - f)b_h\theta_c)$

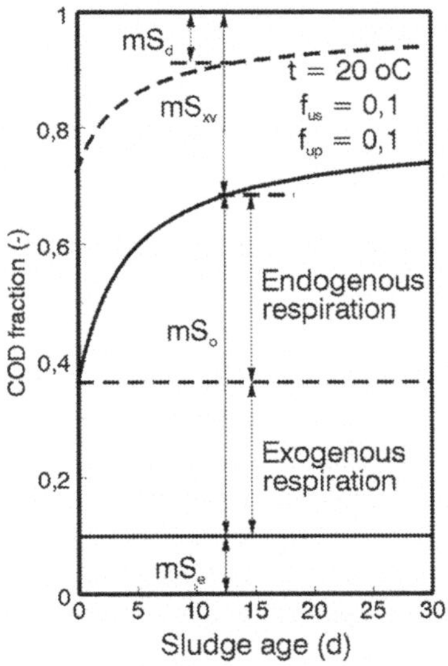

Figure 3.3 The division of the influent COD into three basic fractions mS_{te}, mS_o and mS_{xv} as functions of the sludge age.

mass. These equations are in Table 3.3. For example in Table 3.3, the mX_a indicates the active sludge mass present in the reactor per unit mass applied COD.

3.2.3.2 Sludge production and nutrient demand

The simplified model of activated sludge can be used to predict not only the oxygen consumption and the sludge production, but also the nutrient demand required to produce the sludge. This is exemplified in Figure 3.3, where the three COD fractions (calculated from Equations (3.5), (3.18) and (3.23)) are shown as function of the sludge age.

Equation (3.18) can be used to calculate the sludge production and the nutrient demand for this production as shown in Figure 3.4 for $f_{us} = f_{up} = 0.1$. By considering that the proportion between COD and volatile sludge mass is 1.5 mg COD/mg VSS, the volatile sludge production per unit applied COD can be calculated (first scale

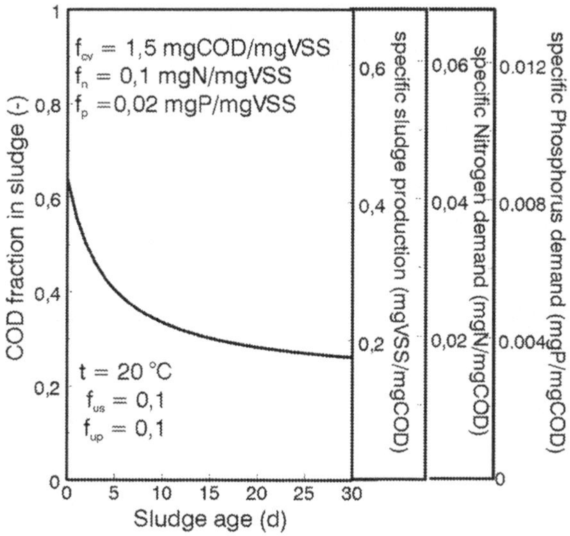

Figure 3.4 Specific sludge production and demand of macro nutrients N and P as functions of the sludge age.

on right hand side of Figure 3.4). Knowing the specific sludge production (kg VSS/kg COD applied) the nutrient demand can now readily be estimated by considering that there exists a constant fraction of the macronutrients N and P in volatile sludge. In Figure 3.4, the specific demands of N and P have been calculated for assumed mass fractions of f_n = 0.1 mg N/mg VSS and f_p = 0.02 mg P/mg VSS.

The actual proportion of N/COD and P/COD in the influent will normally not be an exact match for the demand. Depending on the nature of the wastewater the nutrient content may either be too low (frequent in cases of vegetable wastewaters) or too high (animal wastewaters). If nutrient is lacking it must be added for example as urea and phosphoric acid or as mono or diammonium phosphate (MAP or DAP). If after sludge production there is still a high nutrient concentration, the treatment plant may have to be designed for the removal of nutrients in a tertiary treatment plant. This aspect is discussed in Sections 3.4 and 3.5.

3.2.3.3 *Sludge mass and composition*

The basic equations can be used to calculate the masses of the different fractions that compose the sludge as functions of the sludge age, when the daily load is known. In Figure 3.5, the masses of inert, active, endogenous, organic and total sludge per unit mass of daily applied COD (mX_a, mX_e, mX_i, mX_v and mX_t) are plotted as functions of the sludge age for wastewaters with a f_{us}, value of 0.14 and f_{up} values of 0.02 (settled) and 0.10 (raw wastewater).

The figures show two important aspects: (1) the sludge mass depends strongly on the characteristics of the influent organic material and (2) even though the active sludge mass increases with increasing sludge age, the active sludge fraction decreases. Since the active sludge fraction is an important parameter, it is interesting

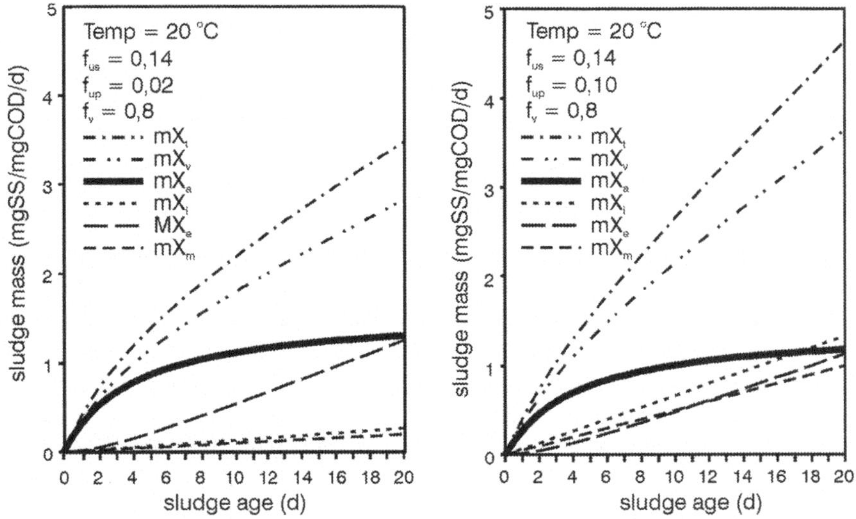

Figure 3.5 Masses of the different sludge fractions per unit of daily applied COD for settled (left) and raw (right) wastewater.

to derive an explicit expression for it. The active sludge fraction can either be defined as a fraction of the organic or a fraction of the total sludge concentration:

$$f_{av} = mX_a/mX_v$$
$$= (1 - f_{us} - f_{up})C_r/[(1 - f_{us} - f_{up})(1 + fb_h\theta_c)C_r + f_{up}\theta_c/f_{cv}] \quad (3.25)$$

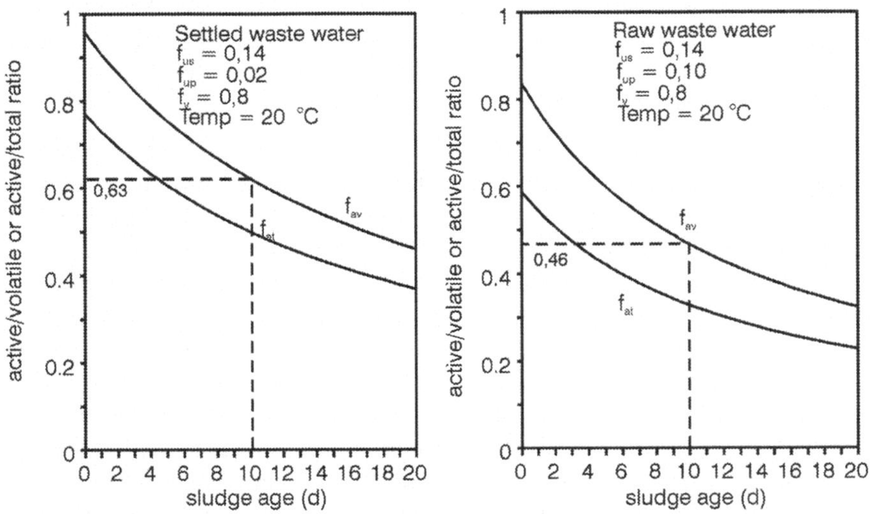

Figure 3.6 Fractions of active (life) sludge as a function of sludge age for settled (left) and raw (right) wastewater.

or

$$f_{at} = mX_a/mX_t$$
$$= (1 - f_{us} - f_{up})C_r/[(1 - f_{us} - f_{up})(1 + fb_h\theta_c)C_r + f_{up}\theta_c/f_{cv}] \cdot f_v = f_{av}/f_v \quad (3.26)$$

where:
f_{av} =proportion of active volatile sludge
f_{at} =proportion of active total sludge.

Figure 3.6 shows values of f_{av} and f_{at} as functions of the sludge age for raw and settled wastewaters. It can be noted in Figure 3.6 that the active sludge fraction depends strongly on the composition of the influent organic material. For example, for the raw wastewater ($f_{up} = 0.10$) the active sludge fraction is $f_{av} = 0.46$ when the sludge age is 10 days. In the case of settled sewage for the same sludge age the active fraction is much higher: $f_{av} = 0.63$. In the case of settled sewage the active fraction of $f_{av} = 0.46$ is found for a sludge age of more than 20 days.

3.2.4 Excess sludge treatment

The activated sludge process removes very efficiently suspended solids, organic material and eventually nutrients from the liquid phase, but at the same time a problem is generated in the form of the excess sludge. The treatment and final disposal of this sludge require a significant fraction of the material and financial resources used at the treatment plant.

The excess sludge from activated sludge process has three undesirable aspects:

(a) biological instability: due to the high fraction of biodegradable organic matter the sludge is putrescible and enters in decomposition within hours after interruption of the aeration.
(b) the hygienic quality of the sludge is very poor: a very large variety of virus, bacteria and other pathogens (protozoa, amoebae and helminth eggs) are present.
(c) The suspended solid concentration in the sludge is low (in the range of 5–50 g/L depending on the nature of the sludge and if a thickening process (settling, flotation) is applied), so that the volume of sludge is large.

The objective of sludge treatment process is to reduce the fraction of biodegradable matter and the pathogen concentration, so that a stable and safe product is obtained, that does not constitute a public health problem. In addition, it is attempted to increase the concentration of solids, thus reducing the volume of treated sludge that needs to be disposed of. Two biological processes are often used for sludge stabilization: aerobic or anaerobic digestion. Both processes also have a positive influence on the hygienic quality of the sludge. The reduction of the water content of sludge is obtained by applying physical processes (filtration, flotation and evaporation), eventually preceded by preparatory processes to accelerate liquid–solid separation.

Aerobic sludge stabilization is mostly applied in small treatment plants in regions with a low temperature, where anaerobic stabilization at environmental temperature is not feasible. Although under those conditions the methane produced in anaerobic digesters may be sufficient to heat the excess sludge to the optimal temperature for anaerobic digestion, this option may be judged to be too complicated for small systems. Aerobic sludge digestion may also be chosen in cases of industrial wastewater treatment plants, where toxicity is likely to be a recurring problem. The design of aerobic digesters is straightforward: the central problem is to produce a sludge with a small active sludge fraction (around 10–20%). The active sludge fraction is reduced due to decay of the active sludge. Normally aerobic digesters are operated as flow-through, completely mixed continuous reactors. Depending on the sludge age in the activated sludge system, the influent organic material composition, the temperature and concentration of the excess sludge, the retention time of aerobic digesters usually is 2–3 weeks. A basis for rational design of aerobic sludge digesters has been presented by Van Haandel *et al.* (1998).

The retention time in anaerobic sludge digesters cannot yet be derived from theory but empiric relationships between the required retention time and the temperature have been proposed by several authors. Araújo *et al.* (1998) taking into consideration several empiric results proposed the following relationship:

$$\theta_{di} = 20 \times 1.1^{(20-t)} + 5 \text{ (considering } 15 < t < 35) \tag{3.27}$$

where:
θ_{di} = retention time in the sludge digester
t = digester temperature (°C).

Van Haandel *et al.* (1998) also established that digestibility of the active sludge is distinctly greater than that of the inactive sludge. Based on experimental data they suggested the following digestion efficiencies in digesters with a sufficient retention time (defined by Equation (3.27)):

$$R_{da} = 0.36 + 0.0067\,t \tag{3.28a}$$

$$R_{dn} = 0.10 + 0.0019\,t \tag{3.28b}$$

where:
R_{da}, R_{dn} = digestion efficiency for active and inactive sludge, respectively.

Hence, for a temperature of 30°C a fraction of $0.36 + 0.0067 \times 30 = 0.56$ of the active sludge can be methanized, but only $0.10 + 0.0019 \times 30 = 0.19$ of the inactive sludge is digestible. Thus, depending on the sludge age a fraction of 25–40% of the volatile sludge may be digested in the anaerobic digester. The fraction of the influent COD wasted as sludge and digested in an anaerobic digester is indicated in Figure 3.3 (interrupted curve).

3.2.5 The activated sludge system under non-ideal conditions

Although the behaviour of the activated sludge system is closely predicted by the simplified or ideal model if the operational conditions are adequate, in practice in several situations a non-ideal behaviour is observed where the removal of the biodegradable organic material may not be complete. The following factors may lead to a non-ideal behaviour:

(1) In practice a system does not operate under steady state conditions. Most wastewaters show strong variations of flow and load over a day. Under these

conditions the oxygen uptake will also be variable, and the aeration capacity must be designed for the maximum demand. It is possible that the metabolic capacity of the sludge is insufficient during periods of maximum load.

(2) Many activated sludge plants are composed of several reactors in series. Since the influent is usually discharged in one reactor, there will be large differences in the availability of substrate and the OUR, so that the behaviour is very different from a completely mixed stirred tank reactor (CSTR).

(3) In practice the final settler is non-ideal in the sense that it has a certain sludge hold up that is not available for metabolic activities. The sludge mass in the reactor varies with time being larger when the influent flow is maximum.

Under non-ideal conditions the velocity of the metabolic processes becomes an important factor, so that it is necessary to formulate a kinetic model of organic material removal. The Monod model discussed above has limited applicability for the metabolism of biodegradable organic material in the activated sludge process. This is due to the fact that much of the organic biodegradable material may be removed from the liquid phase by flocculation, so that there is no direct relationship between the organic material removal and the organic material metabolism. The International Water Association, IWA, (Henze *et al.* 1986) has developed a general model for aerobic treatment systems, which is represented in Figure 3.7.

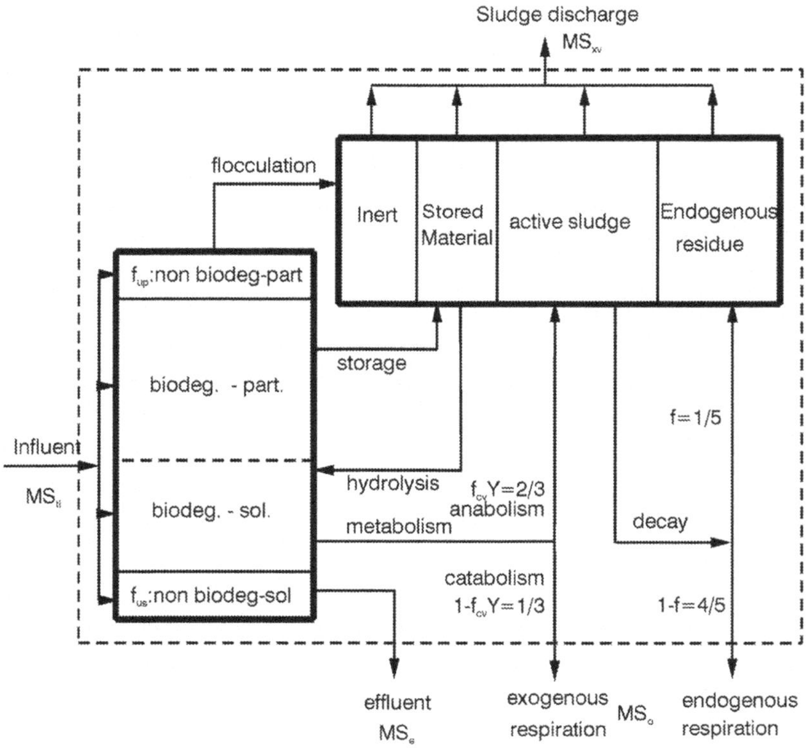

Figure 3.7 Schematic representation of the processes and fluxes in a non-ideal activated sludge process.

The processes that develop in the non-ideal activated sludge process can be resumed as follows:

(1) The biodegradable and soluble material is removed from the liquid phase by metabolism, following Monod kinetics:

$$r_{us} = K_{ms}X_aS_{bs}/(S_{bs} + K_{ss}) \tag{3.29}$$

(2) The biodegradable and particulate organic material cannot be used directly by bacteria. In a preliminary step it is stored onto active sludge and subsequently hydrolysed, producing new biodegradable and soluble material. The kinetics for storage proposed by Dold *et al.* (1980) can be expressed as:

$$r_a = K_aX_aS_{bp}(K_{ap} - S_{ba}/X_a) \tag{3.30}$$

Dold *et al.* (1980) also proposed an expression for hydrolysis kinetics, which is expressed as a modified Monod equation:

$$r_{hi} = K_{mp}X_aS_{pa}/(S_{pa} + K_{sp}X_a) \tag{3.31}$$

(3) The hydrolysed material has now become soluble biodegradable COD and will be utilized together with the incoming soluble biodegradable fraction. As a result of anabolism the active sludge grows at a rate that can be expressed as:

$$r_c = Y_hr_{us} = Y_hK_{ms}X_aS_{bs}/(S_{bs} + K_{ss}) \tag{3.32}$$

(4) Parallel with sludge growth there is also decay that is expressed as a first-order process:

$$r_d = -b_hX_a \tag{3.33}$$

In Table 3.4, the reaction rates of the different parameters involved in organic material removal are resumed. From the above expression it can be seen that several of the differential equations described by the rate expressions are non-linear and have no analytical solution. A numeric solution can be obtained by using computer calculations. Once the values of the kinetic constants in the above model have been established, a computer can simulate the behaviour of the activated sludge system for any specified set of influent flow, and load and operational conditions.

The kinetic constants can only be determined through experimental assays. The calibration of the general model for activated sludge is most conveniently achieved by imposing cyclically varying flow and/or load conditions, and observing the

Table 3.4 Rates of production (or removal) of the parameters involved in kinetics of organic material utilization.

Production rate of soluble biodegradable material	$r_{sbs} = r_{hi} - r_{us}$
Production rate of particulate biodegradable material	$r_{sbp} = -r_a$
Production rate of stored organic material	$r_{spa} = r_a - r_{hi}$
Net growth rate of active sludge	$r_{xa} = r_c - r_d = Y_hr_{us} - r_d$
Production rate of endogenous residue	$r_{xe} = fr_d$
Oxygen uptake rate	$OUR_c = (1 - f_{cv}Y)r_{us} + f_{cv}(1 - f)r_d$

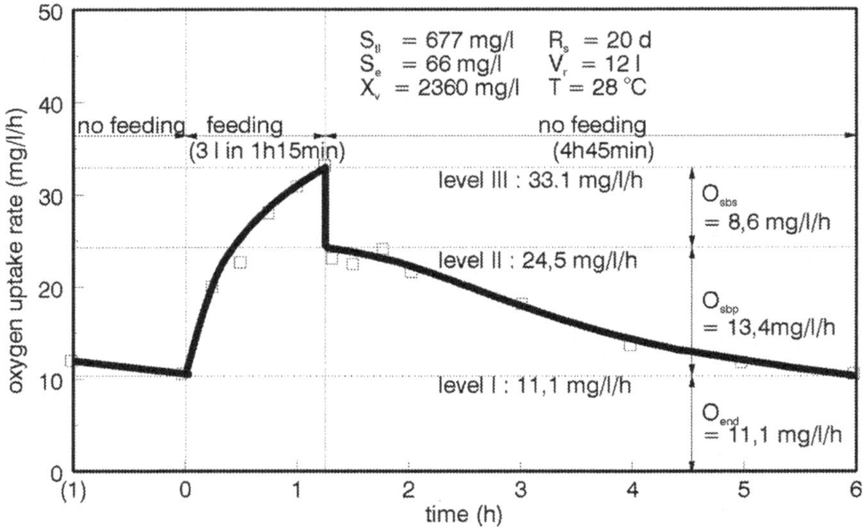

Figure 3.8 OUR profile as a function of time in a cyclically fed aerobic activated sludge system, using domestic sewage as influent.

Table 3.5 Numeric values of the kinetic constants for removal of organic material in the activated sludge process (Dold *et al.* 1980).

Symbol	Value (20°C)	Temperature dependency
K_{ms}	20 mg COD/L	$(1.2)^{(T-20)}$
K_{ss}	5 mg COD/L	–
K_a	0.25 L/mg X_a/L	$(1.1)^{(T-20)}$
K_{mp}	2 mg COD/mg X_a/L	$(1.1)^{(T-20)}$
K_{sp}	0.04 mg COD/mg X_a	$(1.1)^{(T-20)}$
b_h	0.24/day	$(1.04)^{(T-20)}$

behaviour of the activated sludge system. The variable nature of metabolic activity in principle leads to dynamic response of the treatment system, so that the processes vary with time. However, the variation of the concentrations of several parameters like the effluent COD and the sludge concentration is so small that it cannot be used to define the kinetics of metabolism. The parameter that exhibits the largest (and therefore best measurable) variation is the OUR. Figure 3.8 shows the variation of OUR as well as other variables in a bench scale activated sludge process fed with a cyclically varying flow of sewage. The experimental values of influent and effluent COD concentrations (S_{ti} and S_{te}), volatile sludge concentration (X_v) as well as the operational conditions sludge age (θ_c) reactor volume (V_r) and temperature (T) are also indicated. With the basic equations above, a general model can be constructed and the behaviour of the activated sludge process (and particularly the OUR profile) can be simulated. In Figure 3.8, the OUR curve simulates the behaviour and can be compared with experimentally determined OUR values. Now by trial and error using different values for the kinetic constants, the OUR simulation with the best fitting of the simulated OUR profile with the experimental OUR data, the values of the constants can be estimated.

Table 3.5 shows the values of the constants obtained by simulating the experimental results of Figure 3.8. By applying the experimental method at several temperatures and carrying out simulations, the temperature dependency of the constants was determined.

3.2.6 Choice and control of the sludge age

In the previous sections, it has been established that the sludge age is the most important operational parameter of the activated sludge process, so that choosing and maintaining its right value is essential to maintain high quality and low costs. The kinetics of biodegradable organic material showed that at high environmental temperatures ($t = 20°C$) the required sludge age for substantially complete removal of the biodegradable material is very short (1–3 days). At this short sludge age endogenous respiration is incomplete, so that oxygen consumption will be low, but the sludge production will be high, leading to a relatively large sludge digester. The activated sludge reactor will be relatively small since the mass of retained activated sludge is small.

A disadvantage of a very short sludge age is that the predators of free swimming bacteria do not develop, so that the concentration of these bacteria in reactor and in the effluent will be high. The bacteria give a turbid aspect and a relatively high BOD concentration to the effluent. At longer sludge ages (5 days at 20°C) the predators develop and a clear effluent with low turbidity and biochemical oxygen demand (BOD) will be produced.

In regions with a cold-climate-activated sludge plants are often operated at a sludge age that is insufficient for the removal of the biodegradable organic material. Under these conditions, the activated sludge process functions also as a bioflocculator. This operational mode is chosen because it is cheaper to flocculate the biodegradable material and digested it subsequently in an anaerobic reactor (in case of an aerobic digester the advantage does not exist). Thus the disadvantage of having a high sludge production is compensated by the production of methane that can be transformed into electric power in simple stationary explosion motors or gas turbines.

Very often the choice of the sludge age will not be determined by the kinetics of organic material utilization but rather the value required by other processes like nitrogen or phosphorus removal. It can be inferred that tertiary treatment systems will typically produce a high-quality effluent with respect to organic material removal.

Once the sludge age has been chosen, its value must be maintained in the treatment systems, so that a form of control is necessary. Normally the control is done by determining the mass of sludge that is to be wasted as excess sludge. In practice, the sludge discharge is often realized from the return sludge flow, but this is not a recommended procedure. Due to variations in the influent flow, the sludge hold up in the settler and consequently the sludge concentration in the return sludge flow will exhibit large oscillations over a period of 24 h, so that there is no

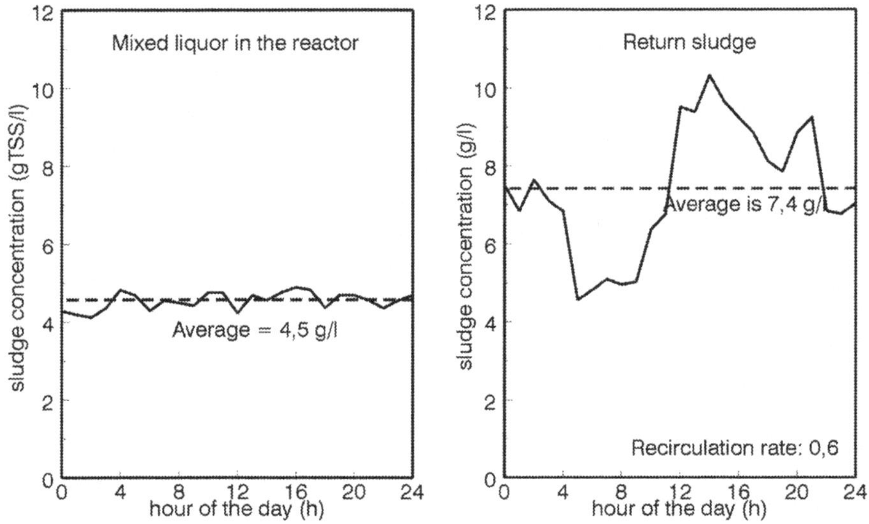

Figure 3.9 Experimental profiles of the mixed liquor concentration in the biological reactor and in the underflow from the settler to the reactor.

clear relationship between the sludge volume and the sludge mass in the return sludge. By contrast, in the biological reactor the sludge concentration varies little over a day. Thus, by using the definition of the sludge age the volume to be wasted from the reactor can be determined directly as $q = V_r/\theta_c$. As an example Figure 3.9 shows the typical profile over a day of the mixed liquor concentration and in the return sludge flow as reported by WRC (1984).

The procedure of discharging from the settler underflow is due to the (mistaken) idea that this sludge (having a larger concentration than mixed liquor) can be thickened more than the reactor contents in a thickener.

Another counterproductive operation procedure often applied in practice is to discharge sludge when the mixed liquor concentration exceeds some prefixed value. In this way it is attempted to maintain a certain desired F/M (food (=substrate) to microorganisms (=sludge) ratio. Under these conditions the sludge age control is always corrective. Instead the sludge age control must be preventative to avoid that the concentration exceeds its maximum value. Sludge must be discharged continuously or cyclically (for example daily) in accordance with the pre-established sludge age.

3.3 ANAEROBIC TREATMENT SYSTEMS

3.3.1 Anaerobic conversion processes

The transformation into biogas of complex macromolecules present in most wastewaters requires the mediation of several groups of microorganisms. Figure 3.10 shows a schematic presentation, suggested by Gujer and Zehnder (1983). Different

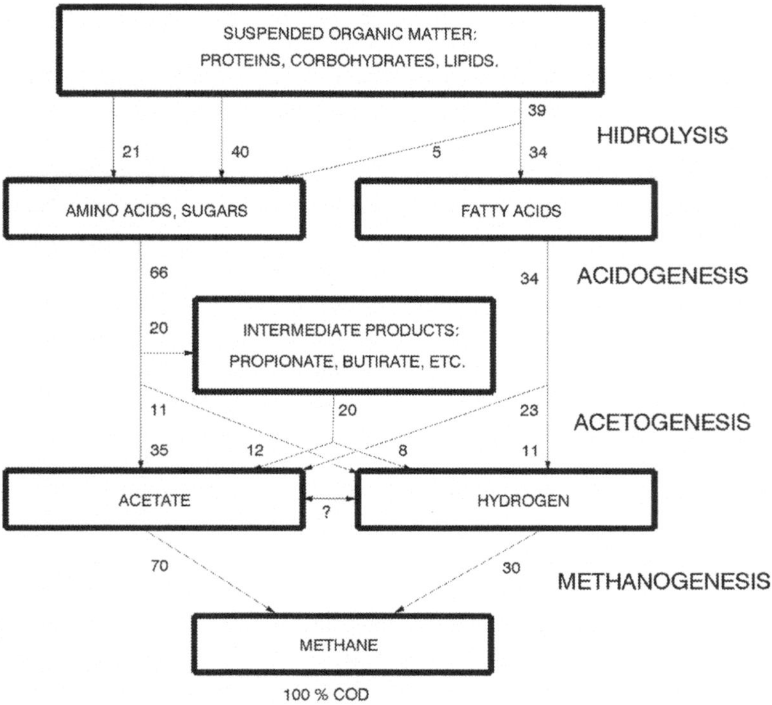

Figure 3.10 Schematic representation of the main conversion processes in anaerobic digestion. (The numbers refer to percentages, expressed as COD, of organic matter destination.)

steps are necessary for the anaerobic digestion of proteins, carbohydrates and lipids. Four different phases can be distinguished in the overall conversion process:

(1) **Hydrolysis**: In this process complex particulate matter is converted into dissolved compounds with a lower molecular weight. The process requires the mediation of exo-enzymes that are excreted by fermentative bacteria. Proteins are degraded via (poly) peptides to amino acids, carbohydrates are transformed into soluble sugars (mono and disaccharides) and lipids are converted to long chain fatty acids and glycerine. In practice, the hydrolysis rate can be limiting for the overall rate of anaerobic digestion. In particular the conversion rate of lipids becomes very low below 18°C (see Section 3.3.2).

(2) **Acidogenesis**: Dissolved compounds, generated in the hydrolysing step, are taken up in the cells of fermentative bacteria and after acidogenesis excreted as simple organic compounds like volatile fatty acids (VFA), alcohols and mineral compounds like CO_2, H_2, NH_3, H_2S, etc. Acidogenic fermentation is carried out by a diverse group of bacteria, most of which are obligate anaerobe. However, some are facultative and can also metabolize organic matter via the oxidative pathway. This is important in anaerobic wastewater treatment, since

dissolved oxygen (DO) otherwise might become toxic for obligate anaerobic organisms, such as methanogens.

(3) **Acetogenesis**: The products of acidogenesis are converted into the final precursors for methane generation: acetate, hydrogen and carbon dioxide. As indicated in Figure 3.10, a fraction of approximately 70% of the COD originally present in the influent is converted into acetic acid and the remainder of the electron donor capacity is concentrated in the formed hydrogen. Naturally the generation of highly reduced material like hydrogen must be accompanied by production of oxidized material like CO_2.

(4) **Methanogenesis**: Methanogenesis may be the rate limiting step in the overall digestion process, especially at high temperatures (>18°C) and when the organic material in the influent is mainly soluble and little hydrolysis is required. Methane is produced from acetate or from the reduction of carbon dioxide by hydrogen using acetotrophic and hydrogenotrophic bacteria, respectively:

$$\text{Acetotrophic methanogenesis:} \quad CH_3COOH \rightarrow CH_4 + CO_2 \tag{3.34}$$

$$\text{Hydrogenotrophic methanogenesis:} \quad 4H_2 + CO_2 \rightarrow CH_4 + 2H_2O \tag{3.35}$$

Different from aerobic treatment where the bacterial mass was modelled as a single bacterial suspension, anaerobic treatment of complex wastewaters, with particulate matter in the influent, is only feasible by the action of a consortium of the four mentioned groups of bacteria that each have their own kinetics and yield coefficients. The bacteria that produce methane from hydrogen and carbon dioxide grow faster than those utilizing acetate (Henzen and Harremoes 1983), so that the acetotrophic methanogens usually are rate limiting for the transformation of acidified wastewaters to biogas.

The different groups of bacteria involved in the conversion of influent organic matter all exert anabolic and catabolic activity. Hence, parallel to the release of the different fermentation products, new biomass is formed associated with the four conversion processes described above. For convenience, the first three processes often are lumped together and denominated *acid fermentation*, while the fourth step is referred to as *methanogenic fermentation*.

The removal of organic matter-COD during the acid fermentation is limited to the release of hydrogen. As shown in Figure 3.10, only 30% of the organic matter is converted into methane via the hydrogenotrophic pathway. Hence, a necessary condition for efficient organic matter removal in an anaerobic treatment system is that a sufficient mass of acetotrophic methanogens develops.

Acid fermentation tends to cause a decrease in the pH because of the production of VFA and other intermediates that dissociate and produce protons. As methanogenesis will only develop well at a neutral pH values, instability may arise, if for some reason the rate of acid removal by methane production falls behind the acid production rate: the net production of acid will tend to cause a decrease in pH, and thus may reduce the methanogenic activity further. In practice, this so called "souring" of the anaerobic reactor contents is the most common cause for operational failure of anaerobic treatment systems. The danger of souring can be avoided, by maintaining the proper balance between acid and methanogenic fermentation which in fact means that both the methanogenic digestion capacity and buffer capacity of the system should be sufficiently high (See Section 3.3.3.2).

3.3.2 Modelling of anaerobic treatment systems

Like in aerobic treatment systems, the main objective of anaerobic treatment systems is to remove, as efficiently as possible, the organic material in wastewaters. The anaerobic treatment generates a division of the digested organic material in three fractions: (1) a solid fraction: the anaerobic sludge, (2) a liquid fraction discharged in the effluent and (3) a gaseous fraction released as biogas. Different from aerobic systems that can be described in detail as shown in the previous sections, modelling of anaerobic treatment systems is still incomplete. This is due to the following factors:

(1) The complex nature of anaerobic digestion and the variability of the composition of influent organic material in wastewaters are the reason that the sludge composition is also variable, that is, the concentration of each of the bacterial groups that are active depend on the influent material composition. If particulate biodegradable material is not hydrolysed in the system, it will become a biodegradable fraction of the solid phase, the sludge, and will be removed when excess sludge is wasted. Indeed it is not unusual that a large fraction (and sometimes the biggest) of the VSS is influent organic material in stead of biological sludge.
(2) While in aerobic system the sludge age is defined at the design stage and the operating sludge mass is constant through daily or even continuous excess sludge discharge, anaerobic systems will be operated at maximum sludge age, because that will lead to the highest organic material removal efficiency. However it is not *a priori* clear what is the value of the maximum sludge age when the anaerobic system is built, since this depends on the sludge generation (hydrolysis efficiency) and the efficiency of the sludge retention device, which in turn depends on the settling properties of the sludge.

Henzen and Harremoes (1983) estimated the most important sludge mass parameters and kinetic constants for acid and methanogenic fermentation from the results of a large number of experimental investigations. The values for a temperature of 35°C are represented in Table 3.6. From the values in the table it follows that it may be expected that a pure culture of acid formers or methanogens will both metabolize a maximum of about 13 mg COD/mg VSS-day. The acid formers grow 0.15 kg VSS per kg COD metabolized substrate, which is complex organic matter, whereas methanogens grow only 0.03 kg VSS per kg COD of methanogenic substrate. Thus, a sludge mass of $0.15 + 0.03 = 0.18$ kg VSS/kg COD will be produced, when 1 kg of COD of complex organic matter is utilized anaerobically. Hence, a combined culture of acid formers and methane producers, generated from a complex organic substrate, would typically be composed of $0.03/(0.03 + 0.15) = 1/6$ of methanogens and 5/6 of acid formers. In this estimate two factors have not been taken into consideration: (1) in fact the biomass production by methanogens will be

Table 3.6 Kinetic constants of anaerobic cultures (After Henzen and Harremoes 1983).

Cultures	μ_{m35} (per day)	μ_{m26} (per day)	Y (mg VSS/ mg COD)	K_m (mg COD/ mg VSS/day)	K_s (mg COD/L)
Acid-producing bacteria	2.0	0.70	0.15	13	200
Methane-producing archaea	0.4	0.14	0.03	13	50
Combined culture	0.4	0.14	0.18	2	–

slightly less, because the influent fraction anabolized by the acid formers does not become available for methanogenesis and (2) decay is not taken in account, because its rate is known to be low. However, these factors only have a very small effect, so that the maximum rate of methane production per unit mass of combined bacterial mass would be only about 1/6 of the one obtained with a pure methanogenic culture, that is 13/6 = 2 mg COD/mg VSS/day. In Table 3.6 the constants of Monod kinetics (K_m and K_s) do not refer to the hydrolysis process, only to acidogenesis, acetogenesis and methanogenesis.

The anaerobic digestion efficiency can be calculated as a function of the sludge age for the constants in Table 3.6 for soluble wastewaters, where no hydrolysis is required. Assuming a temperature dependency of $\mu_{mT} = \mu_{m35} * 1.11^{(T-35)}$ the maximum specific growth constants were calculated for a temperature of 26°C (See Table 3.6) and the efficiency of acid and methanogenic fermentation can be calculated by using the expressions for Monod kinetics. This is shown in Figure 3.11, where the different COD fractions of acid and methanogenic fermentation are plotted as functions of the sludge age. The following situations will occur as the sludge age increases:

(1) The soluble and particulate non-biodegradable influent COD fractions will not be affected by the anaerobic sludge mass. The soluble non-biodegradable material will leave the system dissolved in the effluent, the particulate material will flocculate and leave the system in the excess sludge.
(2) Below the minimum sludge age for acid or methanogenic digestion no biological activity can develop because the rate of wastage exceeds the maximum growth rate of the bacteria. Below the minimum sludge age for acid fermentation all soluble biodegradable COD is in the effluent. The particulate biodegradable material may be flocculated and become part of the solid phase. In this case the reactor acts as a flocculater; not as a biological reactor.
(3) If the sludge age is between the minimum for acid fermentation and methanogenic fermentation, a small fraction of the biodegradable COD is used for production of acidogenic sludge (a Y value of 0.07 mg VSS/mg COD has been adopted in Figure 3.11, assuming partial pre-acidification of the influent) and the rest remains in the effluent. If the decay rate is neglected relative to the growth rate, the minimum sludge age for acid fermentation to develop is estimated at $\theta_{c,min} = 1/\mu_{m25} = 1/0.7 = 1.5$ day. The minimum sludge age for methanogenic fermentation is $\theta_{c,min} = 1/\mu_{m26} = 1/0.14 = 7$ day.

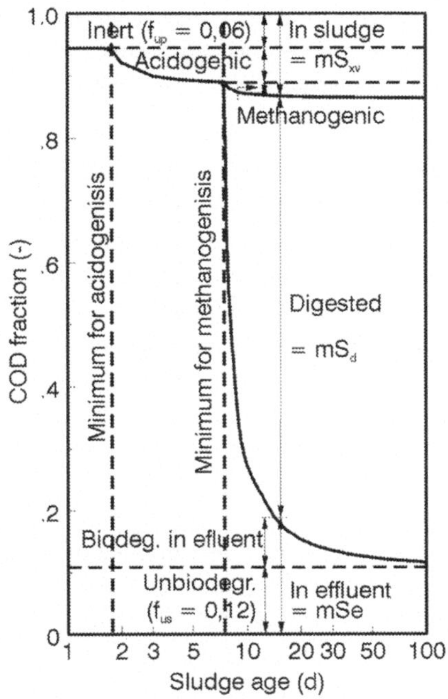

Figure 3.11 Calculated relationships between the division of COD in fractions in the effluent (bottom) digested (middle) and in the sludge (top) as functions of the sludge age for soluble substrates at 26°C (constants of Table 3.6).

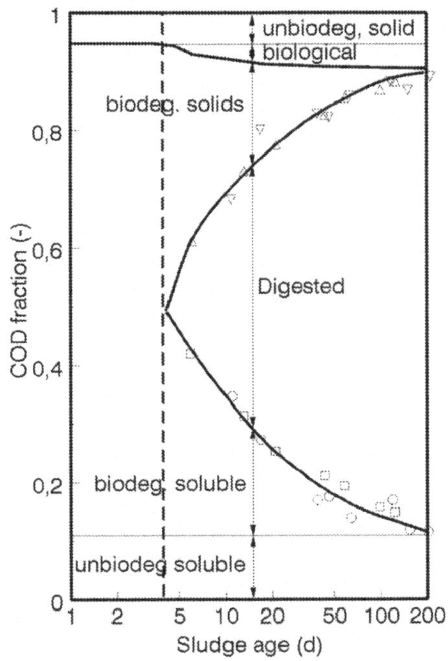

Figure 3.12 Empiric relationship of the COD fractions in the effluent (top) in the excess sludge (bottom) and digested (middle) as functions of the sludge age in anaerobic systems treating raw sewage (predominantly particulate substrate) at 26°C.

(4) For sludge ages longer than the minimum for methanogenesis there is an increasing fraction of the influent COD that is transformed into biogas, but the sludge age has to be long (50–100 days) for a substantially complete digestion of the biodegradable material. In this respect, anaerobic treatment systems are very different from aerobic treatment systems, where the removal of biodegradable material is complete at a sludge age of only 1 day under comparable conditions. In fact aerobic treatment systems are only operated at sludge ages beyond 30 days in cases of toxic wastes.

When raw wastewater with particulate biodegradable material is used as a substrate, the situation becomes more complicated, because now also hydrolysis will take place. The hydrolysis rate is usually the rate determining process and therefore has a very important influence. As a result of incomplete hydrolysis, the influent biodegradable and particulate material will tend to flocculate and become part of the sludge mass, without being metabolized. This is shown in Figure 3.12, where experimental results of the three COD fractions are plotted as a function of the sludge age in the case of raw sewage treatment (Cavalcanti *et al.* 2003). To construct the diagram of Figure 3.12, two different anaerobic treatment systems were operated at their maximum sludge hold up and at different hydraulic retention times. The sludge age was calculated from the sludge mass in the systems and the rate of excess sludge leaving the "full" system. The empiric curve in Figure 3.12 shows some important points that can be resumed as follows:

(1) The experimental data were obtained with different anaerobic systems but the results show that the division of the three fractions is the same for the same sludge age. Thus, the sludge age is the fundamental parameter to describe the behaviour of anaerobic processes (as well as aerobic processes). In other words, the division of influent COD fractions in the solid, liquid and gaseous components tend to be the same in different anaerobic treatment systems, as long as these are operated with the same sludge age. In this respect, anaerobic treatment systems are similar to aerobic systems. However, in practice the sludge age of anaerobic systems is usually not known *a priori* and the liquid retention time is often used as a design parameter, even though different systems with the same hydraulic retention time may have a very different performance.
(2) In addition to biological sludge, there will also be an inert organic fraction due to flocculation of non-biodegradable and particulate organic matter in the influent and, depending on the operational conditions of the treatment systems, there may also be a fraction of biodegradable and particulate organic matter present in the sludge mass. As a consequence, the sludge from an anaerobic wastewaters treatment system will contain only a small biological fraction.
(3) The division of the influent COD fractions over effluent, excess sludge and digested material will be different from that in systems with soluble wastewaters: at short sludge ages the hydrolysis of the particulate biodegradable material is low and due to flocculation of this material the sludge fraction increases with decreasing sludge age, which is contrary to the pre-vision with soluble wastewaters (Figure 3.11).
(4) As the sludge age decreases, the COD fraction in the effluent increases due to an increasing concentration of soluble biodegradable material that is not removed in the anaerobic system (Figure 3.12). The COD fraction in the effluent is similar to that for soluble influents (Figure 3.11).
(5) As the sludge age decreases, the fraction of COD in sludge increases due to the fact that an increasing concentration of particulate biodegradable influent material is flocculated (result of inefficient hydrolysis) and becomes part of the sludge. As in sewage the biodegradable, particulate material is the largest COD fraction, at short sludge ages the biodegradable material usually is the largest component of the sludge mass, so that the excess sludge will be unstable due to the high putrescible fraction.
(6) There is a minimum sludge age below which the anaerobic system fails altogether (i.e. the biological activity ceases because the bacterial populations cannot develop).

3.3.3 Environmental factors

Important environmental factors affecting anaerobic wastewater digestion are temperature, pH, and the concentrations of essential nutrients and of toxic compounds in the influent. An adequate and stable pH in the neutral region is a necessary condition for a proper performance of anaerobic treatment systems. In diluted wastewaters like sewage, normally the alkalinity is sufficient to guarantee a buffer capacity to discharge the effluent at a neutral pH. This may not be the case in more concentrated wastewaters, especially those that are generated as acid waters. In that case alkalinity may be added to and/or acidity subtracted from the influent or the reactor. Nutrients (both macro, N, P and K, and micro) may be present in the wastewater, especially if this is of animal origin. In wastewaters of vegetal origin, the nutrients in the influent may not be sufficient to cover the demand for sludge production. In that case, there will be a need to add these materials to the influent. Toxicity may be a serious problem for anaerobic treatment of industrial wastewaters. Toxic material may be present in the influent or it may be generated in the anaerobic system itself.

3.3.3.1 Influence of temperature on anaerobic digestion

Anaerobic digestion, like other biological processes, strongly depends on temperature. With respect to the conversion rate of digestion processes, there exists a maxim between 30° and 40°C for the mesophilic range and at about 55°C for the thermophilic range. The influence of temperatures on the rate and the extent of anaerobic digestion has been subject to many investigations. Henzen and Harremoes (1983) evaluated the available data. Figure 3.13 shows a graphical representation of their analysis and of some more recent data. From the figure the following conclusions can be drawn: (1) The optimal range is between 30° and 40°C and (2) for temperatures

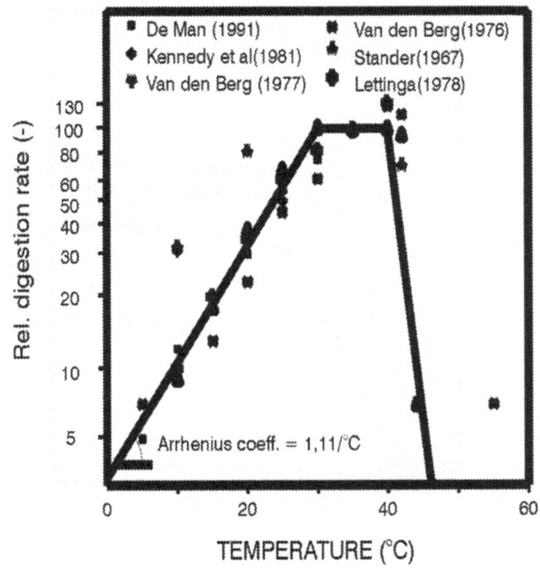

Figure 3.13 Influence of temperature on the rate of anaerobic digestion in the mesophilic range (After Henzen and Harremoes 1983).

below the optimal range the digestion rate decreases by about 11% per °C temperature decrease, or according to the Arrhenius expression:

$$r_t = r_{30}(1.11)^{(t-30)} \tag{3.36}$$

where:

t = temperature in °C;

r_t, r_{30} = digestion rate at temperature t and 30°C, respectively.

Using Equation (3.36), the calculated rates at 20°C and at 10°C amount to about 35°C and 12%, respectively, relative to the rate at 30°C.

The influence of temperature on anaerobic digestion is not limited to the rate of the process. Also the extent of anaerobic digestion is affected, as found by O'Rourke (1968) and Van der Last (1991). Figure 3.14 shows the achieved extent of digestion for settled wastewaters solids (primary sludge) in relation to digestion time at different temperatures, according to the results of O'Rourke (1968). This diagram clearly reveals the strong dependence of solids digestion on the temperature. The decrease of the fraction of organic matter degraded can be attributed to a low rate of hydrolysis. In practical terms this means that suspended organic matter can be removed from the water phase at low temperatures even when it is not metabolized: because it can be entrapped in the sludge bed, consequently becoming part of the sludge mass in the treatment system. For high strength wastewaters, the operational temperature to a certain extent can be considered as a process variable for an anaerobic treatment system, because within limits it can be controlled by using the produced methane to warm up the wastewater. This is not the case for low strength wastewaters like domestic sewage, because the heat obtained from

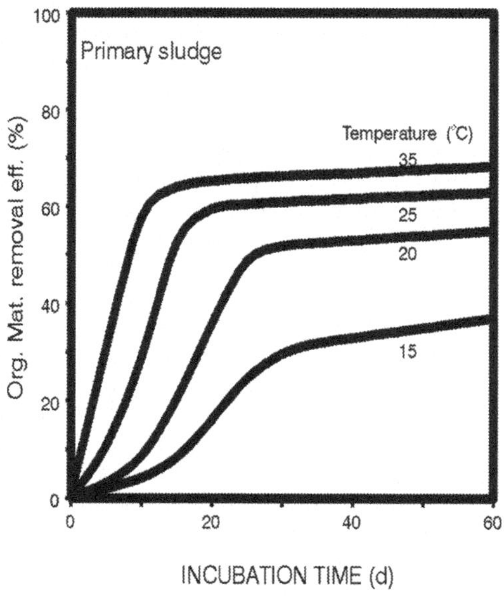

Figure 3.14 Influence of temperature on the extent and the rate of anaerobic digestion of primary wastewaters sludge (After O'Rourke 1968).

combustion of the produced methane is insufficient for a significant temperature increase. For example, if 1 g/L of COD is digested the methane production is ¼ gCH_4/L with a combustion heat of 50.4 KJ/gCH_4. If all produced methane is captured and burnt, the maximum heat production is ¼ × 50.4 = 12.6 KJ/L or 3 kCal/L. With this heat, the temperature of 1 l of water can be increased by 3°C. Thus, if the digested COD concentration is 1 g/L, the generated heat is too small to rise the temperature of the wastewater significantly (at most 3°C). At digestible COD concentrations of 10 g/L or more, at least in principle, there is generation of enough energy to heat the wastewater up to the optimal temperature.

3.3.3.2 pH in the anaerobic reactors

The value and stability of the pH in an anaerobic reactor is extremely important, because methanogenesis only proceeds at a high rate, when the pH is maintained in the neutral range. At pH lower than 6.3, or higher than 7.8 the rate of methanogenis is decreases. Acidogenic populations are significantly less sensitive to low or high pH values, and hence fermentation will prevail over methanogenic fermentation beyond this pH range, which may result in "souring" of the reactor contents.

In an anaerobic digester the most important question with respect to pH and pH stability is, whether the alkalinity present in the sewage plus the alkalinity eventually formed in the digester, are sufficient to maintain the pH in the reactor above a certain minimum value that assures stable operation. Van Haandel and Lettinga (1994) have described a method to calculate the pH in anaerobic reactors treating wastewaters:

$$pH_{eff} = pK_1 - \log f_e \tag{3.37a}$$

with

$$f_e = (b - (b^2 - 4c)^{0.5}) \times 0.5$$

where:

f_e = $10^{(pK1-pH)}$
b = $(K_H + C_{ti} + C_{td} + C_{md} - C_{me} - Alk_e)/Alk_e$
c = $K_H(C_{ti} + C_{td} - Alk_e)/Alk_e^2$
K_H = Henry's constant for CO_2 in water ($pK_H = 1.12 + 0.0138t$ for $0 < t < 35$°C, Loewenthal and Marais 1986)
C_{ti} = total carbonic species concentration in the influent
C_{td} = total generated carbonic species generated in the reactor (≈1/64 of biodegraded COD concentration)
C_{md} = total methane concentration generated in the reactor (=1/64 of biodegraded COD concentration)
C_{me} = total methane concentration dissolved in the effluent (≈1 mmol/L)
Alk_e = effluent alkalinity concentration.

(Concentrations are in mol/L, alkalinity is in eq/L.)

Estimate of required alkalinity addition The validity of Equation (3.37) to calculate pH is limited by the fact that it is presupposed that methanogenesis of biodegradable material will occur. When the calculated pH is low (for example below 6.3) this presupposition is not valid. In that case an alternative approach must be made: It must be calculated what is the required alkalinity to maintain some desired pH value. The alkalinity demand for anaerobic digestion of a wastewater can be determined when the minimum required operational pH is specified.

The value of the pH is set at a determined value, for example $pH_r = 7.0$. The required alkalinity can be determined from:

$$Alk_r = 0.5(b - (b^2 - 4ac)^{0,5})/a \tag{3.37b}$$

where:
$f_r = 10^{(pK1 - pH_r)}$
$a = (1 + f_r)f_r$
$b = (C_{ti} + 2C_{td} - C_{ml})f_r + k_H(1 + f_r)$
$c = k_H(C_{ti} + C_{td})$

The result of Equation (3.37) is the required influent alkalinity to maintain the pH value that has been assumed (pH_r). If the actual influent alkalinity is smaller than the required value, the difference must be added to the influent. Certainly, if the alkalinity is higher than the desired pH, the actual pH will be higher than the chosen pH_r value and may be calculated from Equation (3.36). Figure 3.15 shows the required alkalinity as a function of the digested COD concentration for different pH values in the digester at 25°C and at 35°C. The assumptions made to construct the diagrams are indicated.

Impact of effluent or biogas recycle If the operational pH of anaerobic digester is too low for stable digestion, alkalinity addition is not the only alternative: acidity may be removed by CO_2 subtraction from the liquid or from the gas phase by recirculating

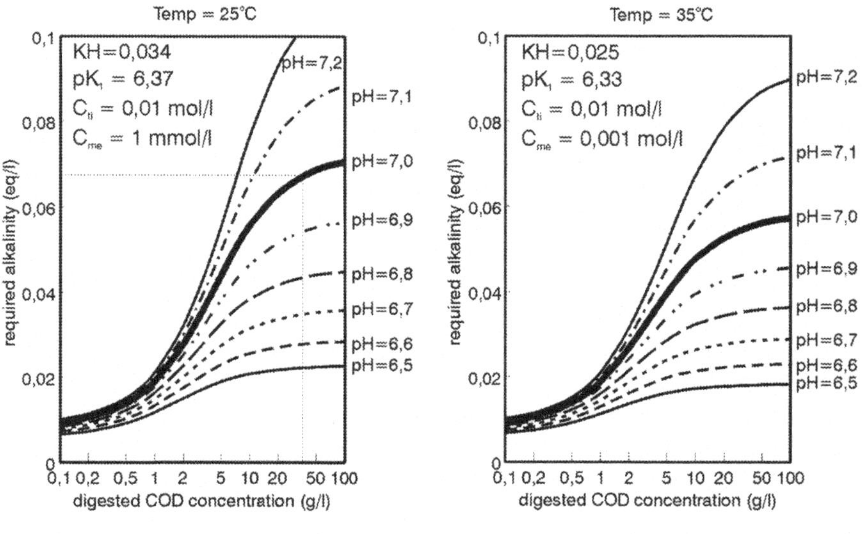

Figure 3.15 Required alkalinity in an anaerobic digester as a function of the digested COD concentration for different pH values in the digester (and effluent) and for temperatures of 25°C (left) and 35°C (right).

effluent or biogas, respectively. Carbon dioxide will desorb from the effluent when this is placed in contact with the atmosphere. Acidity removal by effluent recirculation becomes particularly efficient, when it is mixed with influent with a low pH. In this case the effluent bicarbonate will be transformed into CO_2 and most of the carbonic species concentration will desorb from the supersaturated mixture. CO_2 can also be absorbed from the biogas by contacting it with an alkaline solution, so that (bi)carbonate is formed. It can be demonstrated that normally it is more economical to recirculated effluent than biogas, by considering that the gas phase has a much lower concentration of carbonic species than the effluent, whereas the pumping costs per unit volume are in principle the same for both recirculations (the same head is required). Thus, normally the effluent will be recirculated.

If it is assumed that the removed carbonic species fluxes due to effluent recirculation and biogas recirculation are a factor X and Y, respectively, of the flux discharged in the effluent (and that no methane is lost in the recycles), Equation (3.37a) and (3.37b) may be rewritten to calculate pH and required alkalinity for system with effluent and biogas recirculations. The effect of recirculation is equivalent to a decrease of Henry's constant by a factor (1 + X + Y) that is the apparent Henry constant may be written as:

$$K'_H = K_H/(1 + X + Y) \quad (3.38)$$

where:

K'_H = apparent Henry constant in a digester with recirculation

X = ratio of the carbonic species flux desorbed from recirculated effluent to the carbonic species flux discharged in the effluent

Y = ratio of carbonic species flux absorbed from recirculated biogas to the carbonic species flux discharged in the effluent.

3.3.3.3 *Toxic compounds*

Apart from the hydrogen ion concentration (pH), several other compounds, such as heavy metals and chloro-organic compounds, affect the rate of anaerobic digestion, even at very low concentrations. However, the presence of these compounds at inhibitory concentrations is limited to specific in wastewaters. Potentially, toxic compounds often present are oxygen and sulphide. Some oxygen may be introduced in the influent distribution system, but it will be used for oxidative metabolism in the acidogenesis process. Thus, no DO will be present in the anaerobic reactor, so that its introduction will be of no consequence for the performance of the reactor. Sulphide can be formed in the process due to the occurrence of reduction of sulphate. This aspect will be discussed in depth in Section 3.6 and in Chapter 6.

3.4 NITROGEN REMOVAL IN WASTEWATER SYSTEMS

3.4.1 Forms and reactions of nitrogenous material in wastewaters

In most wastewaters, nitrogenous material is composed of ammonium (gas, NH_3 or ionized, NH_4^+) and organic nitrogen (urea, amino compounds and RNH_2) and amino acids ($R_1NH_2R_2COOH$). In some industrial wastewaters, other compounds like oxidized forms (NO_3^-, NO_2^-) may be present. The sum of ammonium and organic nitrogen concentrations is called the total Kjeldahl nitrogen concentration (TKN).

In aerobic wastewater treatment systems, several processes may occur that can change the nature of nitrogenous compounds. Figure 3.16 shows the most important processes that can occur: (a) ammonification or the inverse, ammonium assimilation, (b) nitrification and (c) denitrification.

(a) *Ammonification/assimilation*
Ammonification is the conversion of organic nitrogen to ammonium, whereas assimilation is the inverse process. Ammonification accompanies the mineralization of organic material during metabolism and occurs in both aerobic and anaerobic processes. Assimilation occurs during sludge synthesis when the bacteria satisfy their demand for nitrogen as a material source. At neutral pH ionized ammonium (NH_4^+) is predominant, and the ammonification and assimilation processes can be expressed as:

$$RNH_2 + H_2O + H^+ \underset{\text{Ammonification}}{\overset{\text{Assimilation}}{\rightleftharpoons}} ROH + NH_4^+ \tag{3.39}$$

(b) *Nitrification*
Nitrification is the biological oxidation of ammonium with nitrate as the final product. The reaction requires the mediation of specific bacteria and is realized in two sequential steps. In the first step, also called nitritation, ammonium is oxidized to nitrite by the action of bacteria of the genera *Nitrosomonas*. The second step is the oxidation of nitrite to nitrate mediated by bacteria of the genera *Nitrobacter*. Both *Nitrosomonas* and *Nitrobacter* develop biochemical activity only in the presence of oxygen, that is, they are strictly aerobic microorganisms.

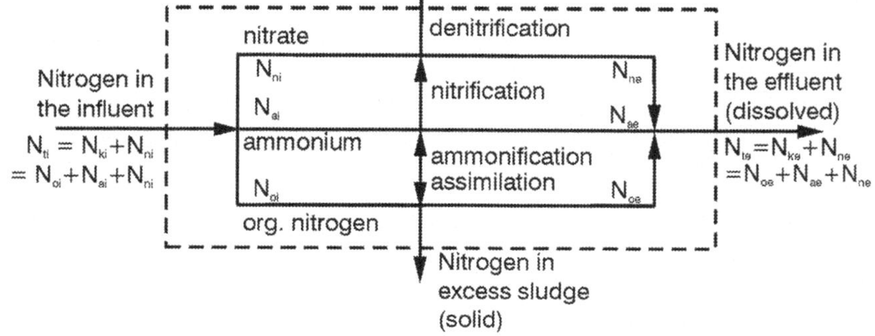

Figure 3.16 Schematic representation of the forms and reactions of nitrogenous material in aerobic wastewater treatment systems.

The two steps can be written as:

$$NH_4^+ + 3/2O_2 \rightarrow NO_2^- + H_2O + 2H^+$$

$$NO_2^- + 1/2O_2 \rightarrow NO_3^-$$

$$NH_4^+ + 2O_2 \rightarrow NO_3^- + H_2O + 2H^+ \quad (3.40)$$

(c) *Denitrification*
In the denitrification process, nitrate is reduced to molecular nitrogen by organic material. This process takes place in an environment without oxygen, when the nitrate substitutes oxygen as oxidant of organic material. To effect denitrification systematically, part of the total reactor volume is kept in an anoxic environment (i.e. without aeration): For a general structural formulae of organic nitrogen $C_xH_yO_z$ the redox reaction can be written as:

Oxidation:

$$C_xH_yO_z + (2x - z)H_2O \rightarrow xCO_2 + (4x + y - 2z)H^+ + (4x + y - 2z)e$$

or:

$$1/(4x + y - 2z)\ C_xH_yO_z + (2x - z)/(4x + y - 2z)\ H_2O \\ \rightarrow 1/(4x + y - 2z)\ CO_2 + H^+ + e$$

Reduction:

$$e + 6/5H^+ + 1/5NO_3^- \rightarrow 1/10N_2 + 1/5H_2O$$

Redox:

$$1/(4x + y - 2z)\ C_xH_yO_z + 1/5H^+ + 1/5NO_3^- \rightarrow x/(4x + y - 2z)\ CO_2 + \\ 1/10N_2 + 1/5(2x + 3y - z)/(4x + y - 2z)\ H_2O$$

or:

$$C_xH_yO_z + (4x + y - 2z)/5H^+ + (4x + y - 2z)/5NO_3^- \\ \rightarrow x\ CO_2 + 1/5(2x + 3y - 2z)H_2O + 1/10\ (4x + y - 2z)N_2 \quad (3.41)$$

The concentration of nitrogenous material in industrial wastewaters depends strongly on the nature of the industrial process: wastes of vegetable products (like beer, paper, canned fruits, etc.) have relatively little nitrogen and the TKN/COD ratio usually is in the range of 0.01–0.04. Wastewaters from industrial, activities of animal processing (slaughter houses, tanneries, milk processing, etc.) have a ratio in the range of 0.10–0.16 gTKN/g COD. Domestic sewage has an intermediate value with ratios in the range of 0.06–0.12 gTKN/g COD, depending on the protein consumption of the contributing population. In many cases, the nitrogenous material in wastewater was originally organic nitrogen, but in the processes of storage and transport, a large fraction of the TKN may already be converted into ammonium. If ammonification does not take place before treatment it will occur readily within the biological treatment system as mineralization of organic material takes place.

3.4.1.1 Mass balance equations

In Figure 3.16, it can be noted that nitrogenous material may leave the treatment system under three different forms: (1) as organic nitrogen in the excess sludge, (2) as dissolved matter in the effluent or (3) as nitrogen gas to the atmosphere after denitrification. Volatilization of ammonium is not considered in Figure 3.16, although this process may occur to some extent, especially in industrial wastes with a high pH. By using the same concepts as developed for organic material, the recovery factor for nitrogen can be expressed as the ratio of the nitrogenous material in the three fluxes leaving the treatment system (solid, liquid and gas) and the flux entering it:

$$B_n = (MN_l + MN_{te} + MN_d)/MN_{ti} \tag{3.42}$$

where:
B_n = recovery factor of nitrogenous material
MN_l = flux of nitrogenous material in the excess sludge
MN_{te} = flux of nitrogenous material in the effluent
MN_d = flux of denitrified nitrogen
MN_{ti} = flux of nitrogenous material in the influent.

Equation (3.42) is only useful when the fluxes are expressed as measurable parameters; only then it can be verified experimentally if the recovery factor approaches its theoretical value of 1.00.

The flux MN_l can be calculated by remembering that a fraction f_n = 0.1 mg N/mg VSS is sludge is nitrogen:

$$MN_l = f_n MX_v/\theta_c \tag{3.43}$$

The fluxes in the influent and effluent can easily be expressed as:

$$MN_{ti} = Q_a(N_{oi} + N_{ai} + N_{ni}) = Q_i N_{ti} \tag{3.44}$$

$$MN_{te} = Q_a(N_{oe} + N_{ae} + N_{ne}) = Q_a N_{te} \tag{3.45}$$

The flux of denitrified nitrogen can be calculated from the difference between the fluxes entering and leaving anoxic reactors in the treatment system.

$$MN_{dk} = Q_k \, \Delta N_{nk} \tag{3.46}$$

where:
MN_{dk} = flux of denitrified nitrogen in anoxic reactor "k"
Q_k = flow passing through reactor "k" (=influent flow + return sludge flow + eventual other recirculation flows)
ΔN_{nk} = variation of nitrate concentration in anoxic reactor "k".

3.4.1.2 Stoichiometry of reactions with nitrogenous material

Oxygen uptake With respect to oxygen uptake only the nitrification and the denitrification processes are of importance. Figure 3.17 shows schematically the electron transfers in these processes. During nitrification the oxidation number of nitrogen increases from its initial value of −3 in ammonium to +5 in nitrate, that is an increase of 8 caused by the loss of 8 electrons accepted by oxygen, so that

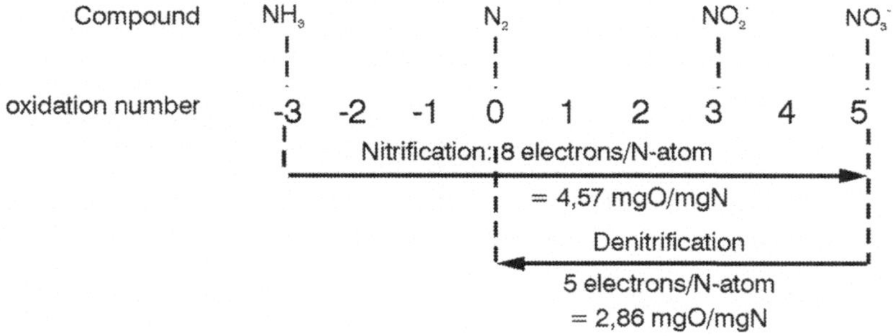

Figure. 3.17 Variation of the oxidation number of nitrogen in the nitrification and the denitrification process.

2 mol of oxygen are required per nitrogen atom. It is concluded that nitrification requires 2 mol (64 g) of oxygen per mol of nitrogen (14 g), that is the oxygen demand is 64/14 = 4.57 mg O/mg N.

In the denitrification process, nitrate (oxidation number +5) is reduced to molecular nitrogen (oxidation number 0), so that a transfer of 5 electrons takes place per denitrified N atom. It is concluded that from the 8 electrons released during nitrification only 5 are recovered during denitrification. Thus, in oxidimetric terms, nitrate has an oxidation equivalent of 5/8 of the oxygen required for nitrification or in other words, 5/8 or 62.5% of the oxygen required for nitrification (4.57 mg O/mg N) can be recovered as "equivalent oxygen" during denitrification, (i.e., $0.625 \times 4.57 = 2.86$ mg O/mg N). Hence, in biological removal by nitrification + denitrification, there is a net oxygen consumption of 4.57–2.86 = 1.71 mg O/mg N.

3.4.1.3 Alkalinity change

The effect of ammonification, nitrification and denitrification on the alkalinity can be evaluated from simple stoichiometric relations. In the three corresponding reaction equations (Equations (3.39) (3.40) and (3.41)) it can be noted that hydrogen ions are involved: during ammonification there is consumption of 1 molH^+ per mol N (14 g N), during nitrification 2 molH^+ are produced per mol N and during denitrification again there is consumption of 1 molH^+ per mol N. Knowing that consumption of 1 mol H^+ is equivalent to production of 1 equivalent alkalinity or 50 g $CaCO_3$, the alkalinity change due to the processes of nitrogenous material is expressed as:

$$(\Delta \text{Alk}/\Delta N)_{\text{am}} = 50/14 = 3.57\,\text{mg CaCO}_3/\text{mg N} \tag{3.47a}$$

$$(\Delta \text{Alk}/\Delta N)_{\text{n}} = -100/14 = -7.14\,\text{mg CaCO}_3/\text{mg N} \tag{3.47b}$$

$$(\Delta \text{Alk}/\Delta N)_{\text{d}} = 50/14 = 3.57\,\text{mg CaCO}_3/\text{mgN} \tag{3.47c}$$

where:
$(\Delta \text{Alk}/\Delta N)$ = alkalinity change/mg N.

Indices am, n and d refer to ammonification, nitrification and denitrification, respectively.

The total alkalinity change of the activated sludge system can now be written as the sum of the alkalinity changes of the three processes:

$$\Delta Alk_t = \Delta Alk_{am} + \Delta Alk_n + \Delta Alk_d = 3.57(N_{oa} - N_{oe} - N_l) - 7.14(N_{ka} - N_{ke} - N_l) + 3.57(N_{na} + N_{ka} - N_{ke} - N_l - N_{ne}) \quad (3.48)$$

Remembering that the TKN concentration, N_k, is the sum of the organic nitrogen (N_o) and ammonium (N_a) concentrations, one has:

$$\Delta Alk_t = -3.57(N_{aa} - N_{na} - N_{ae} + N_{ne}) = 3.57(\Delta N_a - \Delta N_n) \quad (3.49)$$

where:
ΔN_a = variation of the ammonium concentration in the activated sludge system
ΔN_n = variation of the nitrate concentration in the activated sludge system.

In Equation (3.49) all parameters on the right side of the equation can be determined experimentally so that it is possible to calculate theoretically the alkalinity change in an aerobic treatment process on the basis of the change of ammonium and nitrate concentration in the influent and the effluent. The theoretical value can then be compared to the experimentally observed value of alkalinity change. In Figure 3.18, the value of the theoretical alkalinity change is plotted as a function of the experimental alkalinity change. Figure 3.18 refers to aerobic systems with different reactions of nitrogenous material: (a) without nitrification, (b) with nitrification and no denitrification and (c) with nitrification and denitrification (ammonification always occurs). It can be seen that for a wide range of experimental values there was an excellent correlation between experimental and theoretical values. It is concluded that the alkalinity changes in aerobic treatment plants can reliably be predicted from the stoichiometry of nitrogenous material reactions, according to Equation (3.48).

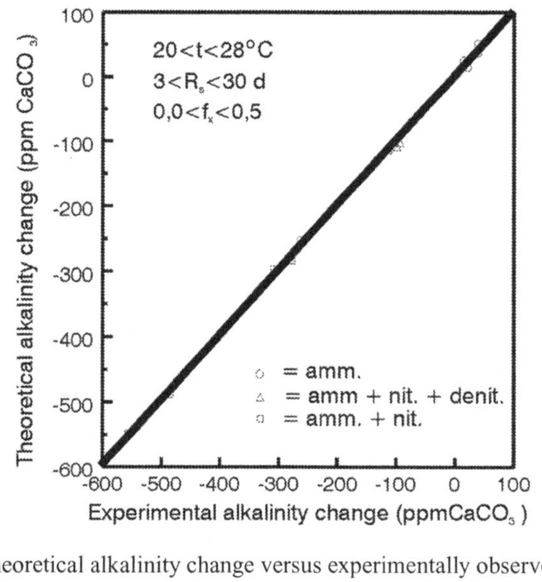

Figure 3.18 Theoretical alkalinity change versus experimentally observed values for different aerobic treatment systems.

The alkalinity change in aerobic wastewater treatment systems is important because it has a direct influence on the pH of the mixed liquor in the biological reactor, which in turn is decisive for the rate of substrate metabolism. Loewenthal and Marais (1976) have described the interrelationship between alkalinity, acidity and pH solutions with weak acid base systems. Van Haandel and Lettinga (1994) have shown that for most wastewaters, this interrelationship is due almost exclusively to equilibrium of the carbonic system CO_2—HCO_3^-—CO_3^{-2}. For this system the alkalinity is defined as:

$$\text{Alk} = [HCO_3^-] + 2[CO_3^=] + [OH^-] - [H^+] \tag{3.50}$$

Now by substituting the dissociation equations for the carbonic system and for water the relationship between pH and alkalinity, it can be written explicitly:

$$\text{Alk} = [CO_2][k_1^*/(H^+) + 2k_1^*k_2^*/(H^+)^2] + k_w^*/(H^+) - (H^+)$$

or

$$\text{Alk} = [CO_2]10^{pH-pk*1}(1 + 2 * 10^{pH-pk*2}) + 10^{pH-pk*w} - 10^{-pH} \tag{3.51}$$

Figure 3.19 shows the pH curve as a function of alkalinity for concentrations of dissolved CO_2 of 0.5 mg/L (saturation value a 20°C) 2 and 10 mg/L (i.e. 4 and 20 times the saturation value, respectively). To construct the diagram a temperature of 20°C and activity coefficients of $f_m = 0.90$ and $f_d = 0.67$ for mono and divalent ions, respectively have been assumed. (These values correspond to an ionic force of 0.01 according to Loewenthal and Marais 1976.)

Figure 3.19 shows that for an alkalinity of more than about 35 ppm $CaCO_3$, the pH value does not vary considerable when the alkalinity increases. For example,

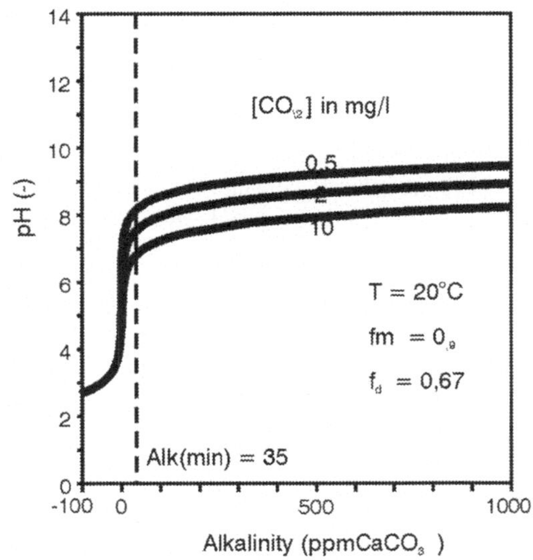

Figure 3.19 pH change as a function of mixed liquor alkalinity.

an increase of 35–500 ppm $CaCO_3$ results in an increase of pH of less than 1 unit. By contrast, when the alkalinity decreases from 35 ppm there is a steep decrease of pH. Reduction from 35–0 ppm leads to a pH of 4.2 approximately.

A low pH value affects the metabolic capacity of microorganisms. In aerobic treatment systems the autotrophic bacteria *Nitrosomonas* and *Nitrobacter*, both active in Nitrification are particularly sensitive to pH and virtually cease activity below pH = 6. Thus, to ensure stable and efficient nitrification, the alkalinity must be less than 35 ppm $CaCO_3$ at any time. It is interesting to note that Haug and McCarty (1971) established experimentally the same minimum alkalinity derived above theoretically. Now by using the expression for alkalinity, change the minimum alkalinity in the effluent may be calculated:

$$\Delta \mathrm{Alk_t} = \mathrm{Alk_e} - \mathrm{Alk_i} = 35 - \mathrm{Alk_i} = 3.57(\Delta N_a - \Delta N_n)$$

or, approximately,

$$\mathrm{Alk_i} > 3.57(\Delta N_n - \Delta N_a + 10) \tag{3.52}$$

where:
$\mathrm{Alk_i}$ = influent alkalinity
$\mathrm{Alk_e}$ = effluent alkalinity.

If the influent alkalinity is smaller than the required minimum, the difference must be added as external alkaline material (not ammonium or urea!).

3.4.2 Nitrification kinetics

Downing *et al.* (1964) were the first to show that nitrification can be modelled with Monod kinetics, by considering that *Nitrosomonas* is the rate limiting step and that nitrite oxidation in many cases is virtually instantaneous. The residual ammonium concentration in a completely mixed steady state aerobic treatment system (and hence in the effluent) can be expressed as:

$$N_{ae} = K_n(b_n + 1/\theta_c)/[\mu_m - (b_n + 1/\theta_c)] \tag{3.53}$$

where:
N_{ae} = effluent ammonium concentration (mg N/L)
μ_m = maximum specific growth rate constant for *Nitrosomonas* (per day)
b_n = decay constant for *Nitrosomonas* (per day)
K_n = Monod half saturation constant (mg N L^{-1}).

Hence the ammonium concentration depends on the three kinetic constants and the sludge age, but not on the influent concentration.

Naturally the ammonium effluent concentration cannot be higher than the ammonium concentration available for nitrification. This condition defines the minimum sludge age for nitrification:

$$N_a < N_p = K_n\,(b_n + 1/\theta_{cn})/(\mu_m - b_n - 1/\theta_{cn})$$

or

$$\theta_{cn} = (1 + K_n/N_p)/[\mu_m - b_n(1 + K_n/N_p)] \tag{3.54}$$

Table 3.7 Typical values as well as atypically low and high values of the kinetic constants for nitrification in aerobic treatment systems.

Kinetic constant	Unit	Low value	Typical value	High value
μ_m	Per day	0.2	0.4	0.8
b_n	Per day	0	0.04	0.1
K_n	mg N/L	0	1	2

where:
N_p = nitrification potential or ammonium concentration available for nitrification in the influent.

For most wastewaters, the available ammonium concentration always will be very much higher than the K_n value, so that $K_n/N_p << 1$ and Equation (3.54) simplifies to:

$$\theta_{cn} \approx 1/(\mu_m - b_n) \tag{3.54}$$

The *Nitrosomonas* concentration can now be calculated from activated sludge theory by using the same expression as the one derived for heterotrophic bacteria using organic material (Equation (3.11)):

$$X_n = [Y_n\theta_c/(1 + b_n\theta_c)]N_p/\theta \tag{3.56}$$

where:
X_n = *Nitrosomonas* concentration (mg VSS/l)
Y_n = yield coefficient for *Nitrosomonas* = 0.1 mg VSS/mg N.

Many research workers have reported experimental values for the kinetic constants. Since these data vary considerably even for comparable environmental conditions, it would appear that their values also depend on the origin of the wastewater. In that case, it is important that the value be known, so that an expedite method is necessary to determine their values experimentally.

In Table 3.7, on the basis of published values, for each of the constants "typical" values are attributed to the kinetic constants, as well as atypically low and high values. The typical value can be found in domestic sewage without industrial contributions. To evaluate the influence of the numeric values of the constants on nitrification efficiency, Figure 3.20A is presented, in which the residual ammonium concentration is depicted as a function of the sludge age for typical values of the constants b_n and K_n (0.04 per day and 1 mg/L, respectively) and a low (0.2 per day) as well as a high value (0.8 per day) for the constant μ_m. It can be noted that the change in μ_m value has a very pronounced influence on the minimum sludge age for nitrification and on the nitrification efficiency. Similarly the influence of variation of the b_n and K_n values is evaluated in Figures 3.20B and C. Clearly, the influence of these constants on nitrification efficiency is much smaller, so that their typical values may be accepted for calculation without the danger of a large error. In that case, only the μ_m value must be determined and the factors that influence it must be evaluated.

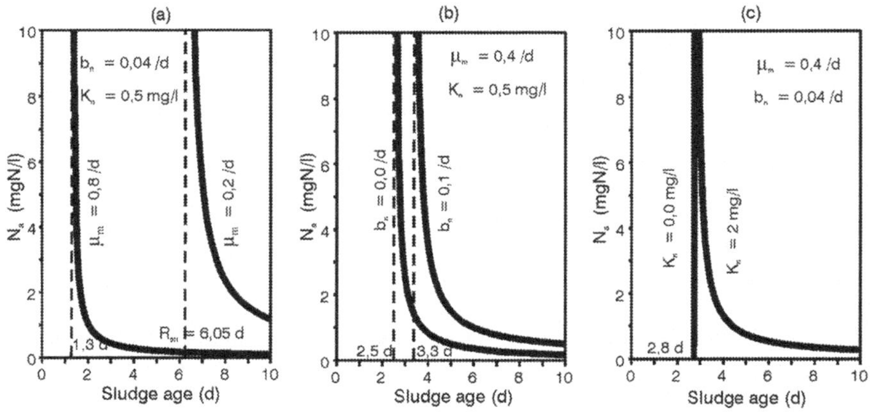

Figure 3.20 Evaluation of the influence of kinetic constants values on nitrification.

3.4.2.1 *Experimental determination of μ_m for nitrification*

A very convenient way to determine the μ_m value is by using respirometry and determining the maximum OUR in a batch of sludge, generated under steady state conditions with the wastewater to be investigated. When a batch of this sludge is aerated without adding substrate, the OUR will decrease and stabilize at a value corresponding to the endogenous respiration rate. Now, when excess ammonium is added (for example as an NH_4Cl solution), the OUR increases to a constant high value, because nitrification will take place at maximum speed. Thus, the nitrification rate can be calculated as a factor 1/4.56 of the OUR increase after ammonium addition. This nitrification rate, r_{mn}, can be linked directly to the *Nitrosomonas* concentration and the μ_m value:

$$r_{mn} = \mu_{mn} X_n / Y_n \tag{3.57}$$

As an example, Figure 3.21 shows the OUR as a function of time in a batch of sludge where 5 mg N/L has been added. The *Nitrosomonas* concentration in the sludge batch (as calculated from Equation (3.55)) was estimated at 30 mg/L. The OUR increase after ammonium addition was from 7 to 18 mg/L/h or 11 × 24 = 264 mg/L/d, whence the nitrification rate is calculated as 264/4.56 = 58 mg N/L/day. For an estimated $Y_n = 0.1$ mg N/mg X_n the μ_m value is estimated 58 × 0.1/30 = 0.19 per day.

3.4.2.2 *Factors influencing the μ_m value*

(a) DO concentration The influence of DO on the μ_m value can be expressed with Monod kinetics. A convenient way to determine the influence of DO experimentally is to carry out respirometric tests at different average DO concentrations and determine the different values of μ for non-limiting ammonium concentrations. The influence of the DO concentration can now be expressed as:

$$\mu_m = \mu_{max}(DO/DO + K_o) \tag{3.58}$$

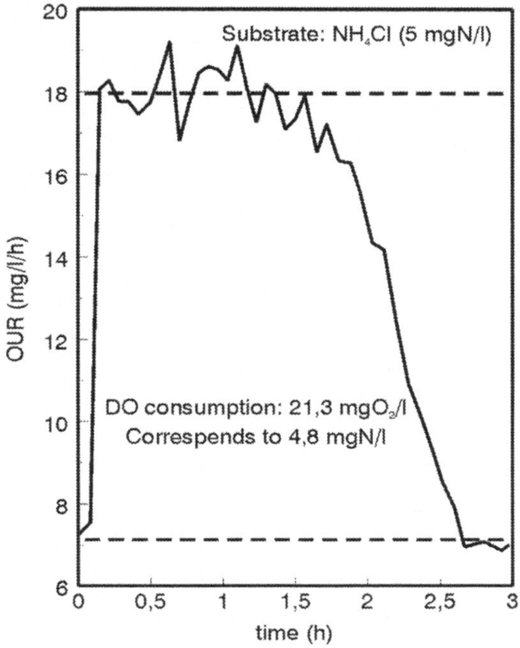

Figure 3.21 OUR as a function of time in a sludge batch after ammonium addition.

where:

μ_m = the maximum specific growth constant for *Nitrosomonas* under non-limiting ammonium and limiting DO concentration

μ_{max} = maximum specific growth constant for *Nitrosomonas* when both ammonium and DO are non-limiting

DO = Dissolved oxygen concentration (mg O/L)

K_o = Monod half saturation constant for DO.

Figure 3.22 shows an example: for several values of DO the maximum specific growth rate for *Nitrosomonas* was determined and plotted as a function of the DO concentration. Then Equation (3.58) was plotted for several K_o values and an assumed value of μ_{max} = 0.4 per day. It can be seen that for K_o = 1 mg/L there is a good correlation between the experimental and theoretical values whence this value is adopted for K_o. It is concluded that under the specific conditions of the test the nitrification rate developed at half its maximum rate when the DO concentration was 1 mg/L. Reported values for K_o vary considerable (0.3–2.0 mg OD/L), which at least in part can be attributed to differences in mixing velocity and other operational conditions.

(b) Temperature Several researchers have presented the influence of temperature on nitrification with the Arrhenius expression as proposed in Section 2.5.7.

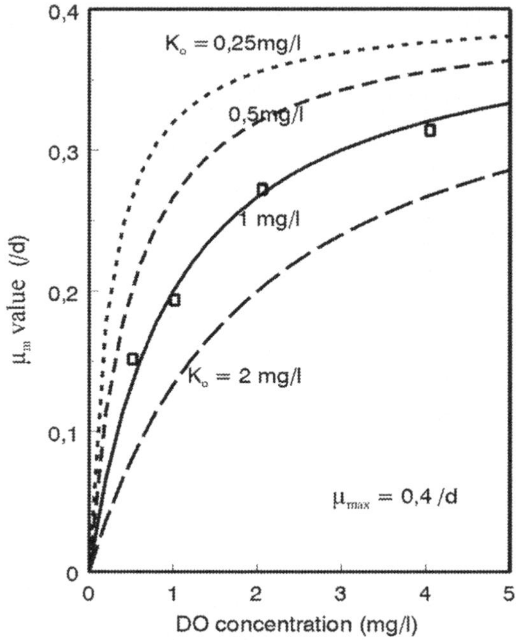

Figure 3.22 Influence of the DO concentration on the apparent μ_m value.

Table 3.8 Values of the temperature dependency (Θ) of on the μ_m value of *Nitrosomonas*.

Θ	Temperature range (°C)	Reference
1.116	19–21	Gujer (1977)
1.123	15–20	Downing *et al.* (1964)
1.123	14–20	Ekama and Marais (1976)
1.130	20–30	Lijklema (1973)

$$\mu_{mT} = \mu_{m20}\Theta^{(t-20)} \tag{3.59}$$

where:

Θ = Arrhenius temperature dependence factor.

Indices *t* and 20 refer to temperatures *T* and 20°C, respectively.

Table 3.8 shows some experimental values of the Θ values, oscillating between 1.11 and 1.13. This means that the μ_m value increases 11–13% per °C which is equivalent to doubling the value every 6–7°C. The repercussion of temperature on the μ_m value is very important as can be seen by the following. In regions with a moderate climate where wastewaters would tend to have a temperature in the range of 8–14°C, the μ_m value for a typical value of $\mu_{m20} = 0.4$ per day at 20°C, would be $\mu_{m14} = 0.2$ to $\mu_{m8} = 0.1$ per day. By using Equation (3.54) the corresponding minimum sludge age for nitrification is calculated as $\theta_{cn} = 6$–14 days. In Europe and the United States aerobic treatment systems commonly are operated at a shorter sludge age than this range, so that nitrification will not develop. In contrast, in tropical regions where wastewater temperatures are high, the development of nitrification is almost unavoidable. For a temperature of 26°C, the typical μ_m value is $\mu_{m26} = 0.8$ per day, and the minimum sludge age for nitrification is only $\theta_{cn} = 1.25$ day. In practice, aerobic system normally will be operated at a sludge age longer than 1.25 day so that at least partial nitrification will develop. If a system is designed for secondary treatment (organic material removal), but nitrification in fact can develop, the system is bound to have operational problems: the heterotrophic and autotrophic bacteria will compete for oxygen and the removal of both will be incomplete. Also in the environment with low DO is it likely that badly settling sludge will develop.

(c) Mixed liquor pH Most research workers indicate that the μ_m value is practically constant in the range $7 < pH < 8.5$. Beyond this range a rapid decrease of the growth rate constant is observed. In practice, many wastewaters have a pH in the neutral range, but due to alkalinity consumption for nitrification, the pH tends to decrease and eventually nitrification will cease or become unstable. To avoid that pH becomes less than 7 a minimum alkalinity of 35 ppm $CaCO_3$ must be maintained. If the natural alkalinity of the wastewater is insufficient, it must be added to the system.

(d) Non-aerated zones In aerobic treatment plants, designed for nitrogen removal, part of the reactor volume is kept under anoxic conditions, that is no oxygen is supplied and nitrate is used as an alternative oxidant of organic material. Since *Nitrosomonas* and *Nitrobacter* only grow under aerobic conditions, but decay under both aerobic and anoxic conditions, the inclusion of anoxic zones is equivalent to a reduction of the maximum specific growth rate constant. For an anoxic sludge mass fraction f_x the apparent constant is given as:

$$\mu'_m = (1 - f_x)\mu_m \tag{3.60}$$

Hence the residual ammonium concentration is now expressed as:

$$N_a = K_n(b_n + 1/\theta_c)/[(1 - f_x)\mu_m - b_n - 1/\theta_c] = K_n(b_n + 1/\theta_c)/[\mu'_m - b_n - 1/\theta_c] \tag{3.61}$$

μ'_m = apparent maximum specific growth rate constant for *Nitrosomonas* in systems with anoxic zones.

Equation (4.37) can also be expressed written in explicit terms for the anoxic sludge mass fraction:

$$f_x = 1 - (1 + K_n/N_a)(b_n + 1)\theta_c/\mu_m \tag{3.62}$$

In nitrogen removing aerobic treatment systems a high nitrification is necessary, since it is the prerequisite for nitrogen removal by denitrification. The maximum anoxic sludge fraction that can be maintained and still guarantee a residual ammonium concentration lower than a particular desired value, N_{ad}, can now be expressed as:

$$f_M = 1 - (1 + K_n/N_{ad})(b_n + 1/\theta_c)/\mu_m \quad (3.63)$$

where:

f_M = maximum sludge fraction than can be maintained under anoxic conditions, allowing the discharge of an ammonium concentration lower than N_{ad}.

N_{ad} = specified maximum ammonium concentration in the effluent.

Apart from the desired ammonium effluent concentration, the maximum anoxic sludge mass fraction depends on the sludge age in the aerobic system and the values of the kinetic constants, particularly the μ_m value. Figure 3.23 shows the maximum fraction f_M as a function of the sludge age for different μ_m values between 0.2 and 0.8 per day.

The numeric value of the maximum sludge mass fraction, f_M, has great practical importance: the larger f_M, the larger the capacity to remove nitrate by denitrification in the anoxic section of the treatment system. Along with the limitation imposed by nitrification, the value f_M is also limited by other considerations: very high anoxic fractions may affect the metabolic and mechanic properties (settleability) of the sludge, so that in practice the anoxic sludge mass is kept below 50% to 60%, even if nitrification is not limiting.

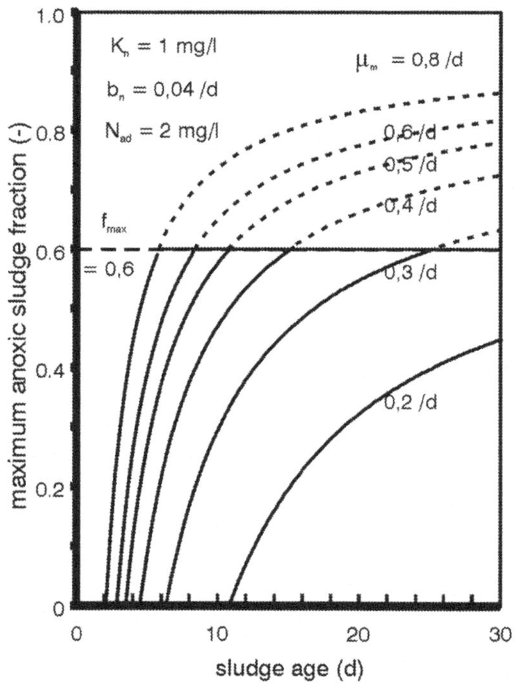

Figure 3.23 Maximum anoxic sludge mass fraction as a function of sludge age for different μ_m values.

3.4.3 Nitrification potential and nitrification capacity

The nitrification potential is the influent TKN concentration that is available for nitrification in aerobic treatment systems. Its value can be expressed as:

$$N_p = N_{ki} - N_l - N_{oe} \tag{3.64}$$

where:
N_p = nitrification potential
N_{ki} = influent TKN concentration
N_l = TKN concentration required for excess sludge production
N_{oe} = effluent organic nitrogen concentration.

The nitrification capacity is the influent TKN concentration that will effectively be nitrified in the treatment system. Hence the nitrification capacity is the difference between nitrification potential and the residual ammonium concentration.

$$N_c = N_p - N_{ae} \tag{3.65}$$

where N_c = nitrification capacity.
Using Equation (3.65) for $N_{ae} = N_{ad}$ and Equation (3.64) for N_p yields:

$$N_c = N_{ki} - N_l - K_n(b_n + 1/\theta_c)/[\mu_m(1 - f_M) - (b_n + 1/\theta_c)] \tag{3.66}$$

In Figure 3.24 the values of N_p and N_c can be observed as functions of the sludge age for the following conditions: (1) Composition of influent organic material

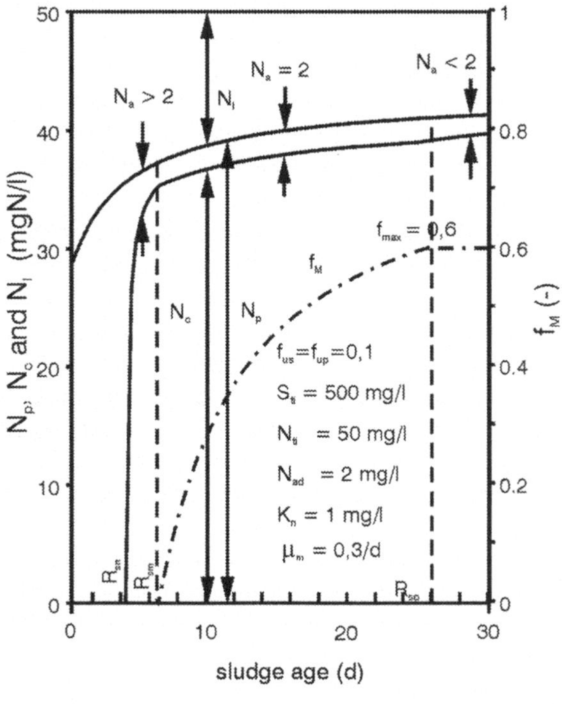

Figure 3.24 Nitrification potential and nitrification capacity as function of sludge age.

(to calculate the value of N_l): $f_{us} = f_{up} = 0.1$; $S_{ti} = 500\,mg/L$ (2) nitrification kinetics: $\mu_m = 0.3$ per day; $K_n = 1.0\,mg/L$; $b_n = 0.04$ per day; $N_{ad} = 2\,mg/L$; $f_{max} = 0.6$. Concentrations of $N_{ti} = 50$ and $N_{ni} = 0\,mg/L$ have been assumed. The minimum sludge ages for the onset of nitrification (θ_{cn}) for efficient nitrification, $f_M = 0$ and $N_{ae} = N_{ad}$ (θ_{cm}) and for the maximum anoxic sludge fraction $f_M = f_{max} = 0.6$ (θ_{co}) have also been indicated.

The fundamental aspect design of nitrogen removing activated sludge processes is to create a system with aerobic zones of sufficient magnitude to have efficient nitrification, whereas in the anoxic zones the produced nitrate is reduced to nitrogen gas. Hence, for efficient nitrogen removal, it is necessary that in the anoxic zone a denitrification capacity is created, that is compatible with the nitrification capacity of the aerobic zone.

3.4.4 Denitrification

3.4.4.1 Necessary conditions for denitrification

The necessary conditions for the denitrification process to develop in aerobic treatment systems are:

(a) **Presence of a suitable bacterial sludge mass**: Most bacteria present in aerobic treatment systems are facultative, that is, they can use nitrate as an oxidant to oxidize organic material when no DO is available. It has been established that bacteria generated in totally aerobic treatment systems start to use nitrate immediately when DO is no longer available and will do so at unchanged rate as long as the anoxic condition persists. It is concluded that there is no need for sludge adaptation to develop denitrification.

(b) **Presence of nitrate and absence of DO in the mixed liquor**: Presence of DO inhibits denitrification as it is the preferred oxidant of most bacteria for organic material. In most wastewaters nitrogen is not present in its oxidized forms (NO_2^-, NO_3^-), so that nitrification normally is a prerequisite for denitrification. The nitrate concentration has little influence on the denitrification rate: at concentrations of more than 0.5 mg N/L denitrification is not limited by nitrate.

(c) **Adequate environmental conditions for the microorganisms**: Temperature and pH are the most important environmental parameters. The denitrification rate increases with increasing temperature up to a maximum of about 40°C. The influence of pH has been described by several authors. The denitrification rate is maximum at pH values around the neutral value. In general it is found that the nitrification rate is much more influenced by pH than the denitrification rate.

(d) **Availability of an electron donor**: The presence of an electron donor is essential for the reduction of nitrate to nitrogen gas. The reductor is biodegradable organic material, which may be introduced from an external source after nitrification or be present as organic material in the influent. The choice of the organic material is of fundamental importance for the configuration of nitrogen removing treatment systems.

3.4.4.2 Configurations of systems with nitrogen removal

Although initially denitrification with an external source of organic material was suggested (Bart *et al.* 1969) modern nitrogen removing plants invariably use the organic material present in the wastewater. In these systems the sludge is placed alternately in an aerobic environment (where nitrification occurs) and in an anoxic environment (where denitrification develops). Normally the alternating environment is established by circulating the sludge between continuously aerated zones and zones where only stirring but no aeration is applied.

The first denitrification system placed in operation by Wuhrmann (1964) was composed of two reactors (Figure 3.25). The first reactor was aerobic and received all the influent as well as the return sludge from the settler. In this reactor, nitrification developed and a large portion of the influent biodegradable organic material was also removed. The nitrified mixed liquor was then introduced in the second reactor which operated in an anoxic environment. Such an anoxic reactor is also called a post-denitrification (Post-D) reactor, because the anoxic environment is established after the mixed liquor is submitted to an aerobic environment.

The denitrification rate in the Wuhrmann system is low because the concentration of biodegradable material is low in the post-D reactor. Thus, in order to obtain considerable denitrification, the anoxic sludge mass fraction must be large, but the magnitude of this fraction is limited because of the need for efficient nitrification in the aerobic reactor (see Equation 3.63).

In the system proposed by Ludzack and Ettinger (1962), the influent organic material is used directly for nitrate reduction. The pre-denitrification (pre-D) has two reactors, the first being anoxic and the second aerobic. The nitrate formed in the second reactor is recirculated to the anoxic reactor via the underflow from the settler as well as a mixed liquor flow directly from the aerobic to the anoxic reactor

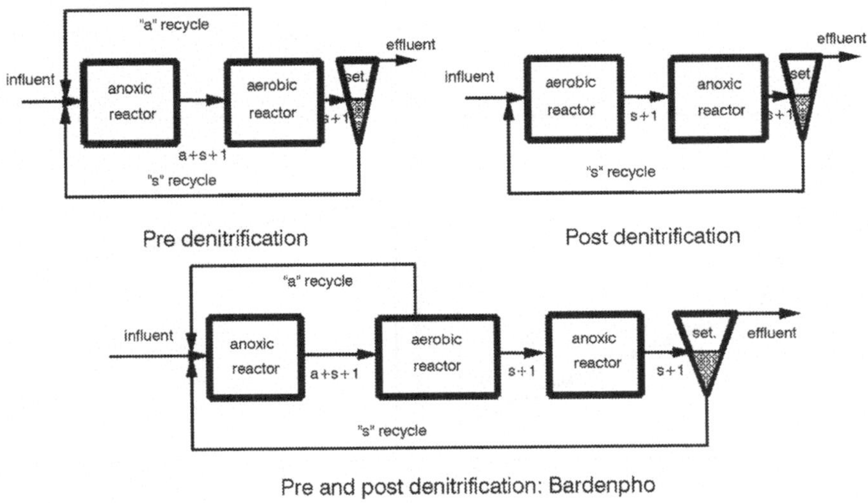

Figure 3.25 Schematic representation of denitrification with internal source of organic material.

as indicated in Figure 3.25. The pre-D system allow a much higher denitrification rate, but has the disadvantage than complete nitrate removal is impossible as a fraction of the mixed liquor passes from the aerobic reactor (with nitrate) through the settler and the liquid phase is discharged without denitrification. The removal efficiency depends on the mixed liquor recirculation flow from the aerobic to the anoxic reactor.

The Bardenpho system proposed by Barnard (1973) has both pre- and post-D reactor (Figure 3.25) and thus allows combining the advantage of pre-D system (high denitrification rate) and post-D system (feasibility of complete denitrification). Sometimes a re-aeration reactor is placed between the post-D reactor and the settler to avoid that the sludge remains in an anoxic environment for a very long time.

3.4.4.3 Denitrification kinetics

It has been amply established that the sludge production and composition in nitrogen removing systems with anoxic zones can be described by the same equations that have been developed for purely aerobic systems. (Marsden and Marais 1976; Sutton *et al.* 1979). The denitrification rate in pre- and post-D reactors can be determined from the nitrate profile as a function of the retention time in anoxic reactors as indicated in Figures (3.26A and 3.26B) for pre- and post-D reactors, respectively.

In pre-D reactors the nitrate concentration decreases rapidly because (a) there is a high concentration of biodegradable material and (b) part of this material is soluble and removed at a very high rate.

The utilization rate of soluble organic material in a pre-D reactor is so high that the duration of the primary phase is only a few minutes, much shorter than the usual retention time of mixed liquor in anoxic zones (a few hours). Thus, for practical purposes the denitrification associated with utilization of soluble organic material can be considered as instantaneous. If in the influent the soluble biodegradable material has a concentration of S_{bsi}, its concentration after mixing with recirculations from the settler (s) and the aerobic reactor (a) will become $S_{bsi}/(1 + a + s)$ (it is assumed

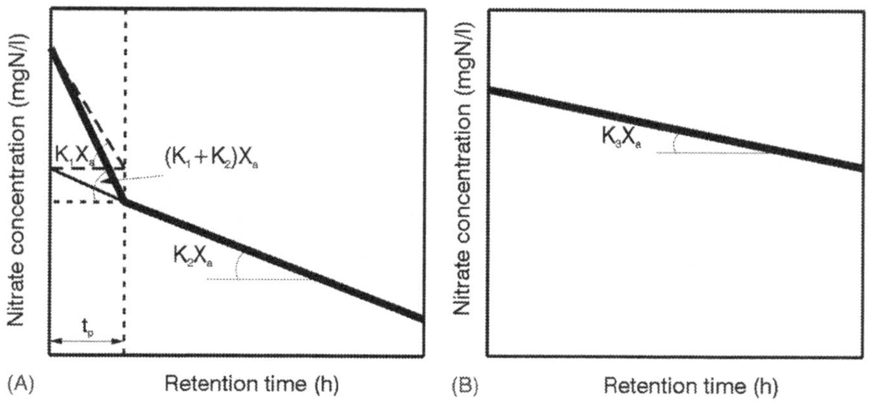

Figure 3.26 Typical profiles of the nitrate concentration in anoxic plug flow reactors Pre-D (A) and Post-D (B).

that in the "a" and "s" recycles no soluble biodegradable material is present). Knowing that upon metabolism a fraction $(1 - f_{cv}Y)$ is oxidized and that nitrate N is equivalent to 2.86 mg O/mg N, the reduction of the nitrate concentration in the pre-D reactor can be expressed as:

$$\Delta N_{ns} = (1 - f_{cv}Y)/2.86 \times S_{bsi}/(1 + a + s) = 0.12S_{bsi} = 0.12f_{bs}S_{bi} = 0.12(1 - f_{us} - f_{up})f_{bs}S_{ti} \quad (3.67)$$

Van Haandel and Marais (1981) showed that after completing the very high denitrification rate in a pre-D reactor due to utilization of soluble material, denitrification proceeds at a slower rate by utilization of particulate biodegradable material, which is proportional to the *active* sludge concentration. Hence:

$$r_{Dp} = (dN_n/dt) = -K_2X_a \quad (3.68)$$

r_{Dp} = denitrification rate associated to the utilization do particulate biodegradable material
K_2 = denitrification constant in a pre-D reactor (mg N/mg X_a/day)

Similarly the denitrification rate in the post-D reactor is described as:

$$r_D = (dN_n/dt) = -K_3X_a \quad (3.69)$$

K_3 = denitrification constant in a post-D reactor (mg N/mg X_a/day)

The numeric values of the constants K_2 and K_3 have been calculated on the basis of experimental data from many reports, all using raw domestic sewage as influent of nitrogen removing systems. It has been shown that in good approximation the constants are given by (Van Haandel and Marais 1981):

$$K_2 = 0.1(1.08)^{T-20}$$
$$K_3 = 0.08(1.03)^{T-20} \quad (3.70)$$

3.4.4.4 Denitrification capacity

It is convenient to introduce the concept of denitrification capacity which is the concentration of nitrate (expressed as mg N/L influent) that can be removed in an anoxic reactor. From the nitrate profiles in anoxic reactors the following expressions can be derived:

(1) for a pre-D reactor with enough sludge mass to assure complete removal of the soluble biodegradable material:

$$D_{c1} = (0.12f_{sb} + K_2C_rf_{x1})S_{bi} \quad (3.71)$$

(2) For a post-D reactor:

$$D_{c3} = K_3C_rf_{x3}S_{bi} \quad (3.72)$$

where:
D_{c1} = denitrification capacity in a pre-D reactor (mg N/L influent)
D_{c3} = denitrification capacity in a post-D reactor (mg N/L influent)
K_2 = denitrification constant in the pre-D reactor

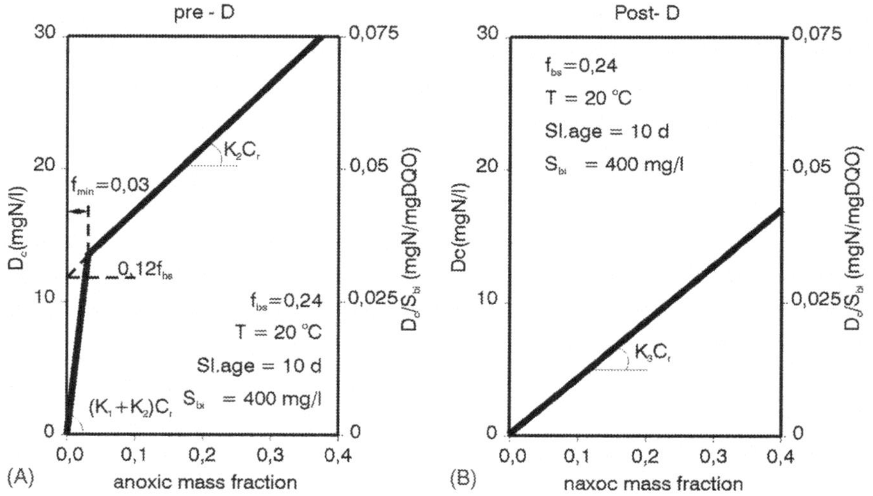

Figure 3.27 Denitrification capacity as a function of the anoxic sludge mass fraction in pre-D and post-D reactors.

K_3 = denitrification constant in the post-D reactor
f_{x1} = sludge mass fraction in the pre-D reactor ($f_{x1} > f_{min}$)
f_{x3} = sludge mass fraction in the post-D reactor
S_{bi} = biodegradable influent COD concentration
f_{sb} = soluble biodegradable COD fraction
$C_r = Y\theta_c/(1 + b_h\theta_c)$

Figures 3.27A and 3.27B show the denitrification capacity of anoxic pre-D and post-D reactors, respectively, as functions of the anoxic sludge mass fraction f_{x1} or f_{x3} and the following conditions: $\theta_c = 10$ d, $S_{bi} = 400$ mg/L; $T = 20$°C; $f_{bs} = 0.24$. The D_c/S_{bi} ratio is also indicated (right-hand scale). The equations show that the denitrification capacity depends on the following factors:

(1) Concentration and composition of influent COD: S_{ti} and f_{us}, f_{up} and f_{sb} fractions.
(2) Sludge age: the value of $C_r = Y\theta_c/(1 + b_h\theta_c)$ increases with increasing sludge age so that D_c also increases.
(3) Temperature: the values of the denitrification constants K_2 and K_3 increase with temperature D_c (b_h also increases and reduces the effect somewhat).
(4) Sludge mass fractions: the larger the f_{x1} and f_{x3}, values, the bigger will be the denitrification capacity D_{c1} and D_{c3}. In practice f_{x1} and f_{x3} are limited by the imposition that nitrification must be efficient which limits the maximum anoxic sludge fraction: $f_{x1} + f_{x3} = f_M$.

3.4.5 Application of the nitrification and denitrification concepts

The concepts of nitrification and denitrification capacity are very convenient to describe the removal of nitrogen in activated sludge system and to optimize these

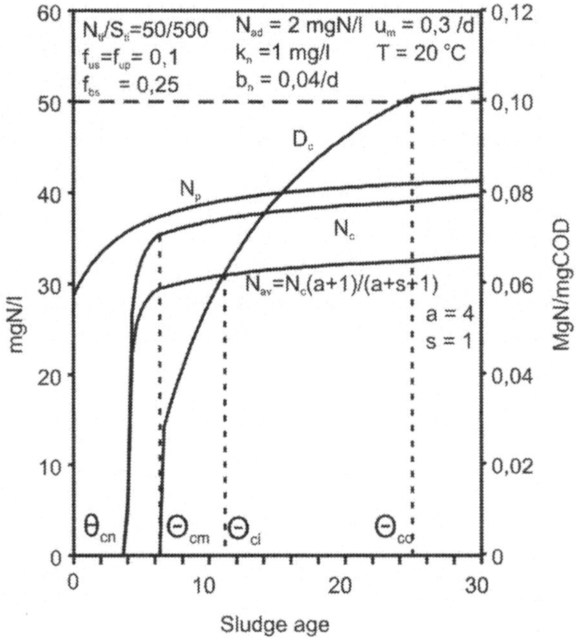

Figure 3.28 Nitrification and denitrification capacities in a pre-D system as a function of the sludge age for the maximum anoxic sludge mass fraction.

systems to maximize nitrogen removal. A numeric example will illustrate the applicability: A pre-D system is operated under the conditions specified in Figure 3.28 and $K_2 = 0.1$ mg N/mg X_a d. With the aid of these data, the nitrification and denitrification capacities can be calculated as a function of the sludge age, using Equation (3.65) for N_c and Equation (3.71) for D_{c1}. In order to be able to calculate the D_{c1} value, it is necessary to determine first the maximum anoxic sludge mass fraction by applying Equation (3.63). In Figure 3.28 the N_c, D_{c1} and f_M curves are shown as functions of the sludge age. It is convenient to introduce another parameter linked to the nitrification capacity: the available nitrate in the anoxic reactor which is the fraction of the nitrification capacity that is recirculated to the anoxic reactor. In the pre-D system the available nitrate is calculated as follows: the total flow entering into the aerobic reactor is $(a + s + 1)$ times the influent flow. With the recirculation ratios "a" from the aerobic reactor and "s" from the settler, a fraction $(a + s)/(a + s + 1)$ of this flow is recirculated to the anoxic fraction and a fraction $1/(a + s + 1)$ is discharged directly with the effluent. Hence a fraction $(a + s)/(a + s + 1)$ of the nitrification capacity is recirculated and the rest is discharged, so that the available nitrate in the anoxic reactor can be expressed as:

$$N_{av} = (a + s)/(a + s + 1)N_c \tag{3.73}$$

where N_{av} = available nitrate in the anoxic reactor (mg N/L influent).

The N_{av} value is also indicated in Figure 3.28 as a function of the sludge age and for $a = 4$ and $s = 1$ (i.e. $N_{av} = (4+1)/(4+1+1)N_c = 0.833N_c$).

In Figure 3.28 the following situations occur with increasing sludge age:

(1) below a minimum value θ_{cn} nitrification is impossible. This minimum sludge age depends on the nitrification kinetics: $\theta_{cn} = 1/(\mu_m - b_n) = 1/(0.3 - 0.04) =$ 3.85 days.

(2) For sludge ages longer than θ_{cn} nitrification is possible. However, to attain the required nitrification efficiency that leaves a residual ammonium concentration that does not exceed a specified value of N_{ad} (=2 mg/L in Figure 3.28), the sludge age must be at least θ_{cm}. This sludge age can be calculated by inserting $f_M = 0.0$ in Equation (3.63):

$$f_M = 0.0 = 1 - (1 + K_n/N_{ad})(1/\theta_{cm} + b_n)/\mu_m$$

or

$$\theta_{cm} = 1/(\mu_m/(1 + K_n/N_{ad}) - b_n) \tag{3.74}$$

For the data in the example: $\theta_{cm} = 1/[0.3/(1 + 1/2) - 0.04)] = 6.25$ days.

(3) At sludge ages longer than θ_{cm} it is possible to maintain efficient nitrification and include an anoxic reactor in the configuration where denitrification can take place. With the aid of the value for f_M (which itself is a function of the sludge age) the denitrification capacity can now be calculated (Equation (3.71)). The nitrification capacity is calculated from Equation (3.63).

(4) For some particular sludge age θ_{co}, the anoxic sludge mass fraction attains it maximum allowable value f_{max}. For an adopted value of $f_{max} = 0.6$ the sludge age is calculated as θ_{co}:

$$f_M = f_{max} = 0.6 = 1 - (1 + K_n/N_{ad})(1/\theta_{co} + b_n)/\mu_m$$

or

$$\theta_{co} = 1/[\mu_m(1 - f_{max})/(1 + K_n/N_{ad}) - b_n] = 1/[0.3(1 - 0.6)/(1 + 1/2) - 0.04] = 25 \text{ days.}$$

(5) When the sludge age $\theta_c > \theta_{co}$, both the nitrification and the denitrification capacity increase marginally with the sludge age.

With the aid of N_c and D_{c1} the nitrogen concentration in the effluent (ammonium and nitrate) can now readily be calculated:

(a) Until $\theta_c = \theta_{cn}$ no nitrification occurs, so that the ammonium effluent concentration is equal to the nitrification potential and no nitrate will be present and no nitrogen removal takes place.

(b) In the range $\theta_{cn} < \theta_c < \theta_m$ nitrification occurs. The ammonium concentration is given by Equation (3.52) The nitrate concentration will be equal to the nitrification capacity (plus eventual nitrate in the influent) and no biological nitrogen removal takes place (no anoxic zone).

(c) For $\theta_c > \theta_{cm}$ the inclusion of an anoxic reactor is feasible. With increasing sludge age the permissible anoxic sludge mass fraction and therefore the denitrification capacity increase. For some sludge age $\theta_c = \theta_{ci}$ the denitrification capacity becomes equal to the available nitrate in the anoxic reactor, so that:

$$D_{c1} = N_{av} \text{ (when } \theta_c = \theta_{ci})$$

or

$$(0.12 f_{bs} + K_2 C_r f_M) S_{bi} = N_c(a + s)/(a + s + 1) \quad (3.75)$$

Graphically in Figure 3.28 the value of $\theta_{ci} = 11$ days is determined. In the range $\theta_{cm} < \theta_c < \theta_{ci}$ the available nitrate N_{av} exceeds the denitrification capacity, D_{c1}. However, D_{c1} represents the maximum removal that can take place in the anoxic reactor. Thus, the anoxic reactor is overloaded with nitrate and the excess nitrate will be returned to the aerobic reactor. It is possible to reduce the recirculation of nitrate without reducing the nitrate removal. For example, for $\theta_c = 10$ days the values of N_c and D_{c1} are 37.2 and 28.9 mg N/L, respectively (Equation (3.65) and (3.71), respectively). Hence, for $D_{c1} = N_{av} = N_c(a + s)/(a + s + 1)$ a value of $(a + s) = 3.4$ is required, which means $s = 1$ (assumed) and $a = 2.4$. Larger $(a + s)$ values will not result in more nitrogen removal.

In the range $\theta_{cm} < \theta_c < \theta_{ci}$ the ammonium concentration is constant: $N_{ad} = 2$ mg N/L ($f_{x1} = f_M$). The concentration of nitrate will be the difference between the nitrification and the denitrification capacities:

$$N_{ne} = N_c - D_{c1}$$

(d) When $\theta_{ci} < \theta_c < \theta_{co}$ then $D_{c1} > N_{av}$ which means that the anoxic reactor is underloaded with nitrate. All recirculated nitrate will be removed in the anoxic reactor and the effluent nitrate concentration will be equal to the fraction of the nitrification capacity that is discharged without recycles: $N_{ne} = N_c/(a + s + 1)$. The ammonium concentration maintains a value of $N_{ad} = 2$ mg/L ($f_{x1} = f_M$ continues). In this sludge age range the performance of the system could be improved by taking some of the pre-D reactor and create a post-D reactor.

(e) When $\theta_c > \theta_{co}$ the anoxic sludge mass fraction is limited by the imposition that always $f_M < f_{max}$ (=0.6). In this range the ammonium concentration will be less than N_{ad}: and can be calculated by using Equation (3.62). Since still $D_{c1} > N_{av}$, the nitrate concentration will continue to be given by $N_{ne} = N_c/(a + s + 1)$.

In Figure 3.29 the different forms under which nitrogen leaves the activated sludge system are depicted as functions of the sludge age for the conditions in Figure 3.28: (1) the nitrogen concentration assimilated in the excess sludge, N_l, (2) the ammonium concentration in the effluent, N_{ae}, (3) the nitrate concentration in the effluent N_{ne} and (4) the concentration of removed nitrogen by nitrification and denitrification, N_d. In the case of the example for a sludge age of $\theta_{ci} = 11$ days (Figure 3.29) it is possible to reduce the influent nitrogen concentration of 50 mg/L to a value of $N_{te} = N_{ad} + N_{ne} = 2.0 + 6.3 = 8.3$ mg/L in the influent ($N_{ne} = N_c/(a + s + 1) = 38/6 = 6.3$ mg/L), with a concentration of $N_l = 10$ mg/L incorporated in the excess sludge. Hence, under these conditions, the nitrogen

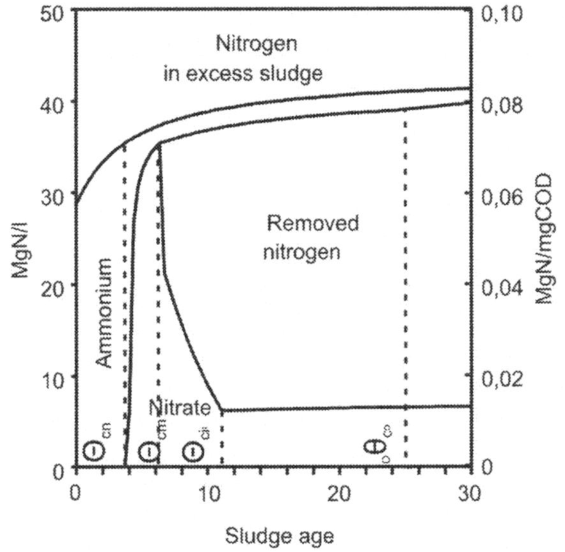

Figure 3.29 Division of nitrogenous material in a pre-D system: fractions in the excess sludge, in the effluent (N_{ne} and N_{ae}) and removed as N_2, as functions of the sludge age.

removal amounts to $N_d = N_{ki} - N_{ad} - N_{ne} - N_l = 50 - 2 - 6.3 - 10 = 31.7$ mg/L, equal to the denitrification capacity for $\theta_{ci} = 11$ days. If the nitrogen concentration in the effluent is to be reduced beyond this minimum (in the example $N_{te} = 8.3$ mg N/L) it is necessary to increase the sludge age and transform the pre-D system in a Bardenpho system.

3.4.6 Optimization of nitrogen removal

The objective of nitrogen removal in activated sludge processes are: (i) to produce an effluent with the lowest possible total nitrogen concentration and (ii) realize this minimization at the lowest construction and operational costs. Before the optimization it is necessary to realize that there are two main restrictions that apply to single sludge nitrogen removing systems:

(1) The mass fraction of sludge in the unaerated zones is limited by two independent criteria: (a) nitrification must be efficient, which requires a minimum sludge mass fraction in the aerobic zone and (b) the anoxic sludge mass fraction must not be excessive (<60%) in order to guarantee adequate properties of the sludge.
(2) The mixed liquor recirculation flows from the aerobic reactor ("*a*") and the settler ("*s*") to the first anoxic reactor are limited by pumping costs. Since the head from the settler to the anoxic zone is always greater than that from the aerobic to the anoxic zone and knowing that the nitrate concentration in the aerobic zone is greater or equal to that in the underflow from the settler, it is concluded that recycling from the aerobic zone is more practical. Hence, the recirculation rate "*s*" must be equal to the minimum required for proper performance of the

settler (generally about $s = 1$). As for the "a" recycle an optimal value is found by considering that in principle a high recirculation rate is favourable because it takes nitrate to the pre-D reactor, where it can be removed at high rate, but at the same time the pumping costs must be considered.

A variable that has not yet been considered is the ratio between the nitrogen concentration and the organic material concentration N_{ti}/S_{ti}. This ratio varies from a low value (0.02–0.04 mg N/mg COD) for wastewaters of vegetal nature to a high value (0.1–0.16 mg N/mg COD) for residues from industries of animal nature. In the case of municipal sewage the N_{ti}/S_{ti} ratio varies between 0.05 and 0.12 mg N/mg COD depending on the social economic habits of the contributing population. A rich population (protein eaters) tends to have a much higher N_{ti}/S_{ti} ratio than a poor population (carbohydrate eaters).

If the N_{ti}/S_{ti} ratio is low, it is possible to create sufficient denitrification capacity to remove all nitrate in the system. However, to attain that objective it is necessary to operate a Bardenpho system with pre- and post-D reactors. If there is complete removal of the nitrate in these reactors, a fraction $a/(a + s + 1)$ of the nitrification capacity will be removed in the pre-D reactor and a fraction $(s + 1)/(a + s + 1)$ will be denitrified in the post-D reactor (Figure 3.29). Hence:

$$D_{c1} = (0.12f_{bs} + K_2C_r f_{x1})S_{bi} = N_c \cdot a/(a + s + 1) \tag{3.76}$$

and

$$D_{c3} = K_3C_r f_{x3}S_{bi} = N_c(s + 1)/(a + s + 1) \tag{3.77}$$

By rearranging these equations one has:

$$f_{x1} = (N_c/S_{bi}) \cdot (a/(a + s + 1) - 0.12f_{bs})/(K_2C_r) \tag{3.78}$$

and

$$f_{x3} = (N_c/S_{bi})(s + 1)/(a + s + 1)/(K_3C_r) \tag{3.79}$$

Now the largest (N_c/S_{bi}) ratio that permits complete denitrification is defined as $(N_c/S_{bi})_o$. This value can be calculated by equating the sum f_{x1} and f_{x3} in Equation (3.78) and (3.79) to the maximum anoxic sludge mass fraction f_M:

$$\begin{aligned} f_M = f_{x1} + f_{x3} = {} & [(N_c/S_{bi})_o(a/(a + s + 1)) - (0.12f_{bs})]/(K_2C_r) \\ & + [(N_c/S_{bi})_o(s + 1)/(a + s + 1)]/K_3C_r \end{aligned}$$

or

$$(N_c/S_{bi})_o = (a + s + 1)(0.12f_{bs} + K_2C_r f_M)/(a + (K_2/K_3)(s + 1)) \tag{3.80}$$

where $(N_c/S_{bi})_o$ = maximum N_c/S_{bi} ratio that allows complete nitrogen removal.

The $(N_c/S_{bi})_o$ ratio is directly linked to the corresponding $(N_{ti}/S_{ti})_o$ ratio:

$$S_{bi} = (1 - f_{us} - f_{up})S_{ti}$$

and

$$N_c = N_{ti} - N_l - N_{ad}$$

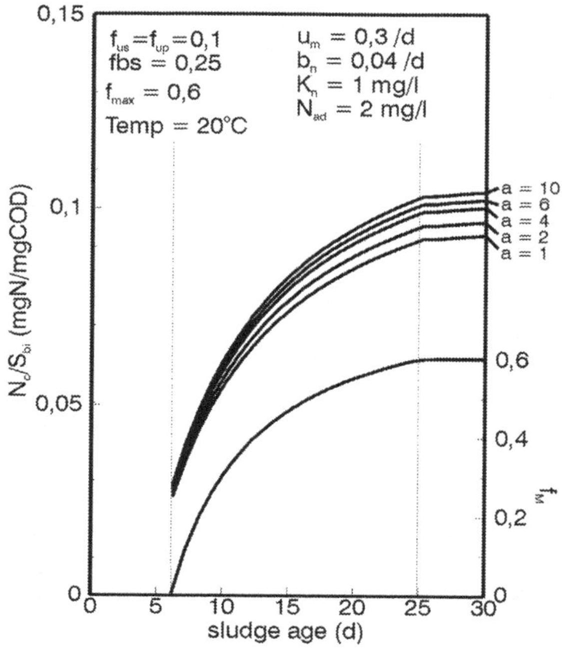

Figure 3.30 Maximum anoxic sludge mass fraction and $(N_c/S_{bi})_o$ ratio as functions of the sludge age for several values of the "*a*" recycle.

Hence:

$$(N_{ti}/S_{ti})_o = (1 - f_{us} - f_{up})(N_c/S_{bi})_o + (N_l + N_{ad})/S_{ti}$$
$$= (1 - f_{us} - f_{up})(0.12f_{bs} + K_2C_r f_M)(a + s + 1)/(a + (K_2/K_3)(s + 1)) + (N_l + N_{ad})/S_{ti} \quad (3.81)$$

Equation (3.81) shows the various factors that influence the TKN/COD ratio in wastewaters that permits complete nitrogen removal:

(1) Composition or the influent organic material (f_{us}, f_{up}, f_{bs})
(2) Kinetic constants of denitrification (K_2 and K_3)
(3) Kinetic constants of nitrification (μ_m, K_n, b_n)
(4) Temperature (influences the kinetic constants (K_1, K_2, K_n, b_n, b_h, μ_m))
(5) Maximum residual ammonium concentration (N_{ad})
(6) Recirculation rates("*a*" and "*s*")
(7) Sludge age (θ_c).

The majority of the above listed factors has values that cannot be changed (Factors 1–4). In principle the maximum residual ammonium concentration (Factor 5) will be set by the standards emitted by environmental authorities. Thus, only the values of the recirculation rates "*a*" and "*s*" and the sludge age (factors 6 and 7) can be chosen by the designer. It has been shown already that the "*s*" recirculation is set by the need to have efficient phase separation in the settler. Thus, the factors to be determined are the "*a*" recirculation and the sludge age. Figure 3.30

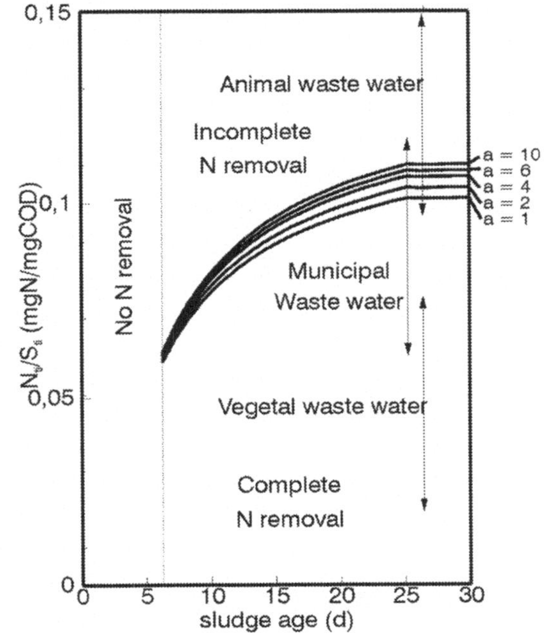

Figure 3.31 Maximum TKN/COD ratio in the influent that allows complete nitrogen removal as a function of the sludge age for several "*a*" recycles.

shows $(N_c/S_{bi})_o$ ratios as a function of sludge age for different "*a*" values: $a = 1, 2, 4, 6$ and 10. To carry out the calculations the following values were assumed: $S_{ti} = 500\,\text{mg/L}$; $f_{us} = 0.14$; $f_{up} = 0.06$; $f_{sb} = 0.24$; $b_n = 0.04$ per day; $K_n = 1\,\text{mg/L}$; $s = 1$; $K_2 = 0.1\,\text{mg N/mg}\,X_a/\text{day}$; $K_3 = 0.08\,\text{mg N/mg}\,X_a/\text{day}$; $\mu_m = 0.3$ per day. Figure 3.31 shows the corresponding $(N_{ti}/S_{ti})_o$ ratios. The usual influent TKN/COD ratio for different wastewaters has also been indicated.

From Figure 3.30 it is noted that the minimum sludge age for complete nitrogen removal increases as the recirculation ratio "*a*" decreases. The selection of the optimal "*a*" value now becomes a question of economics, where the energy costs for pumping are compared to the cost reduction by reducing the sludge age. In practice the "*a*" value will rarely exceed 4.

The diagram in Figure 3.31 indicates that, under the specified conditions, complete nitrogen removal is feasible for many wastewaters. However, it must be stressed that this in part is due to the choice of the constants used in the calculations. Particularly the value of the maximum specific growth rate constant for nitrifiers (μ_m) is important. The chosen value is conservative for the assumed temperature of 20°C, but it can have a much lower value at low temperatures. In that case the TKN/COD ratio for which complete nitrogen removal is feasible will decrease correspondingly.

3.4.7 Anaerobic ammonium oxidation (anammox)

Until the end of the last century, the single biological process recognized for ammonium oxidation was nitrification. Consequently, the oxidation of ammonium had only been investigated under aerobic or oxygen-limited environments. In theory, ammonium could also serve as an electron donor for achieving denitrification, either with nitrite or nitrate as an electron acceptor. Based on thermodynamics, Broda (1977) already predicted the existence of autotrophic microorganisms capable of achieving anammox:

$$NH_4^+ + 3/2O_2 \rightarrow NO_2^- + H_2O + 2H^+ \quad (3.82a)$$

$$NH_4^+ + NO_2^- \rightarrow N_2 + 2H_2O \quad (3.82b)$$

$$2NH_4^+ + 3/2O_2 \rightarrow N_2 + 3H_2O + 2H^+ \quad (3.82)$$

If the basic equation for autotrophic nitrogen removal described above is compared to the one for conventional nitrogen removal with the aid of nitrification and denitrification, there are similarities insofar as stoichiometry is concerned: the oxygen and alkalinity demands for the two nitrogen removing processes are equal (1.71 mg O/mg N and 3.57 mg $CaCO_3$/mg N, respectively). However, the anammox process has one advantage: it does not depend on the presence of organic material as an electron donor, because ammonium is used for the reduction of nitrite. This is a very important advantage, especially when the TKN/COD ratio is high as in the case of animal wastewaters and effluents from anaerobic pretreatment units. In these wastewaters, the high TKN/COD ratio may make complete nitrogen removal impossible, unless external organic material is added.

The first indication of the anammox process was obtained during the experiments on a denitrifying fluidized bed reactor treating effluents from a methanogenic reactor (Mulder *et al.* 1995). There was a simultaneous removal of ammonium and nitrate with a concomitant increase in nitrogen gas production. Nitrogen and redox balances showed that ammonium consumption indeed occurred under anaerobic conditions and that, for every mol of ammonium consumed, 0.6 mol of nitrate was demanded, resulting in the production of 0.8 mol of nitrogen gas. Since then many research workers have dedicated efforts to transform the principle of anaerobic ammonium oxidation into a stable process that can be used in practice. While these efforts have given some results, there is still a long way to go for research and development of the anammox process. Nevertheless the process has so much potential that the principles are briefly discussed here.

3.4.7.1 Microbiology of anammox

Once anammox was defined as an autotrophic respiratory process, in which ammonium is oxidized using nitrite as electron acceptor, a selective enrichment medium was composed for cultivation of the microorganisms responsible for this respiratory process. Since the specific rate of ammonium oxidation in batch incubations was considerably lower than that obtained in continuous experiments (Mulder *et al.* 1995), different reactor configurations were tested for the enrichment of the anammox biomass. Successful enrichments of anammox consortia could be developed in fluidized bed and sequencing batch systems (van de Graaf *et al.* 1996; Strous *et al.* 1998) and more recently, in anoxic fixed-film bioreactors and rotating biological contactors (Furukawa *et al.* 2003; Pynaert *et al.* 2004). All configurations were operated with a medium containing ammonium (5–90 mM), bicarbonate (10–12 mM), minerals and trace elements, and nitrite (5–35 mM) when (partial)-nitrification was not established prior to the anammox reactors.

The enrichment of anammox bacteria in these reactor configurations served as a very influential tool for studying the anammox consortia, especially if one considers the extremely low growth rate of the anammox bacteria ($0.003\,h^{-1}$, Jetten *et al.* 2001). The successfully retained anammox biomass could support stepwise increases in nitrogen loading achieving nitrogen removal rates up to $8.9\,kgN/m^3/day$ (Jetten *et al.* 2001).

All attempts to isolate anammox bacteria by conventional methods have not succeeded so far (Strous *et al.* 1999b). However, first isolates were obtained from the anammox enrichments by physical purification by an optimized Percoll density gradient centrifugation procedure (Strous *et al.* 1999; 2002). Phylogenetic analysis of the 16S rRNA gene of the purified cells revealed that anammox bacteria belong to the planctomycete lineage of descent. Based on this finding, the first discovered anammox planctomycete-like bacterium was named *Candidatus brocadia anammoxidans* (Kuenen and Jetten 2001).

Molecular analysis of several inocula from wastewater treatment systems and some freshwater ecosystems, in which high nitrogen losses had not been deciphered so far (Schmid *et al.* 2000; Egli *et al.* 2001; Fujii *et al.* 2002; Fux *et al.* 2002; Helmer-Madhok *et al.* 2002; Toh and Ashbolt 2002; Toh *et al.* 2002; Dong and Tollner 2003; Pynaert *et al.* 2003; Tal *et al.* 2003), also were shown to contain significant populations of anammox bacteria. Some of these microorganisms were only distantly related to the Brocadia branch, thus creating a high genus level diversity. Therefore, a second group of anammox bacteria was recognized, and provisionally named *Candidatus kuenenia stuttgartiensis* (Jetten *et al.* 2003). Anammox bacteria have also recently discovered in natural ecosystems. *Candidatus scalindua sorokinii* has been detected by newly designed fluorescently labelled 16S rRNA genes probes in the anoxic deep water of the Black Sea (Kuypers *et al.* 2003). Nutrient profiles and ^{15}N tracer studies in marine settings indicate that marine anammox bacteria play a very important role in the oceanic nitrogen cycle, contributing up to 70% of the nitrogen gas production (Dalsgaard and Thamdrup 2002; Dalsgaard *et al.* 2003). Even though *Candidatus S. sorokinii* is quite distantly related to the other two anammox genera, all anammox bacteria have very similar physiology and morphology (Jetten *et al.* 2003).

Anammox bacteria have already been detected in several wastewater treatment facilities in The Netherlands, Germany, Belgium, Switzerland, UK, Australia, Japan and in several natural ecosystems around the world (Schmidt *et al.* 2002; Fujii *et al.* 2002; Dalsgaard *et al.* 2003; Pynaert *et al.* 2004) suggesting that this kind of microorganisms may occur more than originally considered.

3.4.7.2 Biochemistry of anammox bacteria

Microbial incubations of anammox enrichments with ^{15}N-labelled compounds revealed the metabolic pathway of this respiratory process. The experiments showed

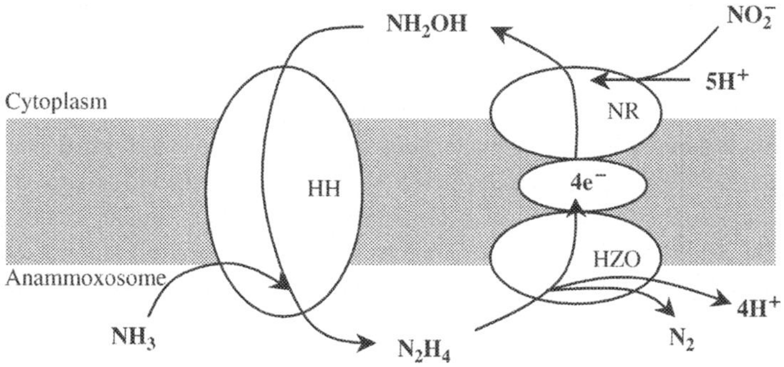

Figure 3.32 Mechanism proposed for anammox. NR: Nitrite reducing enzyme; HH: hydrazine hydrolase; HZO: hydrazine oxidizing enzyme (Schmidt *et al.* 2002).

that the electron acceptor is reduced to hydroxylamine, which subsequently reacts with ammonium, ultimately leading to the formation of nitrogen gas (van de Graaf *et al.* 1997). The study also evidenced the accumulation of hydrazine when hydroxylamine was supplied in excess, thus elucidating a novel metabolic pathway in which hydroxylamine and hydrazine are intermediates. Based on these observations, a hypothesis was proposed for the oxidation of hydrazine to nitrogen gas generating four electrons to support the initial reduction of nitrite to hydroxylamine (Figure 3.32).

The overall nitrogen balance shows a consistent ratio of about 1:1.32 for the conversion of ammonia and nitrite, and a production of nitrate in several anammox incubations at a ratio of 0.26:1 respect to the nitrite converted. The reason for the concomitant production of nitrate during anammox is assumed to be the generation of reducing equivalents demanded for the reduction of CO_2, according to the following stoichiometry (Schmidt *et al.* 2002):

$$NH_4^+ + 1.23NO_2^- + 0.066HCO_3^- + 0.13H^+ \\ \rightarrow 0.26NO_3^- + 1.02N_2 + 0.066CH_2O_{0.5}N_{0.15} + 2.03H_2O$$

Indeed, radioactive bicarbonate was incorporated into biomass confirming the autotrophic profile of anammox bacteria (van de Graaf *et al.* 1997).

The oxidation of hydrazine to nitrogen gas was first reported by Hooper *et al.* (1997). In this report the hydroxylamine oxidoreductase (HAO) enzyme was responsible for the aerobic ammonium oxidation by *Nitrosomonas europaea*. High HAO activity in cell extracts indicated that a similar enzyme is present in *Candidatus B. anammoxidans* (Schalk *et al.* 2000). From these extracts, a homotrimeric, 24 cytochrome-*c*-containing HAO enzyme was purified to homogeneity. Unique peptide sequences of fragments obtained after trypsin digestion were used to locate the *hao* gene in the *Candidatus K. stuttgartiensis* genome. The purified HAO enzyme was also able to catalyse the oxidation of both hydroxylamine and hydrazine in this microorganism. Besides nitrogen gas, NO and N_2O were also detected as end products by the enzyme. The purified enzyme from *Candidatus K. stuttgartiensis* has also been used to obtain polyclonal antibodies for location studies (Lindsay *et al.* 2001). According to immunogold electron microscopic analysis, the enzyme has been found to be present solely inside one of the membrane bounded organelles, named the "anammoxosome", constituting more than 30% of the cell volume (van Niftrik *et al.* 2004). Further experiments revealed that the anammoxosome is surrounded by a membrane almost exclusively composed by unique ladderane lipids (Sinninghe Damsté *et al.* 2002; Schmid *et al.* 2003). The ladderane lipids occur both as ether and ester lipids in all anammox bacteria analysed (Jetten *et al.* 2003). Because anammox bacteria carry out a very slow metabolism, a very dense and impermeable membrane is required to sustain concentration gradients during the anammox reaction. Such membrane characteristics also protect the cells from toxic intermediates. The anammoxosome membrane with a lower degree of rotational freedom and being significantly denser ($1.5\,kg\,dm^{-3}$) and impermeable than a conventional membrane ($1.0\,kg\,dm^{-3}$) is perfectly suited for both of the above mentioned tasks (Sinninghe Damsté *et al.* 2002). The recent findings of hopanoids, which act as rigidifiers in the membranes of the anammox bacteria, gives further support to the necessity of a very dense membrane to limit diffusion of protons and intermediates (Sinninghe Damsté *et al.* 2004; van Niftrik *et al.* 2004).

3.4.7.3 Parameters affecting anammox

The obligate anaerobic nature of anammox bacteria contrast with the capacity of aerobic nitrifying bacteria, which are known to be facultative anaerobes capable of denitrification under oxygen limitation. Certainly, both mixed and pure cultures of *Nitrosomonas eutropha* can produce nitrogen gas from ammonium when either NO or NO_2 is available as a terminal electron acceptor (Schmidt and Bock 1997). The anaerobic ammonium oxidation activity observed in *N. eutropha*, however, is 50-fold slower that typical anammox activities achieved by *Candidatus B. anammoxidans* (Jetten *et al.* 2001). The anammox activity, on the other hand, is extremely sensitive to aerated conditions, so that oxygen concentrations as low as 2 μM can completely inhibit the respiratory process (Jetten *et al.* 1999). Nevertheless, inhibition by oxygen can be reversibly overcome by establishing anaerobic conditions (Strous *et al.* 1997).

The reversibility of oxygen inhibition observed in anammox enrichments has important implications for the relevance of the anammox process. Indeed, to remove ammonium from wastewater, the anammox process has to be supplied with nitrite as the electron acceptor. Thus, complete nitrogen removal can only be achieved when part of the ammonium is firstly oxidized to nitrite in a preceding partial-nitrifying step (Jetten *et al.* 1997). The survival of anammox-bacteria under oxygen limitation allowed the development of the so-called CANON (completely autotrophic nitrogen-removal over nitrite) process, in which oxygen-limited aerobic ammonium-oxidizing bacteria (AOB) and anammox planctomycetes perform two sequential reactions simultaneously in a single vessel (Sliekers *et al.* 2002). Under oxygen limitation, the supplied ammonium is partly oxidized to nitrite. The produced nitrite can then be converted to nitrogen gas with the remaining ammonium by the anammox bacteria. Successful performance of the CANON concept could be established in a sequencing batch (SBR) reactor by carefully introducing limited amounts of oxygen and, within 14 days, a new stable consortium of AOB and anammox planctomycetes was operative (Sliekers *et al.* 2002).

The maximum specific-substrate (ammonium and nitrite) conversion rate of enriched anammox biomass was measured as a function of temperature and pH in batch incubations. The physiological pH and temperature ranges observed were 6.7–8.3 and 20–42°C, respectively (Strous *et al.* 1999). An activation energy for the anammox process was calculated to be 70 kJ/mol from the temperature dependency of the anammox activity. Recent experiments revealed an optimum anammox rate at 15°C and a maximum temperature of 37°C in marine sediments (Dalsgaard and Thamdrup 2002). Thus, anammox activity could be expected in a wide variety of environments including those under relatively cold climates. On the other hand, the fact that anammox activity could contribute up to 70% of the nitrogen gas production in marine sediments (Dalsgaard and Thamdrup 2002; Dalsgaard *et al.* 2003) suggests that the application of the anammox concept for treating effluents with a high content of ammonium and salinity (e.g. discharges from aquaculture activities) may be feasible; a scenario which was shown to occur in a closed recirculated mariculture system (Tal *et al.* 2003).

Several anammox consortia have shown a high affinity for both ammonium and nitrite with typical affinity constants (K_S) below 5 μM (Strous *et al.* 1999; Jetten *et al.* 1999). The anammox process has not been inhibited by ammonium or by the by-product nitrate up to concentrations of at least 1 g N/L. However, in the presence of more than 0.1 g NO_2^--N/L, the process is strongly inhibited (Jetten *et al.* 1999; Strous *et al.* 1999). Prolonged exposure of different consortia, performing anammox activity, to relatively high nitrite concentrations (up to 0.2–0.4 g NO_2^--N/L) collapsed the respiratory process leading to the production of ammonium (Cervantes *et al.* 1999; Strous *et al.* 1999). Addition of either hydrazine or hydroxylamine to nitrite-inhibited anammox cultivations fully restored the metabolic activity, as evidence by the concomitant removal of nitrite and ammonium (Strous *et al.* 1999).

Anammox processes have certainly been successfully applied to wastewaters with a high nitrogen content, but limited amount of organic matter (Jetten *et al.* 1997; Pynaert *et al.* 2004). However, due to its autotrophic nature, anammox has faced difficulties when treating effluents with considerable COD levels. For instance, poor (e.g. 13–22%) ammonium removal could be achieved when evaluating the anammox concept for anaerobic digestion of poultry manure (Dong and Tollner 2003). Previous studies, however, have shown that addition of ammonium as an alternative electron donor to denitrifying reactors avoid accumulation of intermediates (e.g. nitrite and N_2O) due to the coupling between anammox and denitrification (Cervantes *et al.* 1999; 2001). The C/N ratio is a key parameter for allowing both respiratory processes to occur simultaneously. In fact, simultaneous removal of nitrate and ammonium could be achieved at C/N ratios between 0.6 and 1.2 with acetate as a carbon source achieving a maximum nitrogen gas production rate of 30 μg N_2/(g VSS-min). C/N ratios below 0.6 collapsed the nitrogen removal process due to the accumulation of nitrite, consequently leading to the production of ammonium (Cervantes *et al.* 2001).

3.5 BIOLOGICAL PHOSPHORUS REMOVAL

Phosphorus in wastewater is present predominantly in the form of (ortho) phosphates, with a minor fraction of organic phosphate, mainly in proteins. In biological treatment systems most of the organic phosphate is mineralized. The phosphorus concentration in industrial wastewaters depends heavily on the nature of the industry. Wastewaters of vegetable origin have a relatively low *P* concentration with P/COD ratios in the range of 0.002–0.01 mg P/mg COD. In industries with wastewaters of animal origin P/COD ratios of 0.02–0.04 may be encountered. Apart from proteins, an important source of phosphorus in many countries is soap in the form of washing powders. Recently, phosphorus-free soaps have been developed and marketed in some countries.

In the activated sludge process, phosphorus is removed due to excess sludge production: in a conventional aerobic activated sludge process, the phosphorus mass fraction can be estimated at 2–2.5% of the VSS concentration. Thus, the influent phosphorus concentration that is required for sludge production is given as:

$$mP_i = MP_i/MS_{ti} = f_p[(1 - f_{us} - f_{up}) \cdot (1 + fb_h\theta_c)(C_r/\theta_c) + f_{up}/f_{cv}]$$

or

$$P_1 = mP_iS_{ti} = f_p mE_v S_{ti} = f_p[(1 - f_{us} - f_{up})(1 + fb_h C_r/\theta_c) + f_{up}/f_{cv}]S_{ti} \quad (3.83)$$

For usual values of the parameters and variables in Equation (3.83) mP_i is in the range of 0.005–0.007 mg P/mg COD, that is, the removed phosphate concentration may either be smaller or larger than the influent concentration. In the former case it is needed to add phosphorus (e.g. as phosphoric acid) to the influent, while in the latter case there may be a need for additional removal of phosphate to avoid eutrophication of the receiving water body.

Basically two different methods for additional phosphorus removal may be distinguished: (1) precipitation as phosphate with the aid of coagulants like $FeCl_3$ or $Al_2(SO_4)_3$ or as one of the forms of calcium phosphate (p.e. apatite), by lime addition and (2) enhanced biological removal. Enhanced biological phosphorus removal is based on the experimental observation that, under certain specified conditions, the phosphorus mass fraction in active sludge increases and thus a higher phosphorus concentration can be removed in the excess sludge ("luxury *P* uptake" by the active sludge). Biological phosphorus removal by luxury uptake has the advantages that there is no need for the addition of chemicals and that no extra (inorganic) sludge is formed and no additional ions are added to the water (the high salinity may be a factor of importance in case of reuse). For these reasons it is preferable to physical chemical phosphorus removal by precipitation.

The operational conditions for luxury uptake in the activated sludge process have been determined during an extensive investigation by Prof. Marais and his research group at the University of Cape Town (Rabinowitz and Marais, 1980, Siebritz *et al.* 1982, Wentzel *et al.* 1992). This work has resulted in an empiric model that can be used to design activated sludge processes with the objective to remove phosphorus as well as organic material and nitrogen. In this section,

Marais' work and the implications for activated sludge design to obtain a phosphorus-free effluent are discussed. Marais' model is based on the following points:

(1) Research workers on biological phosphorus removal agree that a necessary condition to effect enhanced biological uptake of phosphorus is to create a reactor ahead of the "normal" activated sludge reactors, in which sludge and influent are mixed and kept in an anaerobic condition. The anaerobic environment is characterized by the fact that there is absence of both DO and nitrate. Many authors have observed that in the anaerobic reactor, phosphorus is released from the solid (sludge) phase to the liquid phase, leading to an increase of the phosphate concentration (Fukase *et al.* 1982; Arvin, 1985; Comeau *et al.* 1985; Wentzel *et al.* 1985; Wentzel *et al.*1988). In the subsequent anoxic and aerobic reactors incorporation of phosphorus in the sludge takes place to such an extent that the phosphorus mass fraction in systems with an anaerobic reactor is much higher than in a system without such a reactor (in which it is 0.025 mg P/mg VSS). This phenomenon is called luxury uptake. The higher phosphorus content in the generated sludge naturally will result in a greater phosphorus removal through excess sludge discharge.

(2) In order to create the anaerobic zone, Barnard (1976) suggested a modification of the Bardenpho system and introduced the Phoredox system, by placing an anaerobic reactor upstream of the pre-D reactor and recirculating the return sludge into the anaerobic reactor in which also all the influent is discharged (Figure 3.33). The necessary anaerobic environment for phosphorus removal will be established if nitrate removal in the system is complete or virtually complete. The nitrate eventually introduced together with the return sludge will be removed by reduction with the influent organic material.

(3) Siebritz *et al.* (1982) showed that the exposure of the activated sludge to an anaerobic environmental is a necessary, but not a sufficient condition for luxury uptake. It was established that the *P* sludge mass fraction in the activated sludge was influenced by the concentration of easily biodegradable material in the anaerobic reactor. It was found that a minimum concentration of easily biodegradable material in the aerobic zone is required to trigger off the phenomenon of luxury uptake. The necessity of maintaining a minimum easily biodegradable COD concentration in the anaerobic reactor explains why there may be no *P* release in the pre-D reactor, even if it is in an anaerobic condition: if the nitrate introduced in the pre-D reactor is lower than its denitrification capacity but sufficient for the removal of the easily biodegradable material, then an anaerobic environment is established, but without the necessary easily biodegradable material to trigger off *P* release.

(4) The UCT (University of Cape Town, Figure 3.33) process was introduced with the objective to guarantee a minimum concentration of easily biodegradable material in the aerobic reactor, even if denitrification is incomplete in the treatment system: The "*a*" and "*s*" recycles are introduced in the pre-D reactor, where a very low nitrate concentration is maintained by selecting an adequate "*a*" recycle value. Sludge is introduced into the anaerobic reactor by means of

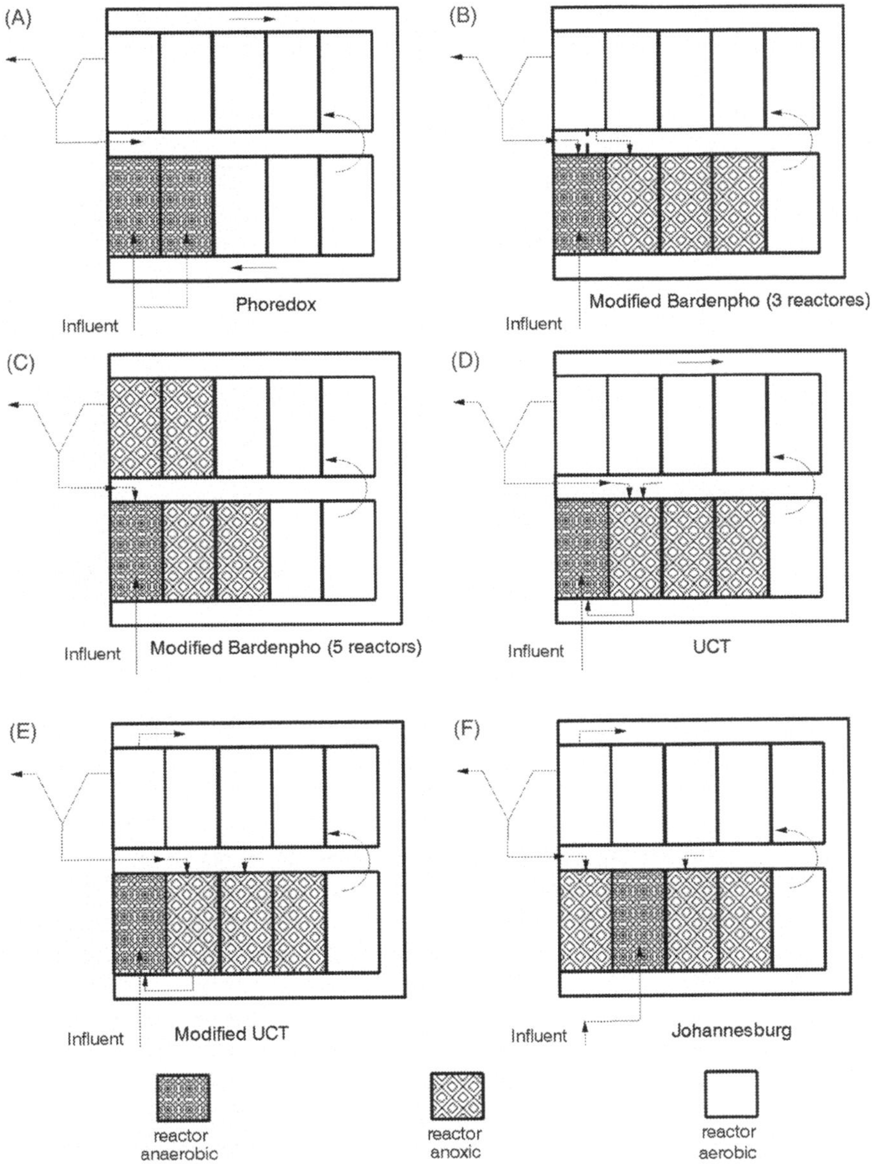

Figure 3.33 Different configurations for activated sludge systems with biological phosphorus and nitrogen removal.

a recycle from the pre-D reactor. Thus, the introduction of nitrate in the anaerobic reactor is minimum.

The fundamental difference between the Phoredox and the UCT system is, that in the latter the easily biodegradable material concentration can be kept at its maximum value in the anaerobic reactor, because in principle no nitrate is introduced in it, so that no organic material is metabolized. By contrast, in the Phoredox system nitrate will be introduced into the anaerobic reactor if denitrification is not complete, because of the underflow recycle from the settler into the anaerobic reactor. The easily biodegradable material concentration in the anaerobic reactor may be estimated as follows:

(a) For the Phoredox system (Figure 3.33A): $S_{bsan} = S_{bsi}/(s + 1) - S_{bni}$ (3.83a)

(b) For the UCT system (Figure 3.33B): $S_{bsan} = S_{bsi}/(r + 1) - S_{bni}$ (3.83b)

In Equations (3.83a) and (3.83b) the concentration S_{bni}, represents the consumption of easily biodegradable material for the reduction of introduced nitrate. Considering that there is a reduction of $(1 - f_{cv}Y)/2.86$ mg N/mg COD utilized and that, as long as easily biodegradable material is present, its utilization will be a fraction $K_1/(K_1 + K_2)$ of the total COD utilization (the remainder, a fraction $K_1/(K_1 + K_2)$ is associated to the utilization of slowly biodegradable material), one has:

(a) for the Phoredox system:

$$S_{bni} = s/(s + 1) \times N_{ne} \times 2.86/(1 - pY) \times K_1/(K_1 + K_2) \quad (3.84a)$$

(b) For the UCT system:

$$S_{bni} = r/(r + 1)N_{n1}2.86(1 - f_{cv}Y)K_1/(K_1 + K_2) \quad (3.84b)$$

where N_{ne} and N_{n1} = nitrate concentration in the settler and the pre-D reactor.

It was experimentally established (Siebritz *et al.* 1982) that release and consequential luxury uptake are only observed if the concentration of easily biodegradable material in the anaerobic reactor is higher than 25 mg/L.

(5) Only after the development of an empiric model for the phenomena of phosphorus release in the anaerobic reactor and luxury uptake in the subsequent anoxic and aerobic zones, a theory was developed to explain biological phosphorus removal. The rational explication can be summarized as follows:
  (a) The presence of the anaerobic reactor ahead of the other reactors originates the development of a bacterial population normally not present in activated sludge. This population is composed of bacteria with a very high *P* mass fraction (35–38% of the dry mass) and for that reason are called poly-P organisms or phosphorus accumulating organisms (PAO) (IWA, 1991). The work of several authors has shown that the *Acinetobacter spp.* genera is predominant in poly-P sludge.
  (b) The poly-P organisms can store organic material (in particular VFA) as polyhydroxybutyrate (PHB) and phosphorus as polyphosphate (Fuhs and Chen, 1975; Buchan, 1981). Under anaerobic conditions the poly-P organisms can absorb VFA as PHB, by using the energy stored in the polyphosphate. In the process, polyphosphate is converted into phosphate and as such released to the liquid phase. Under the subsequent aerobic conditions, PHB is metabolized by the poly-P organisms for catabolic and anabolic processes. Part of the released energy is utilized for the regeneration

of polyphosphate, a process that develops simultaneously with the aerobic metabolism and results in the sorption of phosphate from the liquid phase (luxury uptake).

(6) In processes with an anaerobic reactor two populations develop that co-exists with limited interaction:
   (a) The poly-P sludge obtains its substrate in the anaerobic reactor by absorption of the VFA. It is known that a large fraction of the influent-soluble organic material is converted into VFA in the anaerobic reactor by "normal" sludge.
   (b) The "normal" activated sludge obtains its substrate in the anoxic and aerobic reactors by metabolism of particulate organic material.

   Thus a mixed culture develops partly composed of poly-P organisms and normal sludge. The proportion between these two sludges will determine the extent of P removal.

3.5.1 Process configurations for excess phosphorus removal

Although the mechanism for excess biological phosphorus removal is very recent, several systems have been developed and are applied in practice. The principal difference between the configurations is in the way the anaerobic zone is maintained and protected against introduction of nitrate. Figure 3.33 shows a general configuration that can be converted into several excess phosphorus removing systems.

3.5.1.1 Phoredox or A/O system

The Phoredox system proposed by Barnard (1976) is composed of two reactors in series, the first one (which receives the influent) being anaerobic and the second aerobic. Figure 3.33A is a schematic representation of the Phoredox system. The return sludge is recirculated to the anaerobic reactor and there are no recirculations between reactors. In this variant of the Phoredox process, there is no nitrogen removal nor nitrification, so that there is no need for a long sludge age (for nitrification) or anoxic zones (for denitrification), which makes the system compact. The A/O system has the same basic configuration but by segmenting the anaerobic zone a plug-flow type of flow is created that tends to stimulate the conversion of organic material to volatile acids, thus increasing the P removal capacity. The Phoredox system is applied in countries with a cold or moderate climate principally in Europe and the USA. In regions with a hot climate its applicability is limited, as nitrification is almost unavoidable, so that the anaerobic zone will be "contaminated" with nitrate.

Modified Bardenpho system (3 or 5 reactors) In the modified Bardenpho system (Figure 3.33B and 3.33C) an anaerobic zone is added to the conventional Bardenpho system ahead of the pre-D reactor. The anaerobic reactor receives the influent and return sludge flow. If nitrate removal is not complete, it will be introduced into the anaerobic reactor and, as a consequence, the P removal capacity will decrease.

3.5.1.2 UCT and modified UCT systems

In the UCT system proposed by Rabinowitz and Marais (1980) and represented in Figure 3.33D the introduction of nitrate into the anaerobic reactor is avoided by recycling the return sludge into the anoxic reactor. In the anoxic zone, the nitrate concentration is kept low due to appropriate manipulation of the "*a*" recycle. In such a way that the available nitrate in the reactor is equal to the denitrification capacity, so that, at least in principle, there is no nitrate present. In the modified UCT version (Figure 3.33E) there is an even more stringent nitrate control by dividing the pre-D anoxic reactor in two parts, the first one receiving the return sludge recycle and the second the "*a*" recycle. In both cases, sludge without nitrate is recycled from the first anoxic reactor to the anaerobic reactor. Naturally the limitation of the availability of nitrate in the anoxic reactors will tend to decrease the nitrate removal capacity of the system.

3.5.1.3 Johannesburg system

The particularity of the Johannesburg system (Figure 3.33F) is that the mixed liquor of the aerobic zone flows into the settler and the underflow of the settler flows into an anoxic reactor. As the sludge concentration in the return sludge is a factor $(s + 1)/s$ bigger than the sludge before settling the denitrification rate in the post-D zone will be proportionally higher, so that it becomes easier to have a nitrate-free post-D zone (from where the sludge is recycled to the anaerobic zone) even though there may be nitrate in the effluent.

3.5.2 Modelling of phosphorus removal

In an extensive experimental investigation the research group of Prof. Marais at the UCT developed a quantitative model to describe P removal in activated sludge systems. First, the characteristics of poly-P sludge were established by operating UCT systems with acetate as the sole organic material. On the basis of the experimental data it was established that:

(1) In the anaerobic zone there is a proportionality between the absorbed VFA concentration and the released phosphate concentration. The proportionality constant is $f_{pr} = 0.5$ mg P/mg COD absorbed.
(2) The utilization of any formed PHB in the anaerobic zone is completed in the subsequent aerobic zone, independent of the operational conditions.
(3) The absorption of phosphorus in the anoxic and aerobic zones to produce polyphosphate produces a constant fraction of P in the sludge: independent of operational conditions the P mass fraction is 0.38 m P/mg TSS.

If the poly-P sludge is compared with "normal" sludge, the following important differences come to light:

(1) *Decay of poly-P organisms*: The rate of decay of the poly-P organisms is much smaller than that of "normal" sludge. Experimentally a value for the decay constant of $b_p = 0.04$ per day was established at 20°C, as against $b_h = 0.24$ per day for normal active sludge.

Table 3.9 Differences between poly-P and conventional organisms in activated sludge systems (temperature = 20°C).

Parameter	Symbol	Poly-P organism	Conventional organism
P content (mg P/mg VSS)	f_p	0.38	0.025
Decay constant (per day)	b	0.04	0.24
Endogenous residue (–)	f	0.25	0.20
P in endogenous residue (mg P/mg X_e)	f_{pe}	0.025	0.025
VSS/TSS ratio (mg VSS/mg TSS)	f_v	0.46	0.80
Denitrification constant (mg N/mg X_a day)	K	0	0.10
Phosphate release in anaerobic zone (mg P/mg COD)	f_{pr}	0.5	0

(2) *Sludge masses and concentrations for active and endogenous fractions*: It was determined that 25% of the decayed poly-P mass remains after decay as endogenous residue. However, this residue has a *P* mass fraction of only 2.5% (like normal sludge) instead of 38% for the active poly-P.

(3) *VSS/TSS ratio*: Due to the large inorganic fraction in poly-P organisms the VSS/TSS fraction is much smaller than in normal sludge. In conventional activated sludge systems, the volatile fraction is about $f_v = 0.8$, but in cases of pure poly-P organisms, the proportion is only $f_{vp} = 0.46$ mg VSS/mg TSS. Thus, the sludge production is much larger in P removing systems than in conventional activated sludge systems.

(4) *Denitrification propensity*: Wentzel *et al.* (1986) noted that the rate of denitrification in anoxic reactors with poly-P organisms is so low that for all practical purposes it can be neglected.

Table 3.9 resumes the most important differences between poly-P and conventional activated sludge organisms.

In Figure 3.34, the sludge masses of three different sludges are compared. In Figure 3.34A the sludge masses in conventional activated sludge processes is depicted as a function of sludge age. Figure 3.34B shows the same relationship for a pure poly-P culture and in Figure 3.34C the relationships are represented for a mixed culture, in which 75% of the influent COD is used for conventional sludge generation and the remainder for the generation of poly-P sludge.

The main source for poly-P generation is the VFA concentration both in the influent, and generated in the anaerobic reactor. Wentzel *et al.* (1986) developed a method to calculate the COD fraction that is used by the poly-P organisms. As a first estimate this concentration can be equated to the influent soluble and biodegradable COD concentration, S_{bsi}. Thus, it is tacitly assumed that the entire S_{bsi} is transformed into VFA in the anaerobic reactor. For example, for sewage where the S_{bsi} fraction is often about 25% of the biodegradable influent COD, about 1/4 of the COD would be used for poly-P generation and the remaining 3/4 for conventional sludge generation. This would then lead to the mixed sludge of Figure 3.34C.

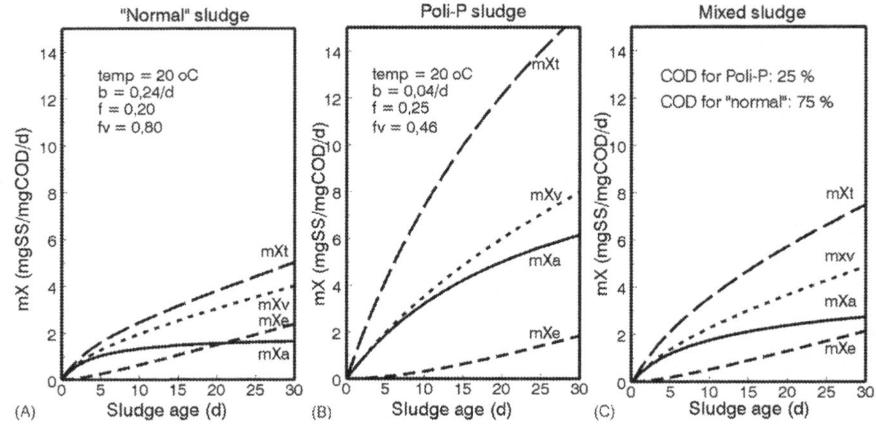

Figure 3.34 Comparison of the composition and mass of sludge in conventional, (A), of pure poly-P cultures (B) and mixed cultures (C).

When the three figures are compared the following points are noticeable:

(1) The concentrations of live organisms in systems with poly-P sludge are very much larger than in a comparable conventional-activated sludge system due to the low decay rate of poly-P organisms.
(2) Also due to the low decay rate of poly-P organisms the endogenous residue concentration is low.
(3) Due to the high inorganic fraction (phosphorus) of poly-P organisms the TSS concentration in systems with these organisms is much higher than in conventional systems.

A problem in the stabilization of the mixed sludge is that in anaerobic digesters the poly-P sludge looses its *P* upon digestion, so that phosphate is released to the liquid phase and eventually will be recycled to the treatment plant. There are two solutions for this problem: (1) precipitate the phosphate in the water from the digester or indeed in the digester by adding $AlSO_4$ or $FeCl_3$ or (2) use aerobic digesters for stabilization making use of the fact that the poly-P sludge decays very slowly, so that there will be little *P* release.

3.6 THE SULPHUR CYCLE

Due to their stability, three oxidation states of sulphur have a major importance in nature: −2 (sulphydryl, R—SH; sulphide, HS^-), 0 (elemental sulphur, S^0) and +6 (sulphate, SO_4^{2-}). Figure 3.35 shows the interrelationships between sulphur compounds with different oxidation states. Sulphur is an essential element; it comprises about 1% dry weight of a bacterial cell, generally is not a limiting growth factor and is present in cysteine, methionine, vitamins and cofactors. Other organic compounds that may also contain sulphur are (poly)peptides, lipids and carbohydrates. The compounds containing *S–S* bonds are light sensitive and radicals are formed

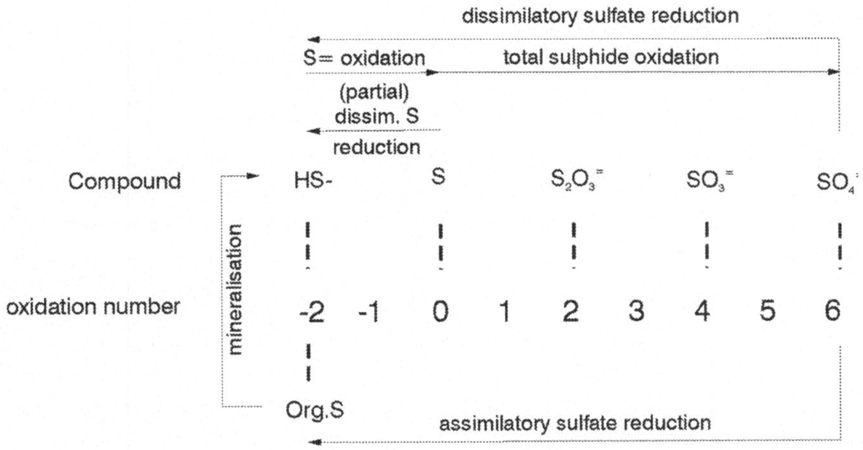

Figure 3.35 Different oxidation–reduction processes that can take place in wastewater treatment plants.

which trigger chain reactions resulting in mixtures of compounds, like polythionates and polysulphides (Steudel, 2000). Sulphides in wastewater may be generated either by industrial processes (tanneries, paper mills, refineries, mines) or as a result from mineralization of organic sulphur compounds (slaughterhouses, tanneries). The presence of sulphate may be a consequence of industrial activities (use of sulphuric acid) or it may be present in natural waters in appreciable quantities.

Under aerobic conditions (in the absence of nitrate) sulphate is energetically the most stable form of sulphur. Similarly, under anaerobic conditions, sulphide is energetically the most stable form of sulphur; in this sense, sulphate and sulphide cannot be further oxidized or reduced, respectively.

Three types of sulphur reduction may be distinguished within the biological sulphur cycle. Assimilatory sulphate reduction, in which sulphate is reduced to sulphide, and incorporated in the biomass of the microorganisms. Dissimilatory sulphate reduction, in this process sulphate is used as terminal electron acceptor leading to sulphide, S^{2-}. And sulphur respiration, a process in which elemental sulphur is used as terminal electron acceptor and reduced to sulphide. On the other hand, sulphur oxidation reactions include oxidation of reduced forms of sulphur, when used as an electron donor by phototrophic bacteria and oxidation of sulphur by chemoautotrophic bacteria that use the energy thus obtained for CO_2 assimilation.

Combination of biological reduction and oxidation processes of sulphur compounds have a very great potential to remove the element from wastewater and contaminated soils, as well as from industrial raw materials like oil and minerals. It has not yet been possible to consolidate the basic knowledge about microbiological processes and biochemistry of reaction of sulphur containing compounds into well-defined treatment systems. Nevertheless, the potential of the biological treatment to remove sulphur compounds is so great that at least the microbiological and biochemical principles are discussed in the following sections.

3.6.1 Sulphate reduction

Sulphate reduction results in the transfer of eight electrons per *S* atom:

$$8H^+ + 8e^- + SO_4^{2-} \rightarrow S^{2-} + 4H_2O \tag{3.85}$$

The dissimilatory sulphate reduction is performed by a unique group of bacteria known as sulphate-reducing bacteria (SRB) that couple the oxidation of electron donors in the form of organic matter to the reduction of sulphate as terminal electron acceptor:

$$\text{Organic matter} + SO_4^{2-} \rightarrow HS^- + HCO_3^- + H_2O \tag{3.86}$$

The main electron donors are fermentation products like hydrogen, acetate, propionate, butyrate and lactate. In this manner, sulphate is used like oxygen in aerobic respiration, with sulphide as product. However, SRB can only develop and use sulphate in the absence of oxygen or nitrate. Growth of SRB is supported by energy generation derived from two major processes: the oxidation of most organic molecules (substrate-level phosphorylation) and the reduction of sulphate (electron transport phosphorylation). The latter is the only energy yielding process if species of SRB grow with hydrogen or acetate (Widdel 1988).

In order to incorporate sulphur into cell components, bacteria, algae, fungi and plants use an assimilatory sulphate reduction system. In this system sulphate has to be reduced first to sulphide and then it is incorporated into aminoacids and other organic sulphur compounds. Although sulphide is the form of sulphur necessary for biosynthesis, organisms take up sulphur in its oxidized form (sulphate); first because sulphate is the most available and stable sulphur compound and second because of sulphide toxicity. Assimilatory sulphate reduction occurs under either aerobic or anaerobic conditions (Widdel 1988).

Dissimilatory sulphate reduction commonly takes place in anaerobic treatment plants, where appropriate conditions for the development of the SRB exist. In many cases, sulphate reduction is an undesirable process because it competes with the methanogens and hence reduces the production of methane (in principle a useful material) and enhances the production of sulphide (a problematic compound that typically requires post treatment). A decrease in methane production due to sulphate reduction can be calculated directly from stoichiometry: oxidimetricly 1 mol of SO_4^{-2} is equivalent to 2 mol of O_2 (i.e. 1 mg SO_4^{-2} can oxidize 0.66 mg COD). On the other hand, 2 mg COD can form 1/2 mg CH_4 through anaerobic digestion. Hence, the reduction of 1 mg of SO_4^{-2} to sulphide, in principle, leads to a decrease of methane production by 0.165 mg CH_4.

Recently, the reduction of sulphate has been used as a first step to remove sulphur from wastewaters. In that case, a complementary step to produce elemental sulphur is required.

3.6.1.1 Microbiology of sulphate reduction

SRB are considered to be strict anaerobes, however their occurrence has been reported in oxic environments, such as the highly oxic chemocline of lake sediments (Sass *et al.* 1997), oxic zones of cyanobacterial mats (Teske *et al.* 1998), and wastewater biofilms grown under oxic conditions (Ramsing *et al.* 1993; Lens *et al.*

1995; Santegoeds *et al.* 1998; Ito *et al.* 2002). Recently, a complete review has been published concerning oxygen respiration by SRB, specifically by *Desulfovibrio* species (Cypionka 2000). It has become clear that the capacity of aerobic respiration is normal among the SRB, however aerobic growth of pure cultures is poor or absent. It appears that the respiration capacity is a protective function.

Morphologically, SRB are diverse and vary from long rods to cocci. Two genera of dissimilatory sulphate reducers are well established: *Desulfovibrio* that comprises non-sporing, curved, motile vibrios or rods, and *Desulfotomaculum* described as spore-forming rods. Other genera of SRB are *Desulfobacter*, *Desulfobacterium*, *Desulfobulbus*, *Desulfoccocus*, *Desulfomonas*, *Desulfonema*, *Desulfosarcina* and *Thermodesulfobacterium* among others. Additionally there exists a mixed group of Archaea (*Acidianus*, *Pyrobaculum*, *Thermodiscus*, *Thermoproteus*) and Eubacteria (*Desulfuromonas*, *Desulfurolobus*) that couple their energy metabolism to the reduction of elemental sulphur. The production of hydrogen sulphide is also the characteristic of this group. Thus, presence of SRB may be easily recognized by the rotten egg smell of gaseous hydrogen sulphide and the black colour of iron sulphides.

SRB are commonly detected in anaerobic reactors and their abundance has been shown to vary depending on the sulphate level. In granular sludge from anaerobic reactors operated under methanogenic conditions *Desulfovibrio* is the most common specie that has been identified by molecular biology techniques (i.e. FISH), moreover the presence of *Desulfobulbus and Desulfobacterium* has also been reported (Godon *et al.* 1997; Merkel *et al.* 1999; Sekiguchi *et al.* 1999).

The SRB conform a broad physiological and ecological grouping, and are the only organisms known to carry out an inorganic fermentation and all dissimilatory reactions of the sulphur cycle (Table 3.10), including the oxidation of sulphur compounds. The molecular phylogenetic studies based on 16S rRNA unrecognized metabolic capabilities of SRB, for example, the capability of reducing iron and uranium. Its application in bioremediation has increased with the isolation of various SRB capable of the complete mineralization of toluene, phenol, *p*- and *m*-cresol, and benzoate coupled to sulphate reduction. (Stackebrandt *et al.* 1995). Most of SRB can grow with sulphite or thiosulphate as electron acceptors; some *Desulfovibrio* strains also can use di-, tri- and tetrathionates as terminal electron acceptors.

The disproportionation of sulphite or thiosulphate is a mechanism where sulphur oxyanions serve as both the electron donor and acceptor. This unique fermentation of inorganic compounds yields sulphide plus sulphate. Thiosulphate is transformed to equal amounts of sulphate and sulphite, while sulphite is disproportionated to three-fourths sulphate and one-fourth sulphide, as can be deducted from Figure 3.35 (Cypionka 1995). Some SRB can use H_2, reduced inorganic sulphur compounds and various organic compounds as electron donors for microaerophilic respiration, with rates comparable to those of aerobic bacteria. Sulphide, sulphite, elemental sulphur and thiosulphate can be oxidized to sulphate. The reduction of molecular O_2 by *Desulfovibrio desulfuricans* (up to 5 mM O_2 added stepwise to keep the concentration low) was coupled to the formation of adenosine triphosphate (ATP), but not to aerobic growth, sulphate was formed during oxidation of sulphite, thiosulphate or elemental sulphur by this specie of SRB (Dannenberg *et al.* 1992).

Table 3.10 The reactions of the sulphur cycle catalysed by SRB.

Reaction	ΔG^0 kJ/mol	Bacteria that use the energy for growth
Complete reduction of sulphur compounds		
$SO_4^{2-} + 4H_2 + 1.5H^+ \rightarrow 0.5HS^- + 0.5H_2S + 4H_2O$	−155	*Desulfotomaculum* *Desulfobacter*
$0.5SO_3^{2-} + 0.5HSO_3^- + 3H_2 + H^+ \rightarrow 0.5HS^- + 0.5H_2S + 3H_2S$	−175	*Desulfobacterium* *Desulfovibrio*
$S_2O_3^{2-} + 4H_2 + H^+ \rightarrow HS^- + H_2S + 3H_2O$	−179	*Thermodiscus*[b]
$S^0 + H_2 \rightarrow 0.5HS^- + 0.5H_2S + 0.5H^+$	−30	*Acidianus*[b] *Desulfovibrio* *Pyrobaculum*[b]
Disproportionation of sulphur compounds		
$S_2O_3^{2-} + H_2O \rightarrow SO_4^{2-} + 0.5HS^- + 0.5H_2S + 0.5H^+$	−25	*Desulfovibrio sulfodesmutans*
$2SO_3^{2-} + 2HSO_3^- \rightarrow 3SO_4^{2-} + 0.5HS^- + 0.5H_2S + 0.5H^+$	−236	*Desulfobacter curvatus*
Oxidation of sulphur compounds		
$0.5HS^- + 0.5H_2S + 2O_2 \rightarrow SO_4^{2-} + 1.5H^+$	−732	*Desulfovibrio desulfuricans* CSN
$0.5HS^- + 0.5H_2S + NO_3^- + 0.5H^+ + H_2O \rightarrow SO_4^{2-} + NH_4^+$	−445	
$S_2O_3^{2-} + 2H_2O + 2O_2 \rightarrow 2SO_4^{2-} + 2H^+$	−738	*Desulfobulbus*
$0.5SO_3^{2-} + 0.5HSO_3^- + 0.5O_2 \rightarrow SO_4^{2-} + 0.5H^+$	−257	*propionicus*[a]

[a]carries out all reactions except $S_2O_3^{2-}$ oxidation.
[b]Organisms are Archaebacteria.
*Modified from Barton (1992), with data of Cypionka (1995) and Kelly (1999).

Depending on the species, organic substrates are oxidized incompletely to acetate (fast-growing species) or completely to CO_2 (slow-growing species). The preferred carbon sources (electron donors) for SRB are always low molecular weight compounds derived from the anaerobic degradation of carbohydrates, proteins and lipids. The incomplete oxidizers grow on a limited numbers of substrates such as lactate, ethanol, pyruvate, propionate and malate, and will only use H_2 and CO_2 for growth in the presence of acetate. The complete oxidizers specialize in the oxidation of fatty acids (acetate, propionate, butyrate) particularly acetate. Some can grow autotrophically with H_2 and CO_2. Table 3.11 shows some of the complete and incomplete reactions carried by SRB and the genera that catalyse those reactions.

The SRB are terminal degraders that oxidize the organic compounds to stable and simple inorganic substances. In addition, SRB are able to use homocyclic aromatic compounds (toluene, benzoate, *p*-cresol and phenol), alkanes, and even chlorinated compounds in mixed or pure cultures (Ensley and Suflita 1995).

3.6.1.2 Biochemistry of sulphate reduction

SRB are loaded with cytochromes, menaquinones, several ferredoxins, flavodoxins and hydrogenases. All are electron carriers, although the function of some is not clear.

Table 3.11 The reactions of the sulphur cycle catalysed by SRB, and some of the genera that catalyse the reaction.

Reaction	ΔG^0 kJ/mole	Genera	μ_{max}[d] (1/day)
$3LA^- + SO_4^{2-} \rightarrow 2HA\text{—}HCO_3^- + H + 2HCO_3^- + HS^- + H^+$	−160	Most *Desulfovibrio* and *Desulfotomaculum*	2.8–4.3[a]
$2LA^- + 3SO_4^{2-} \rightarrow 6HCO_3^- + 3HS^- + H^+$	−255.3	*Desulfobacterium autotrophicum*	0.8–1.0
$4PA^- + 3SO_4^{2-} \rightarrow 4HA^- + 4HCO_3^- + 3HS^- + H^+$	−150.6	*Desulfobulbus propionicus*	0.2–0.6
$4PA^- + 7SO_4^{2-} \rightarrow 12HCO_3^- + 7HS^- + 2H^+$	−331	*Desulfococcus multivorans* *Desulfonema* *Desulfosarcina*	
$4HA^- + SO_4^{2-} \rightarrow 2HCO_3^- + HS^-$	−47.6	Most *Desulfobacter* *Desulfotomaculum acetoxidans*	0.98[b] 0.55[c]

Notes: LA^- $CH_3CHOHCOO^-$ (lactate); PA^- $CH_3CH_2COO^-$ (propionate); HA^- CH_3COO^- (acetate)

[a]*Desulfovibrio desulfuricans*; [b]*Desulfobacter hydrogenophilus*; [c]*Desulfotomaculum acetoxidans*

[d]growing on hydrogen

*Modified from Hao *et al.* (1996) with data of Widdel (1988) and Oude Elferink *et al.* (1994).

The sulphate uptake into the cell for assimilatory purposes is often achieved by primary transport systems and is driven by the hydrolysis of ATP, those systems are unidirectional and prevent the loss of intracellular sulphate. In dissimilatory sulphate reduction, it has been proposed that sulphate transport in SRB is driven by proton symport, which follows the chemiosmotic principles of transport. Sulphide moves across membranes by diffusion and not by an active transport process; an accumulation of sulphate is driven by pre-existing gradients of protons or sodium ions (Barton 1992; Cypionka 1995).

Once in the cytoplasm, the first step in sulphate utilization is the activation by ATP to form adenylyl sulphate (also known as adenosine-5′-phosphosulphate, APS) and pyrophosphate (PP):

$$SO_4^{2-} + ATP \rightarrow APS + PP_i \tag{3.87}$$

The reaction is endergonic (+46 kJ/mol) and is catalysed by ATP sulphurylase (or adenylyl sulphatase). Sulphate is the only electron acceptor that requires activation, this is because sulphate molecule is very stable for APS formation, the equilibrium on the reaction must be shifted by an inorganic pyrophosphatase, and thus a second phosphate ester is cleaved. APS is further reduced to bisulphite and activated protein kinase (AMP) by APS reductase. The reduction of bisulphite to sulphide is carried out by bisulphite reductase. Two mechanisms have been proposed for the reduction of bisulphite to sulphide. One is the direct reduction of bisulphite to sulphite without the formation of any intermediate compound. The other mechanism proposes the formation of trithionate and thiosulphate, with the terminal step being the reduction of thiosulphate to sulphide and bisulphite (Akagi 1995). Bisulphite (HSO_3^-) and not sulphite (SO_3^{2-}), has been determined to be the active species for those reductases that transfer $6e^-$ from reduced ferredoxin or flavodoxin to the sulphur in sulphite with the formation of sulphide (Barton 1992). Up-to-date current evidence supports both mechanisms (Widdel and Hansen 1992; Cypionka 1995).

3.6.1.3 Parameters affecting sulphate reduction

The physicochemical parameters that may affect the growth and activity of SRB have to be taken into account when applying the sulphur cycle to wastewater treatment.

pH Most SRB prefer an environment between 7.5 and 8.0, inhibition has been observed at pH values lower than 5.5 or higher than 9 (Widdel and Hansen 1992). However, sulphate reduction has been reported to occur in reactors treating acid metal containing wastewater at pH as low as 3 (Kaksonen *et al.* 2003).

Temperature Obligate psychrophilic sulphate reducers have not been isolated so far. The optimum temperature for most pure cultures of SRB is in the range of 28–38°C and have an upper temperature limit around 45°C (Widdel and Hansen 1992). Most thermophilic SRB have been isolated from geothermal environments and oil-field waters. The optimum growth temperature for thermophilic eubacterial sulphate reducers of genera *Desulfotomaculum* and *Thermodesulfobacterium* ranges between 54°C and 70°C and the maximum growth temperature vary from 56°C to 85°C. The archaebacterial sulphate reducers of the genus *Archaeoglobus* have an optimum temperature of 83°C and a maximum of 92°C (Stteter 1992).

Sulphide toxicity The metabolic product of sulphate reduction, sulphide, has been reported inhibiting SRB growth and activity, besides to be toxic to other anaerobic bacteria. Sulphide is a diprotic acid and in solution dissociates as follows:

$$H_2S \xrightleftharpoons{pK1} HS^- + H^+ \xrightleftharpoons{pK2} S^{2-} + H^+ \tag{3.88}$$

The p*k* value for the first dissociation reaction is 6.9 (at 30°C). Thus, sulphide in solution is mainly present as hydrogen sulphide (H_2S) and bisulphide (HS^-). The optimum pH for most anaerobic microorganisms is around 7. The exact mechanism of sulphide toxicity is not clear yet. It has been proposed that the neutral undissociated H_2S is the agent of toxicity, because sulphur is membrane permeable only in this form. Once inside the cell it reacts with cell components. Other possible mechanism of inhibition is the combination of any of the sulphide species (H_2S, HS^- and S^{2-}) with the iron of ferredoxin and cytochrome, and other essential iron containing compounds in the cell, causing the electron transport systems to cease activity (Okabe *et al.* 1992).

Cations As a result of closed water loops, certain chemical factories generate sulphate-rich (10 g/L) flows with salinities as high as 105 g/L of total salts. The tanning and seafood industries also generate large quantities of saline wastewater rich in sulphate oxidized sulphur compounds. The presence of sodium and other cation or other salts results in an increase of the ionic strength that causes osmotic stress to the microorganisms and the inhibition of reaction pathways in the substrate degradation process (Pollice *et al.* 2000). The effect of sodium on methanogenic digestion has been studied extensively. Recently, high-rate sulphate reduction (3.7 g SO_4^{2-}/L day) was achieved in upflow anaerobic sludge bed (UASB) reactors inoculated with undapted (to high salt concentration) granular sludge, at salinities exceeding 50 g NaCl/L and 1 g $MgCl_2$/L, using propionate or ethanol as electron donor. In the experiments, it was observed also that granular sludge was adapted to the high salt concentrations relatively fast, around 50 days (Vallero *et al.* 2004). Thus, in practical terms the sodium content of a sulphate-rich wastewater does not generate problems. On the other hand, calcium can cause serious scaling problems at concentrations as low as 400 mg Ca^{2+}, as well as the loss of buffer capacity and decreasing efficiency due to sludge washout. Calcium precipitates as carbonate or with phosphate, which is generally present in wastewater. Eventually a deposition of precipitates within the granule is expected which can completely block substrate transport affecting the activity of the sludge granules (Langerak *et al.* 1998).

Competition with other anaerobic microorganisms As terminal degraders, methanogenic archaea and SRB compete for the same organic substrates. Both types of microorganisms are capable of using acetate and hydrogen as substrates. The SRB may also compete for substrates like propionate and butyrate with acetogenic bacteria. In anaerobic wastewater treatment the outcome of this competition is difficult to predict because, generally, the final result is influenced by the combination of the following parameters: pH, temperature, substrate affinity constant (K_s), specific growth rate (μ_{max}), sulphate concentration, immobilization properties of bacteria and the type of sludge present in the reactor.

Immobilization properties In a high-rate anaerobic reactor, the ability of SRB to immobilize into granules or biofilms becomes important when effluents with high sulphate content are treated. SRB were able to attach and grow in granules when cultivated simultaneously with methanogenic archaea in an upflow anaerobic sludge blanket reactor (Visser *et al.* 1993). The operation of a sulphate reducing reactor for the treatment of acidic wastewater shows the ability of SRB to attach to solid particles (Kaksonen *et al.* 2003).

3.6.2 Sulphide oxidation

In the oxidative part of the sulphur cycle, sulphide and other reduced sulphur compounds (sulphite, thiosulphate, elemental sulphur and tetrathionate) are exclusively oxidized by prokaryotes, being sulphate the major oxidation product, although elemental sulphur may accumulate. The dissimilatory oxidation of reduced inorganic sulphur compounds can take place under aerobic or anaerobic conditions. In the aerobic oxidation of sulphide chemolithoautotrophic microorganisms are involved, most of these live under extreme conditions in hydrothermal vents, at low pH or in

Table 3.12 Examples of some reactions carried out by chemolithotrophic organisms and the free energy changes (Kelly 1999).

Reaction	ΔG° (kJ/mol S-substrate)
$HS^- + 2O_2 \rightarrow SO_4^{2-} + H^+$	−732.7
$S_2O_3^{2-} + 2O_2 + H_2O \rightarrow SO_4^{2-} + 2H^+$	−738.7
$S_4O_6^{2-} + 3.5O_2 + 3H_2O \rightarrow 4SO_4^{2-} + 6H^+$	−1244.6
$S^0 + 1.5O_2 + H_2O \rightarrow SO_4^{2-} + 2H^+$	−507.4
$HS^- + 0.5O_2 \rightarrow S^0 + OH^-$	−145.2
$S_2O_3^{2-} + 0.5O_2 \rightarrow S^0 + SO_4^{2-}$	−231
$5S_2O_3^{2-} + 8NO_3^- + H_2O \rightarrow 10SO_4^{2-} + 2H^+ + 4N_2$	−750.8

hot springs. Phototrophic bacteria carry out the anaerobic oxidation of sulphide under anaerobic conditions. Reactions that take place in the oxidative part of the sulphur cycle are listed in Table 3.12, according to the source the values are definitive, with the caveat that values *in vivo* may deviate in detail from these as consequence of intracellular concentrations of intermediates and cellular pH deviating from those of standard $\Delta G^{0\prime}$ values (Kelly 1999). However, they give an idea of the free energy that the chemolithotrophic bacteria can obtain from the different sulphur compounds.

One interesting biotechnological application of sulphide oxidation is the sulphur removal from effluents (liquids and gases) that contain reduced sulphur compounds and low amounts of organic material. The incomplete reaction to S^0 is preferred because, being insoluble in water, S^0 can be easily separated from the effluent. In terms of free energy changes the complete oxidation to sulphate yields more than two-fold energy that the incomplete oxidation to elemental sulphur. By means of high sulphide loads or low oxygen concentrations, sulphur oxidation is forced towards elemental sulphur production (Stefess *et al.* 1996; Janssen *et al.* 1997).

3.6.2.1 Microbiology of sulphide oxidation

The group of microorganisms that are able to use reduced sulphur compounds comprises a very heterogeneous collection of eubacteria and archaebacteria. Many of these have little or no taxonomic relationship to each other, but are capable to grow on reduced sulphur compounds as electron donors (lithotrophic growth). Under anaerobic conditions, most phototrophic bacteria belonging to Chlorobiaceae, Chloroflexaceae, Chromatiaceae and Ectothiorhodhospiraceae use reduced inorganic sulphur compounds as electron donors during anoxygenic photosynthesis. Principally, sulphide is oxidized via sulphite to sulphate. Elemental sulphur may appear as intermediary storage product inside the cell in *Chromatium*, *Thiocaspa*; or outside the cell as in *Chlorobium* and *Ectothiorhodospira*. These organisms generally use sulphide, the ability to use elemental sulphur is typical for Chlorobiaceae and Chromatiaceae (Trüper and Fischer 1982). The use of phototrophic bacteria for wastewater treatment has been investigated before and it has been said that these systems allow a quick and automatic control in the oxidation rate of sulphide by

Table 3.13 Simplified overview of the physiological types found among sulphur-oxidizing non-phototrophic bacteria (Robertson and Kuenen 1992).

Physiological type	Carbon source		Energy source	
	CO_2	Organic	Inorganic	Organic
Obligate autotroph	+	−	+	−
Facultative autotroph (mixotroph)	+	+	+	+
Chemolithoheterotroph	−	+	+	+
Chemoorganoheterotroph (heterotroph)	−	+	−	+

varying the light intensity when processing a waste stream of variable S^{2-} concentration (Kim *et al.* 1993). Recently, the use of a suspended growth photoreactor fed with dissolved sulphide at laboratory scale was successfully operated for the conversion of sulphide to elemental sulphur, at a loading rate of 4.4 mg/L h (Henshaw *et al.* 1998). However, the main drawback in the application of phototrophic bacteria at large scale is the high energy requirement and the special reactor design to allow the maximum penetration of light, increasing operational costs.

On the other hand, the microorganisms that oxidize sulphur compounds in the presence of oxygen (or nitrate), are classified in four physiological types (Table 3.13) depending on the source of energy and their carbon metabolism. The best known examples of sulphur chemolithotrophic bacteria are the thiobacilli, some *Paracoccus* and *Xanthobacter* species, and the archaea *Sulfolobus* and *Acidianus* (Kelly 1999). Some genera of obligate and facultative autotrophic sulphur oxidizing bacteria are given in Table 3.14 along with the electron donors that they use and optimal pH and temperature.

In biotechnological processes, thiobacilli have been extensively studied mainly for their application in waste effluent treatment (liquid and gas) for sulphur and nitrate removal. In the mining industry, *Acidithiobacillus* and *Acidiphilium* are used in the recovery of metals from poor ores by leaching. All thiobacilli are small (0.3–0.5 × 0.7–4.0 μm) Gram-negative, rod-shaped bacteria, some species are motile by means of polar flagella.

3.6.2.2 Biochemistry of sulphide oxidation

Autotrophic bacteria fix carbon dioxide either via the reductive pentose phosphate cycle or via the reductive tricarboxylic acid cycle. Reductant released from sulphur oxidation is used in lithotrophic bacteria for aerobic respiration and carbon dioxide reduction, while in anaerobic phototrophic bacteria reductant is used mainly for carbon dioxide fixation (Friedrich *et al.* 2001).

Much work has been done in attempts to elucidate the pathways of the oxidation of the inorganic and organic reduced sulphur compounds to sulphate and to establish the mechanisms and efficiency of the coupling of the energy released to the growth of bacteria. However, due to the diversity of bacteria and archaea more than one mechanism of sulphur oxidation has been identified (Kelly 1999). Based

Table 3.14 Survey of the metabolic types found among chemolithotrophic microorganisms.

Species or genera	Electron donor	Product	pH	Temperature (°C)	Denitrifies	
					To NO_2^-	To N_2
Obligate autotrophic						
Thiobacillus thioparus	HS^-, S^0, $S_2O_3^{2-}$	S^0 SO_4^{2-}	6–8	<37	+	−
[b]*Halothiobacillus neapolitanus*	HS^-, S^0, $S_2O_3^{2-}$	S^0 SO_4^{2-}	6–8	<37	−	−
Thiobacillus denitrificans	HS^-, S^0, $S_2O_3^{2-}$	SO_4^{2-}	6–8	<42	+	+
[b]*Acidithiobacillus thiooxidans*	S^0	SO_4^{2-}	2–5			
[b]*Acidithiobacillus ferrooxidans*	HS^-, S^0, $S_2O_3^{2-}$	SO_4^{2-}	1.5–4	<38		
[b]*Thermithiobacillus tepidarious*	HS^-, S^0, $S_2O_3^{2-}$	SO_4^{2-}	5.5–8	20–52	+	−
Thiomicrospira denitrificans	HS^-, S^0, $S_2O_3^{2-}$	SO_4^{2-}	6–8	<37	+	+
Thiovulum	HS^-	SO_4^{2-}	6–8			
[a]*Sulfolobus*	HS^-, S^0	SO_4^{2-}	1–6	60–95		
Facultative autotrophic						
Thiotrix thioparus	HS^-	SO_4^{2-}	6–8	40–80	+	+
Beggiatoa sp (marine)	HS^-, $S_2O_3^{2-}$	SO_4^{2-}	7	<37	−	−
Beggiatoa sp (freshwater)	HS^-, $S_2O_3^{2-}$	SO_4^{2-}	7	<37	+	+
[a]*Sulfolobus acidocaldarius*	HS^-, S^0	SO_4^{2-}	1–6	60–85	−	−
[a]*Acidianus brierleyi*	S^0	SO_4^{2-}	1–5	60–95		

[a]Archaea; [b]Newly designated genera (Kelly and Wood 2000) formerly named as *Thiobacillus*
*Modified from Robertson and Kuenen (1992).

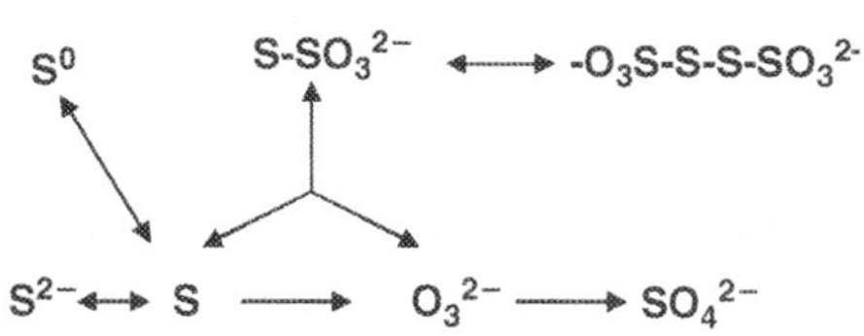

Figure 3.36 Reactions in the oxidation of organic sulphur compounds (Suzuki 1999).

on enzymatic reactions, a general scheme for the oxidation of inorganic sulphur compounds has been proposed (Figure 3.36) in which the main oxidation pathway is the oxidation of S^{2-} to SO_4^{-2} (Suzuki 1999).

Based on physiological and biochemical data, at least two major pathways have been proposed to exist for different sulphur-oxidizing bacteria: The sulphur oxidation pathway and the tetrathionate pathway that involves polythionates.

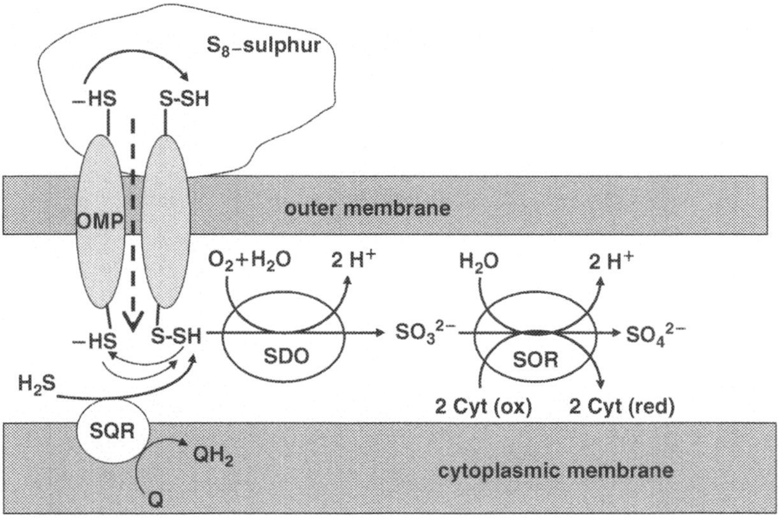

Figure 3.37 Biochemical model proposed for sulphur oxidation in *Acidithiobacillus* and *Acidiphilum*. Elemental sulphur (S_8) is mobilized as persulphide sulphur by special outer-membrane proteins (OMP) and oxidized by periplasmic sulphur dioxygenase (SDO). The produced sulphite is oxidized to sulphate by sulphite: acceptor oxydoreductase (SOR) which probably uses cytochromes as electron acceptors (Cyt.) Free sulphide is oxidized by a separate dehydrogenase (SQR), which uses quinones (Q) as electron acceptors (after Rohwerdert and Sand 2003).

The oxidation to elemental sulphur is considered a bottleneck in the sulphur cycle because it is an important reaction in biotechnological processes such as the biomining or the desulphurization of wastewaters. Recently, Rohwerdert and Sand (2003) proposed a biochemical sulphur oxidation mechanism for the meso-acidophilic bacteria *Acidithiobacillus* and *Acidiphilium* that catalyse the oxidation of inorganic sulphur compounds under acidic conditions (Figure 3.37). For these genera the lowest pH after growth on sulphur compounds reaches 1–3, and some species oxidize ferrous iron or use natural and synthetic metal sulphides for energy generation (Kelly and Wood 2000). Taking into account that sulphur consists of a stable octasulphane ring system (S_8) that forms orthorhombic crystals with extremely poor water solubility, an activation reaction is postulated. This activation occurs most probably between elemental sulphur and the thiol groups of especial outer-membrane proteins to form persulphides. The persulphide sulphur is oxidized by periplasmic glutathion-dependent sulphur dioxygenase to sulphite that is further oxidized to sulphate by sulphite: acceptor oxidoreductase. This complex is membrane associated and most probably coupled to *c*-type cytochromes as electron acceptors (de Jong *et al.* 2000).

The thiol-bearing membrane proteins have not yet been identified, but a hydrogen sulphide-binding protein, has been purified from the membrane fraction of *Acidithiobacillus ferrooxidans* AP19-3 (Sugio *et al.* 1991). In the proposed model, free sulphide is oxidized to elemental sulphur by a separate sulphide:quinone oxidoreductase (SQR), located at the periplasmic site of the cytoplasmic membrane. Recently, SQR activity was observed in *Acidithiobacillus ferrooxidans* sulphur-

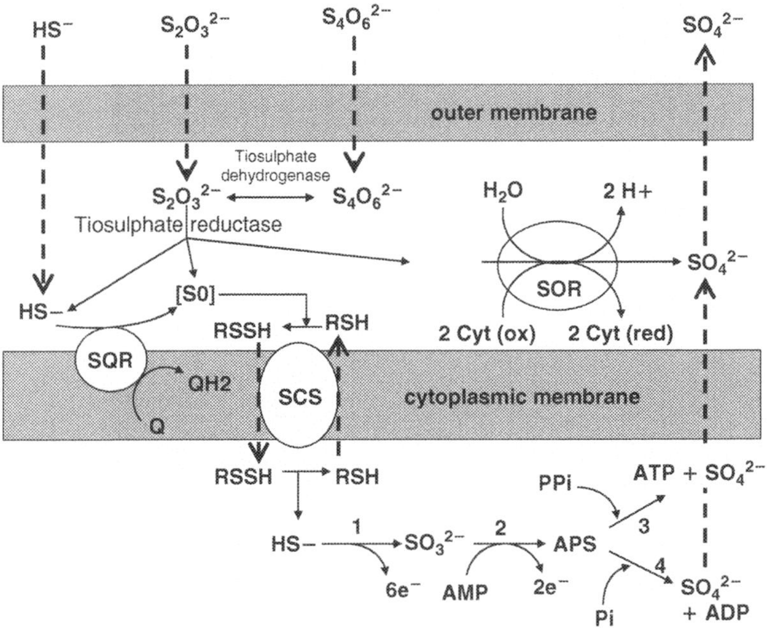

Figure 3.38 Biochemical model proposed for the thiosulphate cleavage pathway. The sulphane sulphur accumulates as S^0. Sulphane sulphur is converted into sulphate. Produced sulphite is oxidized to sulphate by sulphite:acceptor oxydoreductase (SOR) which probably uses cytochromes as electron acceptors (Cyt.) or in the cytoplasm by the APS route. Enzymes required are as indicated: 1, siroheme sulphite reductase; 2, APS reductase; 3, ATP sulphurylase; 4, APAT (APS:phosphate adenylyltransferase). SCS: sulphide carrier system sys transports periplasmic polysulphanes or polysulphides to the cytoplasm (modified from Brüser *et al.* 2000b).

growing cells (Wakai *et al.* 2004). The enzyme has only been isolated and purified to homogeneity from *Rhodobacter capsulatus* and *Oscillatoria limnetica* which are neutrophilic sulphur bacteria. It has been proposed that the electrons from sulphide reduce the two electron carrier flavin-adenine dinucleotide (FAD), and the two electron carrier quinone becomes reduced (Griesbeck *et al.* 2002).

Thiosulphate oxidation is more complicated, there are at least three possible pathways, one is the pathway starting with the cleavage of the sulphur–sulphur bond by thiosulphate reductase (or rhodanase), followed by sulphur-oxidizing enzyme. The second is the pathway involving tetrathionate hydrolase and thiosulphate-oxidizing enzyme (Suzuki 1999). A third pathway consists of a multienzyme complex (Kelly *et al.* 1997).

The cleavage of the sulphur–sulphur bond in thiosulphate has been well studied and established for anaerobic sulphur oxidation in *Allochromatium vinosum* and *Thiobacillus denitrificans* (Figure 3.38).

In the pathway a thiosulphate reductase (or rhodanase) located in the periplasm is responsible for the thiosulphate cleaving step into sulphane sulphur and sulphone sulphur that are further oxidized by different pathways. The sulphane sulphur accumulates as S^0 before further oxidation, whereas the sulphone sulphur is rapidly converted into sulphate, in this conversion sulphite is an intermediary. For sulphite oxidation, two pathways may play a role: (1) Direct oxidation of sulphite to sulphate in the periplasm by the action of sulphite:acceptor oxydoreductase (Kappler *et al.* 2000) and (2) Indirect AMP-dependent oxidation of sulphite to sulphate via the intermediate APS. In the second pathway, APS is formed from sulphite and AMP by the enzyme APS reductase acting in reverse. Sulphate is released in the second step either by ATP sulphurylase or by APS:phosphate adenylyltransferase (APAT), this enzyme has been recently isolated from *Thiobacillus denitrificans* (Brüser *et al.* 2000a).

The third pathway is the thiosulphate-oxidizing enzyme system that has been analysed in detail for *Paracoccus versutus* (formerly *Thiobacillus versutus*) and it is known as The Thiosulphate-Oxidizing Multienzyme System (TOMES). This system is the best studied and most widely distributed pathway for the oxidation of thiosulphate to sulphate. It is also referred to as the Sox system because of the gene region coding for sulphur oxidation. In some organisms, the TOMES pathway is also employed for the oxidation of sulphide, sulphur, sulphite and tetrathionate (Rother *et al.* 2001). The different Sox systems are located exclusively in the periplasm and are present in phototrophic, lithotrophic and methylotrophic bacteria that oxidize reduced sulphur compounds to sulphate (Friedrich *et al.* 2001). Sox systems use a small, monohaem, *c*-type cytochrome as the direct electron acceptor. The Sox system of *Paracoccus versutus* is composed of enzyme A (16,000 Da), enzyme B (60,000 Da), cytochrome c_{551} (260,000 Da) which is intimately associated with sulphite dehydrogenase (44,000 Da), and homodimeric cytochrome $c_{552.5}$, this system catalyses the following reaction stoichiometrically (Kelly *et al.* 1997):

$$^{-}S{-}SO_3^- + 5H_2O \rightarrow 2SO_4^{2-} + 8e^- + 10H^+ \tag{3.89}$$

The total process is carried out in an integrated way, and no free intermediaries have been identified. In the Sox system of *Paracoccus pantotrophus* it has been suggested that the initial reaction in the pathway is the oxidative formation of a disulphide linkage between the sulphane sulphur of thiosulphate and the SoxY cysteine. The reaction is catalysed by a haemoprotein SoxAX complex. The terminal sulphone group of the resultant adduct is realeased as sulphate by SoxB in a hydrolytic reaction. The sulphane atom is then oxidized to a sulfone by the SoxCD complex and released from SoxY as sulphate by the activity of SoxB (Bamford *et al.* 2002).

In the tetrathionate pathway, which is not completely understood, thiosulphate can be oxidized to sulphate via polythionates as intermediates according to the following equations (Friedrich *et al.* 2001):

$$2\ ^{-}S{-}SO_3^- \rightarrow\ ^{-}O_3S{-}S{-}S{-}SO_3^- + 2e^- \tag{3.90}$$

(thiosulphate dehydrogenase)

Tetrathionate is then hydrolyzed to thiosulphate, sulphur and sulphate:

$$^{-}O_3S{-}S{-}S{-}SO_3^- + H_2O \rightarrow\ ^{-}S{-}SO_3^- + SO_4^{2-} + S^0 + 2H^+ \tag{3.91}$$

(tetrathionate hydrolase)

If such reactions represent the main stream degradation route for these compounds, the implication is that the first metabolically oxidizable intermediate would be elemental sulphur, which can be oxidized to sulphite and then to sulphate by a pathway that is not known yet (Kelly 1999). If sulphite accumulates, it can react with the tetrathionate and form thiosulphate and trithionate, which can be hydrolyzed to thiosulphate and sulphate (Suzuki 1999):

$$^{-}O_3S{-}S{-}SO_3^- + H_2O \rightarrow {}^{-}S{-}SO_3^- + SO_4^- + 2H^+ \qquad (3.92)$$

From all the information presented above, it can be concluded that chemolithotrophic microorganisms have not only one option for the oxidation of reduced sulphur compounds.

3.6.2.3 Parameters affecting sulphide oxidation

Environmental parameters Chemolithotrophic bacteria have been found in nature growing at a broad range of pH (from 9 to 1), and at temperatures between 4 and 95°C. In biotechnological applications, the variation of these parameters may affect the outcome of competition for a substrate if a particular microbial community is used. Variations in pH may be used for the selection of acidophilic species or halophilic species within a community. This is because the oxidation of reduced sulphur compounds can be carried out in a wide pH range but the bacteria can only grow under a narrow range of pH.

Culture conditions Obligate chemolithoautotrophs are difficult to cultivate on solid media, as they are very sensitive to organic matter including the small quantities of sugar present as impurities in polysaccharide-based gelling agents such as agar or agarose. In the particular case of acidophiles, attempts to use highly purified agars have not been successful, probably because some of the sugar molecules in the gelling agent are released due to acid-hydrolysis at low pH (Rawlings 2002).

Nutrient availability One of the most important parameters affecting the selection of populations in natural or laboratory environments is the ratio of inorganic substrate to organic nutrients. If the available substrate is predominantly inorganic, obligate autotrophs will normally tend to dominate over facultatively mixotrophic species. On mixed substrates, facultative autotrophs or chemolithoheterotrophs will predominate, depending on the ratio between the two types of substrate. On abundant organic substrates, the sulphide-oxidizing heterotrophs will tend to dominate (Robertson and Kuenen 1992). In general, the obligate species exhibit much greater flexibility in responding to variable supplies of their organic substrate, also frequently showing higher affinities for the substrate, ensuring their survival in fluctuating environments (Wood *et al.* 2004).

REFERENCES

Akagi, J.M. (1995) Respiratory sulphate reduction, In: *Sulphate-reducing Bacteria* (ed. L.L. Barton), Plenum Press, New York, pp. 89–111.

Araújo, L.d.S., Catunda, P.F.C. and Van Haandel, A.C. (1998) Biological sludge stabilisation Part 2 influence of the composition of waste activated sludge on anaerobic stabilisation. *Wat. S.A.* **24**, 231–236.

Arvin, E. (1985) Biological removal of phosphorus from waste waters. *CRC. Critical Rev. Environ. Cont.*, **15**, 24–94.

Bamford, V.A., Bruno, S., Rasmussen, T., Appia-Ayme, C., Chessman, M.R., Berks, B.C. and Hemmings, M. (2002) *EMBO J.* **21**, 5599–5610.

Barnard, J.L. (1991) *Entwurf einer Belebungsanlage für nitrifikation/denitrifikarion, Institut für Siedlungswasserwirtschaft,* 50, Univ. de Braunschweig, Alemanha, 9–18 (in German).

Barnard, J.L. (1976) "A Review of biological phosphorus removal in the activated sludge process". *Wat. S.A.,* **2**(3), 163.

Bart, Brenner and Lewis (1969). Chemical–Biological control of nitrogen and phosphorus in waste–water effluent. *J.Wat.Pollut. Contr. Fed.* **40**, 2040–2054.

Barton, L.L. (1992) Sulphur metabolism. *Encycloped. Microbiol.* **4**. 13–150.

Broda, E. (1977) Two kinds of lithotrophs missing in nature. *Z. Allg. Mikrobiol.* **17**, 491–493.

Brüser. T., Selmer, T. and Dahl, C. (2000a) "ADP sulfurylase" from *Thiobacillus denitrificans* is an adenylylsulfate:phosphate adenylyltransferase and belongs to a new family of nucleotidyltransferases. *J. Biol. Chem.* **275**, 1691–1698.

Brüser. T., Lens, P.N.L. and Trüper, H.G. (2000b) The biological sulphur cycle. In: *Environmental Technologies to Treat Sulphur Production. Principles and Engineering* (Eds. P. Lens and L. Hulshoff Pol) IWA Publishing, London pp. 33–85.

Buchan, L. (1981) The location and nature of accumulated phosporus in seven sludges from activated sludge plants which exhibited enhanced phosphorus removal, *Wat. S.A* **7**, 1–7.

Cervantes F.J., de la Rosa, D.A. and Gómez, J. (2001) Nitrogen removal from wastewaters at low C/N ratios with ammonium and acetate as electron donors. *Biores. Technol.* **79**, 165–170.

Cervantes F.J., Monroy, O. and Gómez, J. (1999) Influence of ammonium on the performance of a denitrifying culture under heterotrophic conditions. *Appl. Biochem. Biotechnol.* **81**, 13–21.

Comeau Y., Hall K.J. and Oldham (1985) A biochemical model for biological enhanced phosphorus removal. *Wat. Res.* **20**(12), 1511–1521.

Coura Dias M., Catunda P.F.C. and Van Haandel A.C. (1983) O sistema de lodo ativado. Parte I: O estado estacionário. *Anais do Congresso Brasileiro de Engenharia Sanitária*, Camburiu.

Cypionka, H. (1995) Solute transport and cell energetics. In: *Sulfate-reducing Bacteria* (ed. L.L. Barton) Plenum Press, New York pp. 151–184.

Cypionka, H. (2000) Oxygen respiration by *Desulfovibrio* species. *Annu. Rev. Microbiol.* **54**, 827–848.

Dalsgaard, T. and Thamdrup, B. (2002) Factors controlling anaerobic ammonium oxidation with nitrite in marine sediments. *Appl. Environ. Microbiol.* **68**, 3802–3808.

Dalsgaard, T., Canfield, D.E., Petersen, J., Thamdrup, B. and Acuña-González, J. (2003) Anammox is a significant pathway of N_2 production in the anoxic water column of Golfo Dulce, Costa Rica. *Nature* **422**, 606–608.

Dannenberg, S., Kroder, M., Dilling, W. and Cypionka, H. (1992) Oxidation of H_2, organic compounds and inorganic sulphur compounds coupled to reduction of O_2 or nitrate by sulfate-reducing bacteria. *Arch. Microbiol.* **158**, 93–99.

de Jong, G.A.H., Tang, J.A., Bos, P., de Vries, S. and Kuenen, J.G. (2000) Purification and characterization of a sulfite:cytochrome *c* oxidoreductase from *Thiobacillus acidophilus*. *J. Mol. Catal. B* **8**, 61–67.

De Man A.W.A., van der Last, A.R.M. and Lettinga, G. (1988) The use of EGSB and UASB anaerobic systems for low strength soluble and complex wastewaters at temperatures ranging from 8°C to 30°C. In: Hall ER and Hobson PN (Ed), *Proceedings of the Fifth International Symposium on Anaerobic Digestion* (pp. 735–738), Bologna, Italy.

Dold, P.L., Ekama, G.A. and Marais, G.v.R. (1980) A general model for the activated sludge process. *Prog.Wat.Tech.* **12**, 47–77.

Dombroski, S.A.G. (2003) Nitrificação e desnitrificação de água residuária de coqueria utilizando fonte interna de carbono. PhD Thesis, EP-USP, São Paulo, Brazil.

Dong, X. and Tollner, E.W. (2003) Evaluation of anammox and denitrification during anaerobic digestion of poultry manure. *Biores. Technol.* **86**, 139–145.

Downing, A.L., Painter, H.A. and Knowles, G. (1964) "Nitrification in the Activated Sludge Process". *J.Proc.Inst.Sew.Purif.* **64**(2), 130–158.

Egli, K., Franger, U., Alvarez, P.J.J., Siegrist, H., Vandermeer, J.R. and Zehnder, A.J.B. (2001) Enrichment and characterisation of an anammox bacterium from a rotating biological contactor treating ammonium-rich leachate. *Arch. Microbiol.* **175**, 198–207.

Ensley, B.D. and Suflita, J.M. (1995) Metabolism of environmental contaminants by mixed and pure cultures of sulfate reducing bacteria. In: *Sulfate-reducing Bacteria* (ed. L.L. Barton), Plenum Press, New York pp. 293–332.

Folsom, B, Schieche, D., DiGrazia, P., Werner, J. and Palmer, S. (1999). Microbial desulfurization of alkylated dibenzothiophenes from a hydrodesulfurized middle distillate by *Rhodococcus erythropolis* I–19. *Appl. Environ. Microbiol.* **65**, 4967–4972.

Friedrich, C.G., Rother, D., Bardischewsky, F., Quentmeier, A. and Fischer, J. (2001) Oxidation of reduced inorganic sulphur compounds by bacteria: emergence of a common mechanism? *Appl. Environ. Microbiol.* **67**, 2873–2882.

Fuhs, G.W. and Chen, M. (1975) Microbiological bases for phosphate removal in the activated sludge process for the treatment of waste water. *Microbiol. Ecol.* **2**, 119–138.

Fujii, T., Sugino, H., Rouse, J.D. and Furukawa, K. (2002) Characterization of the microbial community in an anaerobic ammonium-oxidizing biofilm cultured on a nonwoven biomass carrier. *J. Biosci. Bioeng.* **94**, 412–418.

Fukase T., Shibata M. and Mijayi (1982) "Studies on the mechanism of biological phosphorus removal". *Japan. J. Water Pollut.Res.* **5**, 309–317.

Furukawa, K., Rouse, J.D., Yoshida, N. and Hatanaka, H. (2003) Mass cultivation of anaerobic ammonium-oxidizing using a novel nonwoven biomass carrier. *J. Chem. Eng. Japan* **36**, 1163–1169.

Fux, C., Boehler, M., Huber, P., Brunner, I. and Siegrist, H.R. (2002) Biological treatment of ammonium rich wastewater by partial nitrification and subsequent anaerobic ammonium oxidation (anammox) in a pilot plant. *J. Biotechnol.* **99**, 295–306.

Godon, J.J., Zumstein, E., Dabert, P., Habouzit, F. and Molleta, R. (1997). Molecular microbial diversity of an anaerobic digestor as determined by small-subunit rDNA sequence analysis. *Appl. Environ. Microbiol.* **63,** 2802–2813.

Griesbeck, C., Schütz, M., Schödl, L., Bathe, S., Nausch, L., Mederer, N., Vielreicher, M. and Hauska, G. (2002) Mechanism of sulphide-quinone reductase investigated using site-directed mutagenesis and sulphur analysis. *Biochemistry* **41**, 11552–11565.

Grossman, M.J., Lee, M.K., Prince, R.C., garret, K.K., George, G.N. and Pickering, I.J. (1999) Microbial desulfurization of a crude oil middle-destillate fraction: analysis of the extent of sulphur removal and the effect of removal on remaining sulfur. *Appl. Environ. Microbiol.* **65**, 181–188.

Gujer, W. (1977) Design of nitrifying activated sludge process with the aid of dynamic simulation. *Prog.Wat.Tech.* **9**, 323–336.

Gujer, W. and Zehnder, A.J.B. (1983) Conversion processes in anaerobic digestion. *Wat.Sci.Techn.* **15**, 127.

Hao, J.O., Chen, J.M., Huang, L. and Buglass, R.L. (1996) Sulfate-reducing bacteria. *Crit. Rev. Environ. Sci. Technol.* **26**, 155–187.

Harms, G., Zengler, K., Rabus, R., Aeckersberg, F., Minz, D., Roselló-Mora, R. and Widdel, F. (1999) Anaerobic oxidation of *o*-xylene, *m*-xylene, and homologous alkylbenzenes by new types of sulfate-reducing bacteria. *Appl. Environ. Microbiol.* **65**, 999–1004.

Haug, R.T. and McCarty, P.L. (1971) *Nitrification with the Submerged Filter*. Report prepared by the Department of Civil Engineering, Stanford University for the US Environmental Protection Agency, Research Grant No. 17010 EPM, August. 1971.

Helmer-Madhok, C., Schmid, M., Filipov, E., Gaul, T., Hippen, A., Rosenwinkel, K.H., Seyfried, C.F., Wagner, M. and Kunst, S. (2002) Deammonification in biofilm systems: population structure and function. *Wat. Sci. Technol.* **46**, 223–231.

Henshaw, P.F., Bewtra, J.K. and Biswas, N. (1998) Hydrogen sulphide conversion to elemental sulphur in a suspended-growth continuous stirred tank reactor using *Chlorobium limicola*. **32**, 1769–1778.

Henze, M., Grady, C.P.L., Gujer, W., Marais, G.v.R. and Matsuo, T. (1986) 'Activated sludge model NO. 1" *Scientific and Technical reports NO. 1, IAWPRC,* London, UK.

Henzen, M. and Harremoes, P.(1983) Anaerobic treatment of waste water in fixed film reactors – a literature review. *Wat.Sci.Techn.* **15**, 1.

Hooper, A.B., Vannelli, T., Bergmann, D.J. and Arciero, D.M. (1997) Enzymology of the oxidation of ammonia to nitrite by bacteria. *Anton. van Leeuwen.* **71**, 59–67.

Ito, T., Nielsen, J.L., Okabe, S., Watanabe, Y. and Nielsen, P.H. (2002) Phylogenetic identification and substrate uptake patterns of sulfate-reducing bacteria inhabiting an oxic-anoxic sewer biofilm determined by combining microautoradiography and fluorescent in situ hybridization. *Appl. Environ. Microbiol.* **68**, 356–364.

Janssen, A.J.H., Ma, S.C., Lens, P. and Lettinga, G. (1997) Performance of a sulphide-oxidizing expanded-bed reactor supplied with dissolved oxygen. *Biotechnol. Bioeng.* **53**, 32–40.

Jetten, M.S.M., Horn, S.J. and van Loosdrecht, M.C.M. (1997) Towards a more sustainable municipal wastewater treatment system. *Water Sci. Technol.* **35**(9), 171–180.

Jetten, M.S.M., Sliekers, O., Kuypers, M., Dalsgaard, T., van Niftrik, L., Cirpus, I., van de Pas-Schoonen, K., Lavik, G., Thamdrup, B., Le Paslier, D., Op den Camp, H.J.M., Hulth, S., Nielsen, L.P., Abma, W., Third, K., Engström, P., Kuenen, J.G., Jorgensen, B.B., Canfield, D.E., Sinninghe Damsté, J.S., Revsbech, N.P., Fuerst, J., Weissenbach, J., Wagner, M., Schmidt, I., Schmid, M. and Strous, M. (2003) Anaerobic ammonium oxidation by marine and freshwater planctomycete-like bacteria. *Appl. Microbiol. Biotechnol.* **63**, 107–114.

Jetten, M.S.M., Strous, M., van de Pas-Schoonen, K.T., Schalk, J., van Dongen, U.G.J.M., van de Graaf, A.A., Logemann, S., Muyzer, G., van Loosdrecht, M.C.M. and Kuenen, J.G. (1999) The anaerobic oxidation of ammonium. *FEMS Microbiol. Rev.* **22**, 421–437.

Jetten, M.S.M., Wagner, M., Fuert, J., van Loosdrecht, M., Kuenen, G. and Strous, M. (2001) Microbiology and application of the anaerobic ammonium oxidation (anammox) process. *Curr. Opin. Microbiol.* **12**, 283–288.

Kabe, T., Ishihara, A. and Tajima, H: (1992) Hydrodesulfurization of sulfur-containing polyaromatic compounds in light oil. *Ind. Eng. Che. Res.* **31**, 1577–1580.

Kaksonen, A.H., Riekkola-Vanhanen, M.-L. and Puhakka, J.A. (2003) Optimization of metal sulphide precipitation in fluidized-bed treatment of acidic wastewater. *Wat. Res.* **37**, 255–266.

Kappler, U., Bennett, B., Rethmeier, J., Schwarz, G., Deutzmann, R., McEwan, A.G. and Dahl, C. (2000) Sulfite:cytochrome *c* oxydoreductase from *Thiobacillus novellas* purification, characterization and molecular biology of a heterodimeric member of the sulfite oxidase family. *J. Biol. Chem.* **275**, 13202–13212.

Kato MT, Florencio, L. and Arantes, R.F.M. (2003) Post-treatment of UASB effluent in an expanded granular sludge bed reactor type using flocculent sludge. *Water Sci. Technol.* **48,** 279–284.

Kaufman, E., Harkins, J.B. and Borole, A.P. (1998) Comparison of batch stirred and electrospray reactors for biodesulfurization of dibenzothiophene in crude oil and carbon feedstocks. *Appl. Biochem. Biotechnol.* **73**, 127–144.

Kelly, D.P. (1999) Thermodynamic aspects of energy conservation by chemolithotrophic sulphur bacteria in relation to the sulphur oxidation pathways. *Arh. Microbiol.* **171**, 219–229.

Kelly, D.P. and Wood, A.P. (2000) Reclassification of some species of *Thiobacillus* to the newly designated genera *Acidithiobacillus* gen. nov., *Holothiobacillus* gen. nov. and *Thermithiobacillus* gen.nov. *Int. J. Syst. Evol. Microbiol.* **50**, 511–516.

Kelly, D., Shergill, J., Lu, W., Wood, A. (1997) Oxidative metabolism of inorganic sulphur compounds by bacteria. *Anton. van Leeuwen.* **71**, 95–107.

Kim, B.W., Kim, E.H., Lee, S.C. and Chang, H.N. (1993) Model-based control of feed rate and illuminance in a photosynthetic fed-batch reactor for H_2S removal. *Bioprocess Eng.* **8**, 263–269.

Kuenen, J.G. and Jetten, M.S.M. (2001) Extraordinary anaerobic ammonium oxidising bacteria. *Am. Soc. Microbiol. News* **67**, 456–463.

Kuypers, M.M.M., Sliekers, A.O., Lavik, G., Schmid, M., Jorgensen, B.B., Kuenen, J.G., Sinninghe Damsté, J.S., Strous, M. and Jetten, M.S.M. (2003) Anaerobic ammonium oxidation by anammox bacteria in the Black Sea. *Nature* **422**, 608–611.

Langerak, van E.P.A., Gonzalez-Gil, G., van Aelst, A., van Lier, J.B., Hamelers, H.V.M. and Lettinga, G. (1998). Effects of high calcium concentrations on the development of methanogenic sludge in upflow anaerobic sludge bed (UASB) reactors. *Wat. Res.* **32**, 1255–1263.

Lens, P.N., de Poorter, M.-P., Cronenberg, C.C. and Verstraete, W.H. (1995) Sulfate reducing and methane producing bacteria in aerobic wastewater treatment systems. *Wat. Res.* **29**, 871–880.

Lettinga, G., van Velsen, A.F.M., Hobma, S.W., de Zeeuw, W. and Klapwijk, A. (1980) Use of the upflow sludge blanket (USB) reactor concept for biological wastewater treatment, especially for anaerobic treatment. *Biotechnol. Bioeng.* **22,** 699–734.

Lijklema, L. (1973). Model for nitrification in the activated sludge process. *Envir. Sci. and Tech.* **7**(5), 428.

Lindsay, M.R., Web, R.I., Strous, M., Jetten, M., Butler, M.K. and Fuerst, J.A. (2001) Cell compartmentalization in planctomycetes: novel types of structural organization for the bacterial cell. *Arch. Microbiol.* **175**, 413–429.

Loewenthal, R.E. and Marais, G.v.R. (1976). *Carbonate Chemistry of Aquatic Systems: Theory and Application.* Ann Arbor Science, Michigan, USA.

Ludzack, F.J. and Ettinger, M.G. (1962). Controlling Operation to Minimize Activated Sludge Effluent Nitrogen. *J.Wat.Pollut.Control Fed.* **34**, 920–931.

Macedo Alves R.F.C. (1990) Tratamento biológico da lixívia MSc Thesis – University of Paríba – Campina Grande, Brazil.

Marais, G.v.R. and Ekama, G.A. (1976). The activated sludge process: steady state behaviour. *Wat. S.A.* **2**(4), 163–200.

Marsden, M.G. and Marais, G.v.R. (1976). *The Role of the Primary Anoxic Reactor in Denitrification and Biological Phosphorus Removal.* Research Report. No.19, Department of Civil Engineering, University of Cape Town.

Merkel, W., Manz, W., Szewzyk, U. And Krauth, K. (1999). Population dynamics in anaerobic wastewater reactors: modelling and *in situ* characterization. *Wat. Res.* **33**, 2392–2402.

Mulder, A., van de Graaf, A.A., Robertson, L.A. and Kuenen, J.G. (1995) Anaerobic ammonium oxidation discovered in a denitrifying fluidized bed reactor. *FEMS. Mircobiol. Ecol.* **16**, 177–184.

Okabe, S., Nielsen, P.H. and Characklis, W.G. (1992) Factors affecting microbial sulfate reduction by *Desulfovibrio desulfuricans* in continuous culture: limiting nutrients and sulphide concentration. *Biotech.Bioeng.* **40**, 725–734.

O'Rourke J.T. (1968) *Kinetics of anaerobic treatment at reduced temperature.* PhD Thesis, Stanford University Stanford, CA, USA.

Oude Elferink, S.J.W.H., Visser, A., Hulshoff Pol, L.W. and Stams, A.J.M. (1994). Sulfate reduction in methanogenic bioreactors. *FEMS Microbiol. Rev.* **15**, 119–136.

Pollice, A., Rozzi, A., Tomei, M.C., Di Pinto, A.C. and Limoni N. (2000). Monitoring the inhibitory effect of NaCl on anaerobic wastewater treatment process by the rantox biosensor. *Environ. Technol.* **21**, 535–544.

Pynaert, K., Smets, B.F., Wyffels, S., Beheydt, D., Siciliano, S.D., Verstraete, W. (2003) Characterisation of an autotrophic nitrogen-removing biofilm from a highly loaded lab-scale rotating biological contactor. *Appl. Environ. Microbiol.* **69**, 3626–3635.

Pynaert, K., Smets, B.F., Beheydt, D. and Verstraete, W. (2004) Start-up of autotrophic nitrogen removal reactors via sequential biocatalyst addition. *Environ. Sci. Technol.* **38**, 1228–1235.

Rabinowitz B., and Marais, G.v.R. (1980) *Chemical and Biological Phospohrus Removal in the Activated Sludge Process*, Research Report W32, Department of Civil Engineering, University of Cape Town, Rondebosch, Africa do Sul.

Ramsing, N.B., Kühl, M. and Jørgensen, B.B. (1993) distribution of sulfate reducing bacteria, O_2 and H_2S in photosynthetic biofilms determined by oligonucleotide probe and microelectrodes. *Appl. Environ. Microbiol.* **59**, 3840–3849.

Rawlings, D.E. (2002) Heavy metal mining using microbes. *Annu. Rev. Microbiol.* **56**, 65–91.

Robertson, L.A. and Kuenen, J.G. (1992) The colorless sulphur bacteria, In: *The Prokaryotes, 2nd edn. A Handbook on the Biology of Bacteria: Ecophysiology, Isolation, Identification, Applications*. Vol. 1 (eds. A. Balows, H.G. Trüper, M. Dworkin, W. Harder and K.H. Schleifer). Springer-Verlag pp. 385–413.

Rohwerdert, T. and Sand, W. (2003) The Sulfane sulphur of persulfides is the actual substrate of the sulfur-oxidizing enzymes from *Acidithiobacillus* and *Acidiphilium* spp. *Microbiology* **149**, 1699–1709.

Rother, D., Henrich, H.-J., Quentmeier, A., Bardischewsky, F. and Friedrich, C.G. (2001) Novel genes of the *sox* gene cluster, mutagenesis of the flavoprotein SoxF, and evidence for a general sulfur-oxidizing system in *Paracoccus pantotrophus* GB17. *J. Bacteriol.* **183**, 4499–4508.

Santegoeds, C.M., Ferdelman, T.G., Muyzer, G. and de Beer D. (1998) Structural and functional dynamics of sulfate-reducing populations in bacterial biofilms. *Appl. Environ. Microbiol.* **64**, 3731–3739.

Sass, H., Cypionka, H. and Babenzien, H.-D. (1997) Vertical distribution of sulfate-reducing bacteria at the oxic-anoxic interface in sediments of the oligotrophic Lake Stechlin. *FEMS Microbiol. Ecol.* **22**, 245–255.

Schalk, J., Devries, S., Kuenen, J.G. and Jetten, M.S.M. (2000) A novel hydroxylamine oxidoreductase involved in the anammox process. *Biochemistry* **39**, 5405–5412.

Schmid, M., Twachtmann, U., Klein, M., Strous, M., Juretschko, S., Jetten, M.S.M., Metzger, J., Schleifer, K.H., Wagner, M. (2000) Molecular evidence for genus level diversity of bacteria capable of catalysing anaerobic ammonium oxidation. *Sys. Appl. Microbiol.* **23**, 93–106.

Schmid, M., Walsh, K., Webb, R., Rijpstra, W.I.C., van de Pas-Schoonen, K.T., Verbruggen, M.J., Hill, T., Moffett, B., Fuerst, J., Schouten, S., Sinninge Damsté, J.S., Harris, J., Shaw, P., Jetten, M.S.M. and Strous, M. (2003) Candidatus "*Scalindua brodae*", sp. Nov., Candidatus "*Scalindua wagneri*", sp. Nov., two new species of anaerobic ammonium oxidizing bacteria. *Syst. Appl. Microbiol.* **26**, 529–538.

Schmidt, I. and Bock, E. (1997) Anaerobic ammonium oxidation with nitrogen dioxide by *Nitrosomonas eutropha. Arch. Microbiol.* **167**, 106–111.

Schmidt, I., Sliekers, O., Schmid, M., Cirpus, I., Strous, M., Bock, E., Kuenen, J.G. and Jetten, M.S.M. (2002) Aerobic and anaerobic ammonia oxidizing bacteria – competitors or natural partners? *FEMS Microbiol. Ecol.* **39**, 175–181.

Sekiguchi, Y., Kamagata, Y., Nakamura, K., Ohashi, A. and Harada, H. (1999). Fluorescence in situ hybridization using 16S rRNA-targeted oligonucleotides reveals localization of methanogens and selected uncultured bacteria in mesophilic and thermophilic sludge granules. *Appl. Environ. Microbiol.* **65**, 1280–1288.

Siebritz, I.P, Ekama, G.A. and Marais, G.v.R. (1982), A Parametric Model for Biological Excess Phosphorus Removal. Wat.Sci.Tech. **15** (3/4), 127–52.

Sinninghe Damsté, J.S., Strous, M., Rijpstra, W.I.C., Hopmans, E.C., Geenevasen, J.A.J., van Niftrik, L.A. and Jetten, M.S.M. (2002) Linearly concatenated cyclobutane lipids from a dense bacterial membrane. *Nature* **419**, 708–712.

Sinninghe Damsté, J.S., Rijpstra, W.I.C., Schouten, S., Fuerst, J.A., Jetten, M.S.M. and Strous, M. (2004) The occurrence of hopanoids in planctomycetes: implications for the secondary biomarker record. *Org. Geochem.* **35**, 561–566.

Sliekers, A.O., Derwort, N., Campos Gomez, J.L., Strous, M., Kuenen, J.G. and Jetten M.S.M. (2002) Completely autotrophic nitrogen removal over nitrite in a single reactor. *Wat. Res.* **36**, 2475–2482.

Speece, R.E. (1983) Anaerobic biotechnology for industrial wastewaters. *Environ. Sci. Technol.* **17**, 416A–427A.

Stackebrandt, E., Stahl, D.A., Devereux, R. (1995) Taxonomic relationships, In: *Sulfate-reducing Bacteria.* (ed. L.L. Barton) Plenum Press. New York pp. 49–87.

Stefess, G.C., Torremans, R.A.M., de Schrijver, R., Robertson, L.A. and Kuenen, J.G. (1996) Quantitative measurement of sulphur formation by steady-state and transient-state continuous cultures of autotrophic *Thiobacillus* species.

Steudel, R. (2000) The chemical sulphur cycle, In: *Environmental Technologies to Treat Sulphur Production. Principles and Engineering.* (eds. P. Lens and L. Hulshoff Pol), IWA Publishing, London pp. 1–31.

Strous, M., van Gerven, E., Kuenen, J.G. and Jetten, M. (1997) Effects of aerobic and microaerobic conditions on anaerobic ammonium-oxidizing (anammox) sludge. *Appl. Environ. Microbiol.* **63**, 2446–2448.

Strous, M., Heijnen, J.J., Kuenen, J.G. and Jetten, M. (1998) The sequencing batch reactor as a powerful tool to study very slowly growing micro-organisms. *Appl. Microbiol. Biotechnol.* **50**, 589–596.

Strous, M., Fuerst, J.A., Kramer, E.H.M., Logemann, S., Muyzer, G., van de Pas-Schoonen, Webb, R., Kuenen, J.G. and Jetten, M.S.M. (1999a) Missing lithotroph identified as new planctomycete. *Nature* **400**, 446–449.

Strous, M., Kuenen, J.G. and Jetten, M.S.M. (1999b) Key physiology of anaerobic ammonium oxidation. *Appl. Environ. Microbiol.* **65**, 3248–3250.

Strous, M., Kuenen, J.G., Fuerst, J.A., Wagner, M. and Jetten, M. (2002) The anammox case – a new experimental manifesto for microbiological eco-physiology. *Anton. van Leeuwen.* **81**, 693–702.

Stteter, K.O. (1992) The genus *Archaeglobus*, In: *The Prokaryotes, 2nd edn. A Handbook on the Biology of Bacteria: Ecophysiology, Isolation, Identification, Applications.* Vol. 1 (eds. A. Balows, H.G. Trüper, M. Dworkin, W. Harder and K.H. Schleifer). Springer-Verlag pp. 707–711.

Sugio, T., Suzuki, H., Oto, A., Inagaki, K., Tanaka, H. and Tano, T. (1991) Purification and some properties of a hydrogen sulphide binding protein that is involved in sulphur oxidation of *Thiobacillus ferrooxidans*. *Agric. Biol. Chem.* **55**, 2091–2097.

Sutton, P.M., Jank, B.E., Monaghan, B.A. and Murphy, K.L. (1979). *Single Sludge Nitrogen Removal Systems*, Research Report No. 88, Environmental Protection Service, Canada.

Suzuki, I. (1999) Oxidation of inorganic sulphur compounds: chemical and enzymatic reactions. *Can. J. Microbiol.* **45**, 91–105.

Tal, Y., Watts, J.E.M., Schreier, S.B., Sowers, K.R. and Schreier, H.J. (2003) Characterisation of the microbial community and nitrogen transformation processes associated with moving bed bioreactors in a closed recirculated mariculture system. *Aquaculture* **215**, 187–202.

Teske, A., Ramsing, N.B., Habichi, K., Fukui, M., Küver, J., Jørgensen, B.B. and Cohen, Y. (1998) Sulfate-reducing bacteria and their activities in cyanobacterial mats of solar lake (Sinai, Egypt). *Appl. Environ. Microbiol.* **64**, 2943–2951.

Toh, S.K. and Ashbolt, N.J. (2002) Adaptation of anaerobic ammonium oxidising consortium to synthetic coke-ovens wastewater. *Appl. Microbiol. Biotechnol.* **59**, 344–352.

Toh, S.K., Webb, R.I. and Ashbolt, N.J. (2002) Enrichment of autotrophic anaerobic ammonium-oxidising consortia from various wastewaters. *Microbiol. Ecol.* **43**, 154–167.

Trüper, H.G. and Fischer, U. (1982) Anaerobic oxidation of sulphur compounds as electron donors for bacterial photosynthesis. *Phil. Trans. R. Soc. Lond.* B **298**, 529–542.

van de Graaf, A.A., de Bruijn, P., Robertson, L.A., Jetten, M.S.M. and Kuenen, J.G. (1996) Autotrophic growth of anaerobic ammonium-oxidizing micro-organisms in a fluidized bed reactor. *Microbiology* **142**, 2187–2196.

van de Graaf, A.A., de Bruijn, P., Robertson, L.A., Jetten, M.S.M. and Kuenen, J.G. (1997) Metabolic pathway of anaerobic ammonium oxidation on the basis of N-15 studies in a fluidized bed reactor. *Microbiology* **143**, 2415–2421.

van de Graaf, A.A., Mulder, A., de Bruijn, P., Jetten, M.S.M., Robertson, L.A. and Kuenen, J.G. (1995) Anaerobic oxidation of ammonium is a biologically mediated process. *Appl. Environ. Microbiol.* **61**, 1246–1251.

van der Last, A.R.M. and Lettinga, G. (1991) Anaerobic Treatment of Domestic Sewage Ander Moderate Climatic (Dutch) Conditions using upflow reactors at increased superficial velocities. *In Proceedings of the Sixth International Symposium on Anaerobic Digestion* (pp. 153–154), São Paulo, Brazil.

Vallero, M.V.G., Sipma, J., Lettinga, G. and Lens, P.N.L. (2004). High-rate sulfate reduction at high salinity (up to 90 mS cm^{-1}) in mesophilic UASB reactors. *Biotechnol. Bioeng.* **86**, 226–235.

Van Haandel, A.C. and Lettinga, G. (1994) *Anaerobic Sewage Treatment in Regions with a Hot Climate.* Ed John Wiley & Sons, Chichester, UK.

Van Haandel, A.C. and Marais, G.v.R (1981) *Nitrification and Denitrification Kinetics in the Activated Sludge Process*, Research Report No. W41, Department of Civil Engineering, University of Cape Town.

Van Haandel, A.C. and Marais, G.v.R (1999) *O Comportamento do sistema de lodo ativado: Teoria e aplicações para operação e projetos* ed. *Epgraf Campina Grande* P.B. (in Portugese).

Van Haandel, A.C., Catunda, P.F.C and Araújo L.d.S. (1998) Biological sludge stabilisation Part 1: kinetics of aerobic sludge stabilisation. *Wat. S.A.* **24**, 223–230.

Van Lier, J., Rebac, S. and Lettinga, G. (1997) High-rate anaerobic wastewater treatment ander psychrophilic and thermophilic conditions. *Wat. Sci. Technol.* **36**, 199–206.

van Niftrik, L.A., Fuerst, J.A., Sinninghe Damsté, J.S., Kuenen, J.G., Jetten, M.S.M. and Strous, M. (2004) The anammoxosome: an intracytoplasmic compartment in anammox bacteria. *FEMS Microbiol. Lett.* **233**, 7–13.

Visser, A., Alpheenar, P.A., Gao, Y., van rossum, G. and Lettinga, G. (1993) Granulation and immobilization of methanogenic and sulfat-reducing bacteria in high rate anaerobic reactors. *Appl. Microbiol. Biotechnol.* **40**, 575–581.

Wakai, S., Kikumoto, M., Kanao, T. and Kamimura, K. (2004) Involvement of sulphide: quinone oxidoreductase in sulphur oxidation of an acidophilic iron-oxidizing bacterium, *Acidithiobacillus ferrooxidans* NASF-1. *Biosci. Biotechnol. Biochem.* **68**, 2519–2528.

Wentzel, M.C., Dold, P.L. Ekama, G.A. amd Marais G.v.R. (1985) Kinetics of biological phosphorus release. *Wat. Sci.Tech.* **17**(11/12), 57–71.

Wentzel, C.M., Ekama, G.A. and Marais, G.v.R. (1992) Processes and modelling nitrification denitrification biological excess phosporus removal systems – a review. *Wat.Sci. Tech.* **25**(6), 59,92.

Wentzel, M.C., Loewenthal, R.E., Ekama, G.A. and Marais, G.v.R. (1988) Enhanced polyphosphate organism cultures in activated sludge systems: Part II experimental behaviour. *Wat. S.A.* **15**(2), 71–88.

Wentzel, M.C., Lötter, L.H., Loewenthal, R.E. and Marais, G.v.R. (1986) Metabolic behaviour of *Acinetobacter* spp. in enhanced biological phosphorus removal: a biochemical model. *Wat. S.A.* **12**(4), 209–224.

Widdel, F. (1988) Microbiology and ecology of sulfate- and sulfur-reducing bacteria, In: *Biology of Anaerobic Microorganisms* (ed. A.J.B. Zehnder) John Wiley & Sons pp. 469–585.

Widdel, F. and Bak, F. (1992) Gram-negative mesophilic sulfate-reducing bacteria, In: *The Prokaryotes, 2nd. Ed. A Handbook on the Biology of Bacteria: Ecophysiology, Isolation, Identification, Applications*. Vol. 1 (eds. A. Balows, H.G. Trüper, M. Dworkin, W. Harder and K.H. Schleifer), Springer-Verlag, pp. 3352–3378.

Widdel, F. and Hansen, T.A. (1992) The dissimilatory sulfate and sulphur bacteria. In: *The Prokaryotes, 2nd. Ed. A Handbook on the Biology of Bacteria: Ecophysiology, Isolation, Identification, Applications*. Vol. 1 (eds. A. Balows, H.G. Trüper, M. Dworkin, W. Harder, and K.H. Schleifer), Springer-Verlag pp. 583–624.

Wood, A.P., Aurikko, J.P., Kelly, D.P. (2004) A challenge for 21st century molecular biology and biochemistry: what are the causes of obligate autotrophy and methanotrophy? *FEMS Microbiol.Rev.* **28**, 335–352.

WRC (1984) *Theory, Design and Operation of Biological Nutrient Removal Activated Sludge Processes* Water Research Commission, P.O. Box 824, Pretoria 0001, SA.

Wuhrmann, K. (1964). Nitrogen Removal in Sewage Treatment Processes. *Verh.Int.Verh. Theor. Angew. Limnol.* **15**, (in German).

Young, J.C. and McCarty, P.L. (1969) The Anaerobic filter for waste treatment. *J. Wat. Poll. Cont. Fed.* **41**, 160–165.

4

Biological wastewater treatment systems

G. Buitrón, E. Razo-Flores, M. Meraz and F. Alatriste-Mondragon

4.1 INTRODUCTION

Different types of biological treatment systems are used in the field of environmental engineering. The biochemical reactions leading to the oxidation of organic matter are conducted in reactors that can be classified as aerobic or anaerobic, suspended growth or biofilm, with mechanical or without mechanical mixing, etc. In order to design an appropriate reactor for a given wastewater treatment system, both the microbial kinetics of substrate removal and the fundamental properties of different reactors have to be understood.

The presence or absence of oxygen, the physicochemical characteristics of the wastewater, the wastewater strength, the efficiency of treatment required, the reactor configuration required and costs are some of the main factors influencing the selection of a biological reactor. In this chapter, the reactors most often used and with a potential application for the treatment of industrial wastewaters are discussed.

4.2 AEROBIC SYSTEMS

4.2.1 Activated sludge systems

The activated sludge process, in any of its variants, is the most utilized biological process for industrial wastewater treatment. The activated sludge process is capable of converting most organic wastes to stable inorganic forms and cellular mass. In this process much of the soluble and colloidal organic material remaining after primary sedimentation is metabolized by a diverse group of microorganisms to carbon dioxide and water. At the same time, a fraction is converted to a cellular mass that can be separated from the waste flow by gravity sedimentation.

Activated sludge is a heterogeneous microbial culture composed mostly of bacteria, protozoa, rotifers, and fungi. However, it is the bacteria which are responsible for assimilating most of the organic material, whereas the protozoa and rotifers are important in removing the dispersed bacteria that otherwise would escape in the plant effluent.

The utilization of substrate (organic material) by a bacterial cell can be described as a three-step process: (1) the substrate molecule contacts the cell wall, (2) the substrate molecule is transported into the cell, and (3) metabolism of the substrate molecule by the cell takes place. However, as bacteria require substrate in the soluble form, colloidal or sterically incompatible molecules, which cannot be readily transported into a cell, have to be first adsorbed to the cell surface and then broken down or transformed externally to transportable fractions by exoenzymes or wall-bound enzymes.

To produce a high-quality effluent, the biomass (after removing the organic material from the wastewater) must be separated from the liquid stream. This is accomplished in the secondary clarifier and is effective only if the microbial species present readily agglomerate. Secondary clarification is almost always the effluent quality-limiting step. The soluble biological oxygen demand (BOD_5) of the effluent is generally below 5 mg/L, but biomass solids carryover may produce an effluent BOD_5 of 20 mg/L or greater.

Biological flocculation has been found to be governed by the physiological state of the microorganisms and does not occur until depletion of the substrate by the microorganisms or endogenous growth phase. Biological flocculation is proposed to result from the interaction of exocellular polymers which accumulate at the cell surface during endogenous growth. Cells are bridged into three-dimensional matrices as a result of physical and electrostatic bonding of these polymers to the surface of the cells. After separating the liquid phase from the solid phase, excess biomass, resulting from synthesis during substrate utilization, is wasted and the remainder returned to the aeration tank. Thus, a relatively constant mass of microorganisms is maintained in the system, and performance of the process depends on the recycle of sufficient biomass. If biomass separation and concentration fails, the entire process fails.

The flow scheme for a typical activated sludge plant is presented in Figure 4.1. In general, this process may be considered to be one that involves: (1) wastewater aeration in a reactor containing a microbial suspension, (2) solids–liquid separation with return of the sludge to the reactor, (3) discharge of a clarified effluent, and (4) discharge of excess biomass.

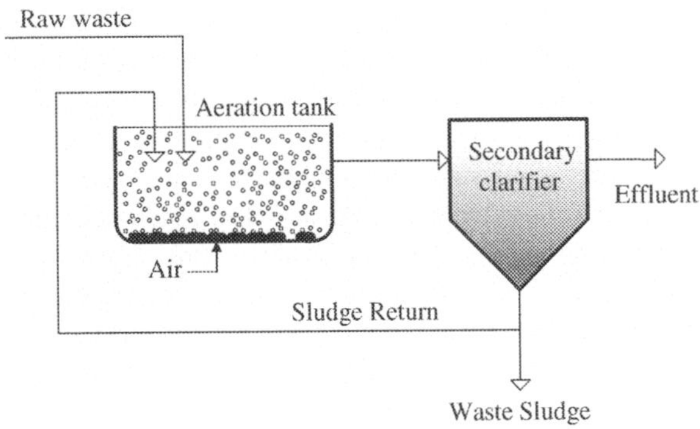

Figure 4.1 Typical activated sludge process.

With a typical municipal wastewater, a well-designed activated sludge process should achieve a carbonaceous, soluble BOD_5 effluent quality of 5 mg/L or less. Similarly, with clarifiers designed to maximize solids removal at peak flows and adequate process control, the average suspended solids (SS) in the effluent should not exceed 15–25 mg/L. On a practical basis, an effluent with 20/20 mg/L BOD_5 and SS should be attained, assuming proper operation.

Of the total oxygen demand exerted by the wastewater, there is often a sizeable fraction associated with the oxidation of ammonia to nitrate. The autotrophic bacteria *Nitrosomonas* sp. and *Nitrobacter* sp. are responsible for this two-stage conversion. Being autotrophic, these nitrifying organisms must reduce oxidized carbon compounds in the wastewater, such as CO_2 and its related ionic species, for cell growth. As a result, this characteristic markedly affects the ability of the nitrifying organisms to compete in a mixed culture.

The nitrifying bacteria obtain their energy by oxidizing ammonia to nitrite and then to nitrate. As very little energy is obtained from these oxidation reactions, and because energy is needed to reduce CO_2 to cellular carbon, the population size of nitrifiers in activated sludge is relatively small. When compared to the normal bacteria in activated sludge, the nitrifying bacteria have a slower reproduction rate.

As the mean cell retention time (MCRT) is increased, nitrification generally takes place. The longer MCRT prevents nitrifying organisms from being lost from the system when carbonaceous wasting occurs or, more accurately, the longer MCRT permits the build-up of an adequate population of nitrifiers. Due to the longer MCRT required for nitrification, some systems are designed to achieve nitrification in the second stage of a two-stage activated sludge system.

4.2.1.1 Modes of operation

The process design for activated sludge requires many decisions, the first of which must be the type to be employed. Economics will guide that choice, but many other factors relating to the characteristics of the wastewater and the objectives of the entire treatment system must be considered as well.

The principal types of biological reactors (aeration basins) are plug flow, complete mix, and arbitrary flow. In a plug-flow reactor the particles pass through the tank, and are discharged in the same sequence in which they enter. This type of flow is achieved in a long, narrow basin. In a complete-mix reactor, the entering particles are dispersed immediately throughout the entire basin. Complete mix is achieved in circular or square basins. Arbitrary-flow reactors exhibit partial mixing somewhere between the plug-flow and complete-mix reactors.

In a biological reactor with solids recycle the MCRT, or θ_c, in the system is obtained from the mass of solids maintained in the reactor divided by the mass of solids produced or removed per day from the system (Equation (2.73), see Section 2.7). Some other useful process design equations are:

$$\frac{\mathrm{F}}{\mathrm{M}} = \frac{QS_o}{VX} = \frac{1}{\theta}\frac{S_o}{X} \tag{4.1}$$

$$q = \frac{(\mathrm{F}/\mathrm{M})E}{100} \tag{4.2}$$

$$E = \frac{(S_o - S)}{S_o}100 \tag{4.3}$$

where F/M: food to microorganism ratio, kg BOD_5 applied per day per kg of mixed liquor volatile suspended solids (MLVSS) in the aeration basin, per day; E: process efficiency, percent; X: active volatile SS in the reactor, g/m^3; V: volume of the reactor, m^3; Q: flow rate, m^3/day; S_o: influent substrate concentration (BOD_5 or chemical oxygen demand (COD)), g/m^3; S: effluent substrate concentration (BOD_5 or COD), g/m^3; q: specific substrate utilization rate, per day.

Although the activated sludge system is primarily designed to remove carbonaceous BOD, with sufficient operational control using suitable modifications it can also achieve nitrification, denitrification, and phosphorus control. In general, high-rate processes have a high sludge loading, a short MCRT, a high sludge activity, and a short retention time. Although large quantities of BOD can be removed per unit volume of reactor, there is a relatively high concentration of organic matter remaining in the final effluent. In contrast, low-rate processes have a low sludge loading, a long MCRT, a low sludge activity, and a long retention time. The sludge is in the endogenous respiration phase so food is limited, resulting in a low residual concentration of organic matter in the final effluent, and as the rate of microbial decay is high compared with the rate of microbial growth, there is little excess sludge produced. Conventional operation falls between these two extremes. There exist several variations of the activated sludge process as shown in Table 4.1.

Once a type has been chosen, to perform the actual process design, the selection of the MCRT, the reactor volume with its associated mixed liquor suspended solids (MLSS) concentration, the recycle ratio, the F/M ratio, and the wastage flow ratio must be evaluated. The volume of the reactor and the flow rate applied determine the hydraulic retention time (HRT) or θ. In addition the mixing and

Table 4.1 Design parameters of various activated sludge process modifications.

Process modification/ type of reactor	MCRT (day)	F/M (kg BOD_5/ kg MLSS/day)	Aerator loading (kg BOD_5/m^3/day)	MLSS (g/L)	HRT (h)	Recirculation ratio (Q_r/Q)
Conventional/plug flow	3–15	0.2–0.4	0.3–0.7	1.0–3.0	4–8	0.25–0.75
Tapered aeration/plug flow	5–15	0.2–0.4	0.3–0.6	1.5–3.0	4–8	0.25–0.50
Step-feed aeration/plug flow	3–15	0.2–0.4	0.6–1.0	1.5–4.0	3–5	0.25–0.75
Complete-mix aeration/ complete mix	3–15	0.2–0.6	0.3–2.0	1.5–6.0	3–5	0.25–1.00
High-rate aeration/plug flow	0.5–2	1.5–2.0	1.2–15	0.2–1.0	1.5–3	1.0–2.0
Extended aeration/plug flow	20–40	0.04–0.15	0.1–0.3	3.0–6.0	18–36	0.5–2.0
SBR/batch	10–30	0.05–0.3	0.2–0.7	1.5–5.0	4–40	Not applicable
Contact stabilization/ plug flow	5–15	0.2–0.6	1.0–1.2 (contact tank) (stabilization	1.0–4.0 (contact tank) 4.0–10.0 (stabilization tank)	0.5–1.0 3.0–6.0 tank)	0.5–1.0
Kraus process/plug flow	5–15	0.3–1.0	0.5–1.5	2.0–3.0	4–8	0.5–1.0
High-purity oxygen/plug flow	8–20	0.25–1.0	1.6–3.3	6.0–8.0	2–5	0.25–0.5
Oxidation ditch/plug flow	15–30	0.04–0.1	0.1–0.3	3.0–5.0	15–30	75–150

Adapted from Qasim (1999) and Metcalf and Eddy (2003).

oxygen requirements must be determined so that a suitable aeration system may be designed (Qasim 1999; Metcalf and Eddy 2003).

By using different combinations of the main operating parameters, various different rates, and degrees of treatment are possible. The major advantage of the activated sludge process over other treatment processes is just this flexibility in design, allowing operation over a wide range of loadings to suit specific treatment objectives. The processes differ in terms of their aeration configuration, aeration equipment design, design, MCRT, operating mode, and ability to remove nitrogen and some are proprietary. The high-rate aeration, contact stabilization, and high-purity oxygen processes are used primarily for BOD removal only, are designed for relatively short MCRT, and require less space than other processes. Where nitrification is not needed to meet treatment discharge limits, the three processes cited above are particularly attractive for large municipalities where space is limited. The conventional plug-flow, step-feed, and complete-mix processes are used for both BOD removal and nitrification and are applied over a wide range of MCRT, depending on the wastewater temperature and treatment needs. The Kraus process is seldom used, but it is included to show how oxidized nitrogen can be used to help BOD degradation in the first pass of a plug-flow aeration tank.

In contrast to the processes described above, conventional extended aeration, and the oxidation ditch are processes that represent a different approach to biological wastewater treatment. The processes employ a much simpler system by generally eliminating primary treatment and anaerobic digestion from the overall treatment system. Larger aeration tanks with longer MCRT exceeding 20 days, are used. The process approach is attractive for smaller communities where space is not an issue and less complex operation is preferred. The large aeration tank volume provides good equalization at high flow and loading occurrences, and a high-quality effluent is produced. With the exception of the conventional extended aeration process, the systems are operated usually to promote denitrification in addition to nitrification. The aeration and mixing of the channel-flow processes (oxidation ditch) require much less energy for mixing than need for aeration so that aeration equipment design is based on meeting oxygen requirements instead of tank mixing. Less energy is required in comparison to conventional extended aeration processes. In the past, the oxidation ditch and extended aeration processes where thought to need long MCRT to provide well-stabilized biosolids for reuse. However, with stricter regulations governing biosolids stabilization, separate aerobic digestion facilities are used to meet the requirements for reuse (Metcalf and Eddy 2003). The sequencing batch reactors (SBR) are attractive to small communities because of the simplicity of operation and relatively low cost. Sequentially operated processes are also adaptable to nitrogen removal.

4.2.1.2 Factors affecting the efficiency of activated sludge systems

The efficiency of the activated sludge process in achieving carbonaceous BOD removal and nitrification will depend on many factors, including: how readily organic material and ammonia can be metabolized by the microorganisms; the

MCRT and food to microorganisms ratio (F/M); how readily organic material can be oxidized or used for cell synthesis; the numbers and types of active microorganisms present in the aeration tank; the HRT; environmental factors such as dissolved oxygen (DO) concentration, nutrients, pH, temperature, and presence of toxic materials; adequacy of the original engineering design for mixing, recycling and wasting system, and aeration capacity; proper maintenance of plant equipment; and adequate training of treatment plant staff, including laboratory, maintenance, operations, and management.

An operational problem frequently present in activated sludge for the solids separation is the sludge bulking. Sludge bulking is the rising of sludge, poor settling, or foaming. The main causes of sludge bulking are (1) characteristics of wastewater, (2) design limitations, and (3) plant operation. Fluctuations in flow and strength, pH, temperature, nutrients, and nature of wastes are related to wastewater characteristics. The design limitations may include an insufficient capacity of aeration, mixing, and return sludge. The plant operation factors constitute low DO, nutrient limitations, low F/M ratio, excessive aeration, and *Nocardia* growth. Sludge bulking, troubleshooting, and control measures are discussed in Qasim (1999).

4.2.1.3 Industrial wastewaters treated by the activated sludge process

Table 4.2 presents a summary of the performances obtained for several industrial wastewaters treated by activated sludge systems.

4.2.2 Sequencing batch reactors

4.2.2.1 Description of the process

The term SBR is given to wastewater treatment systems, operated on a sequence of fill and draw cycles. It can work with suspended or attached biomass. The unit operations involved in an SBR are equivalent to those of conventional-activated sludge systems. Therefore, aeration and sedimentation are performed. The difference between the systems is that, in conventional systems, these two processes take place simultaneously in two different tanks, whereas in SBR systems, they occur sequentially in the same tank. The SBR system is time oriented, where flow, mixing, aeration, and the reactor volume are variable according to some predetermined periodical operational strategy.

Usually an SBR-type bioreactor operates under five well-defined phases: fill, react, settle, draw, and idle (Figure 4.2). The duration of these phases is usually determined by an expert operator based on his/her experience and exhaustive testing in the laboratory with a pilot plant. The settle and draw phases are fixed in duration by the characteristics and constraints of the activated sludge and the reactor itself. In contrast, adequately controlling the fill and reaction times can improve the overall efficiency of the plant (Betancur *et al.* 2004). The avoidance of lengthy periods of starvation, for example, can be of great importance regarding this issue, as well as the nature of the specific growth rate of biomass. The SBR system can be fully automated using the oxygen uptake rate to determine the

Table 4.2 Treatment performance for selected industrial wastewaters.

Wastewater	Influent		Effluent		F/M		MCRT (day)	MLVSS (mg/L)	HRT (day)	SVI (mL/g)
	BOD (mg/L)	COD (mg/L)	BOD (mg/L)	COD (mg/L)	BOD (per day)	COD (per day)				
Pharmaceutical	2950	5840	65	712	0.11	0.19		4970	5.4	
	3290	5780	23	561	0.11	0.18		5540	5.4	
Coke and byproducts chemical plant	1880	1950	65	263	0.18	0.21		2430	4.1	42.4
Diversified chemical industry	725	1487	6	257	0.41	0.71		2874	0.61	119
Cellulose	1250	3455	58	1015	0.51	1.03		3280	0.75	
Tannery	1020	2720	31	213	0.18	0.45	16	1900	3	
	1160	4360	54	561	0.15	0.49	20	2650	3	
Alkyl-amine manufacturing	893	1289	12	47	0.146	0.21		1977	3.1	133
ABS	1070	4560	68	510	0.24	0.94	6	2930	1.5	23
Complex organic	1630	1660	111	415					2.58	98.5
Viscose rayon	478	904	36	215	0.30	0.47		2759	0.57	117
Polyester and nylon fibers	207	543	10	107	0.18	0.40		1689	0.664	116
	208	559	4	71	0.20	0.48		1433	0.712	144
Protein processing	3178	5335	10	362	0.054	0.08		2818	21	180
	3178	5335	5.3	245	0.100	0.16		2451	12.7	215
Tobacco processing	2420	4270	139	546				3840		61

Propylene oxide	532	1124	49	289	0.20	0.31		2969	1	51
	645	1085	99	346	0.19	0.25		2491	1.4	32
Paper mill	375	692	8	79	0.111	0.19	18.9	1414	2.38	63
	380	686	7	75	0.277	0.45	5.2	748	1.83	504
Synthetic fuel	990		15		0.32			3100	1	
	790		18		0.21			2500	1.5	
Vegetable	3474	6302	76	332	0.57	1.00		1740	3.5	49.2
High-strength chemical	3110	3910	174	450					2.25	176
Tetraethyl lead	41	92	1.2	26					0.045	36
Organic chemicals	453	1097	3	178	0.10	0.21		2160	2.02	111
Hardboard mill	3725	5827	58	643	0.30	0.42		2793	4.45	250
High-strength, saline organic chemicals	3030	8750	830	6125	0.75	0.95		233	17.4	
	3171	8597	82	3311	0.24	0.41		328	40	
Coke	1618	2291	52	434	0.23	0.27		3060	2.3	152
Textile dye	393	951	20	261	0.12	0.23	18.8	2620	1.21	185
	393	951	14	173	0.22	0.44	13.4	2440	0.75	350
Kraft and semi-chemical pulp and paper	308	1153	7	575	0.28	0.36		2800	0.58	83

Adapted from Eckenfelder and Musterman (1995).

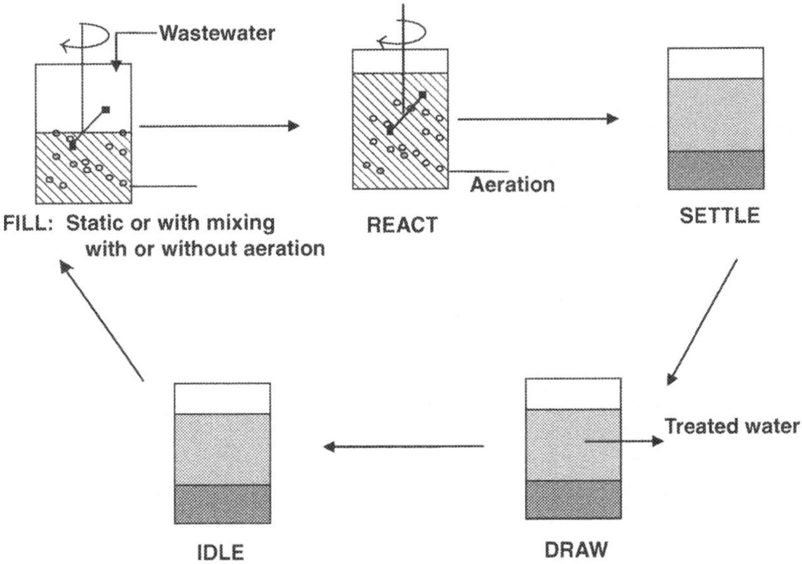

Figure 4.2 Schematic operation on an SBR during a cycle.

end of the reaction phase (Vargas *et al.* 2001; Buitrón *et al.* 2005) and an optical or ultra sound device to detect settling.

In this type of reactors, as a consequence of the substrate concentration variations in each cycle (decreasing in time), the growth rate of the microorganisms changes from high to low. Thus, a selection of a microbial community with a vast metabolic range takes place in which the microbial species can differ greatly in growth rate and yield. This is particularly important when toxic compounds are to be degraded, since the specialized microorganisms can be selected.

In discontinuous processes there is a generation of more active microorganisms than in the continuous processes. This is due to the fact that microorganisms that are periodically exposed to high concentrations of substrate are more active than those that grow under extended periods of low concentrations of substrate. When the substrate concentration is elevated, the cells can accumulate enzymes that increase the reaction rate. Similarly, the overall enzyme activity of cell grown under starvation conditions (i.e. maintenance levels) is decreased.

The SBR presents a better maintenance of a robust population capable of meeting desired effluent limits when transient or shock load conditions occur since they present the ability to adjust the time and magnitude of energy input, the fraction of each tank's volume used and the number of tanks placed into operation to meet actual loading conditions. Industrial effluents are characterized by such variations. Thus, the use of a discontinuous system will generate effluents with better quality, allowing the water to be reused.

Experimentally, it has been demonstrated that repeating shifting of aerobic microorganisms between zones with high and low substrate concentrations, is a very effective method of controlling the excess growth of filamentous microorganisms.

Table 4.3 Effect of cyclic exposure of microorganisms to different process conditions on bioreactor performance.

Factor varied cyclically	Effects achieved
High and low concentrations of readily biodegradable substrates	Growth rate differentials that suppress excess growth of filamentous bacteria (Chiesa and Irvine 1985). Minimizes sensitivity to shock loads and general variations in environmental factors and influent constituents (Buitrón *et al.* 2005).
High substrate concentration followed by an extended period of starvation	Accumulation of exopolymeric substances and suppression of excess growth of filamentous bacteria (Chiesa *et al.* 1985). Enrichment of floc-forming bacteria with the physiological characteristics needed to meet treatment objectives.
Anaerobic and anoxic conditions	Enrichment of both nitrifiers and denitrifiers for nitrogen removal (Keller *et al.* 2001).
Anaerobic and aerobic conditions	Enrichment of Bio-P bacteria for phosphorous removal (González-Martínez and Wilderer 1991).

Adapted from Wilderer *et al.* (2001).

This is exactly what happens in a batch process. Frequent but controlled shifting of microorganisms between aerobic, anoxic, and anaerobic periods also permits the establishment of microbial communities capable of carrying out nitrification, denitrification, and biological phosphorous removal (Metcalf and Eddy 2003).

4.2.2.2 Applicability

SBR systems are typically used at flow rates of 220 L/s (19,000 m^3/day) or less. The more sophisticated operation required at larger SBR plants tends to discourage the use of SBR reactors for large flow rates. As these systems have a relatively small footprint, they are useful for areas where the available land is limited. In addition, cycles within the system can be easily modified for nutrient removal in the future, if it becomes necessary. This makes SBR systems extremely flexible to adapt to regulatory changes for effluent parameters such as nutrient removal. They are also very cost effective if treatment beyond just biological treatment is required, such as filtration.

The SBR provides benefits beyond the simple flexibility of varying the reaction period to improve contaminant removal (Irvine *et al.* 1997). By controlling the cycle times, flow rates, nutrient and oxygen availability, the SBR has the ability to apply environmental pressures on a microbial consortium (Table 4.3). The benefits of the SBR have been shown for nutrient removal (Vallés-Morales *et al.* 2004), the control of filamentous bacteria (Wanner 1992) and the removal of specific organic compounds present in industrial wastewaters (Vargas *et al.* 2000). The use of alternated anaerobic/aerobic environments to treat recalcitrant compounds has been explored (Buitrón *et al.* 2003a; Melgoza *et al.* 2004). During the past several years,

Table 4.4 Selected applications of SBR for the treatment of industrial wastewaters.

Type of wastewater	Characteristics	References
Water from road and rail car clearing installations	*Flow*: 190–320 m^3/day *Influent*: COD 3.6 to 10.5 g/L; BTEX 1.9–12.7 mg/L; AOX 1.3–4.3 mg/L; COD removal 85–90% at 0.25 kg COD/kg SS-day; 7.5–10 g/L MLSS; AOX and BTEX removal >99% Aerated fill 1 h; reaction 18 h; settle 4 h; decant 1 h.	Zilverentant (1997)
Refinery wastewater	*Flow*: 3000 m^3/day *Influent*: total COD 1.4 g/L; phenol 5 mg/L; MCRT 40 days; HRT 53 h; 4.25 g MLSS/L; F/M ratio: 0.08 Effluent total COD 150 mg/L; SS 30 mg/L	Hudson *et al.* (2001)
Winery industry	*Flow*: 8 m^3/day; 1 cycle per day; removal 93% total COD; sludge production 0.21 kg SS/kg of total COD	Torrijos and Moletta (1997)
Piggery wastewater	*Flow*: 150 m^3/day *Influent*: total COD 4.6 g/L; TNK 594 mg/L; TSS 2 g/L; total P 82 mg/L; removal of COD, N and $P > 98\%$; 24 h per cycle: five modules of intermittent conditions: anoxic–anaerobic 2 h; aerobic 2 h and one module of settling phase of 4 h	Tilche *et al.* (2001)

BTEX: benzene, toluene, ethyl-benzene and xylene; TSS: total suspended solids.

a large number of applications based on SBR with biofilm (SBBR) have been developed (Wilderer *et al.* 2001), the common objective of them all was to control the composition and the activity of the bacterial communities. Several types of supports such as activated carbon (Kolb and Wilderer 1995), zeolites (Zwerger *et al.* 2000) or volcanic rocks (Buitrón and Ortiz 1998; Buitrón *et al.* 2003b) have been utilized (a bibliographic review of this technology used to treat different kinds of industrial effluents is presented by Mace and Mata-Alvarez (2002)).

The implementation of control strategies through the use of high-performance controllers, low-cost software sensors, and supervision systems will reduce to a minimum the fill and reaction times (Vargas *et al.* 2000; Betancur *et al.* 2004). With the automation of the overall process, the efficiency and load capacity of the reactor can be greatly improved with respect to a typical operation, therefore, resulting in a lower operational cost, besides the enhancement of the water quality that can be obtained. Table 4.4 presents the operational characteristics for some applications of SBR for the treatment of industrial wastewaters.

4.2.2.3 Design

The design of an SBR plant should be based on the results of pilot studies whenever possible. For industrial wastewater facilities, treatability studies should almost always be performed on bench or pilot-scale. Design procedures can be found in Ketchum (1997), EPA (1999), and Wilderer *et al.* (2001).

4.2.2.4 Limitations of the process

A higher level of sophistication is required (compared to conventional systems), especially for larger systems, of timing units and controls; higher level of maintenance (compared to conventional systems) associated with more sophisticated controls, automated switches, and automated valves; potential of discharging floating or settled sludge during the draw or decant phase with some SBR configurations; potential plugging of aeration devices during selected operating cycles, depending on the aeration system used by the manufacturer; potential requirement for equalization after the SBR, depending on the downstream processes.

4.2.3 Biofilm processes

In common biological treatment systems like the activated sludge process or the SBR discussed previously, microorganisms are mixed with the waste material. The microorganisms, which are suspended, decompose the waste material and convert it to microbial biomass and stable inorganic products. The solid phase (sludge) is present as a suspension in the liquid phase (the treated wastewater). A second step of treatment is needed in these systems to separate the microbial biomass from the treated wastewater.

Biofilm systems distinguish from other biological waste treatments by the fact that the solid phase and the liquid phase are separated. In these processes the microbial biomass is static (immobilized to a support material), while the treated fluid is mobile. This arrangement creates a separation between the microbial biomass and treated water, although such separation is not complete and certain portion of the biomass is suspended.

The immobilization of microorganisms to the support material can be divided into two main immobilization processes: (1) the self-attachment of microorganisms to the support material, which is defined as attached growth, (2) the artificial immobilization of microorganisms to the support material (e.g. entrapment of microorganisms within polymer beads, microencapsulation, etc.). This section will center on the attached growth systems. For the treatment of wastewater, a variety of types of attached growth processes exists, namely trickling filters, rotating biological contactors, submerged biofilters, combined (i.e. hybrid) systems (biofilm and suspended biomass), and fluidized bed (FB) reactors. As there are extensive reports for the first three processes (see e.g. Vesilind 2003; Rother and Cornel 2004), in this section we will only focus on the two later systems.

Biofilm processes present several advantages over the suspended biomass systems as: (a) with some variants a higher biomass concentration is feasible, leading to an efficient treatment with a more compact system; (b) higher metabolic activity due to a high concentration of nutrients around the attached biomass and also because of a physiological difference between attached and suspended microorganisms; (c) greater resistance to toxicity due to physiological differences and also by the protective effect of the extracellular matrix to the attached biofilm; and (d) better biomass properties resulting in denser biofilms (Cohen 2001). An important drawback of biofilm processes is that the sludge age cannot be easily controlled as in suspended growth systems.

4.2.3.1 Moving bed reactor

The group of processes using suspended and attached biomass in the same reactor is referred to as combined systems. They result from the combination of the activated sludge process and the submerged biofilters (fixed bed biofilters). In such systems, a carrier material (polystyrene, activated carbon, polyethylene, polypropylene, or polyurethane), used to increase the specific surface area available for bacterial growth, is placed in the tank and maintained in suspension by aeration and/or by mechanical mixing. This system is called moving bed reactor (Rusten *et al.* 1992).

In general, the moving bed reactors combine the advantages of suspended and attached biomass systems. The major advantage of the combined systems, as compared to conventional-activated sludge systems, is a lower space requirement because of the use of biofilms. Also, it presents the advantage, compared to submerged biofilters, of low headloss and no backwash requirement. The entire reactor volume of the reactor is efficiently used and its capacity, based on total volume required, is about the same as in most submerged biofilter systems. This process is often used to restore overloaded activated sludge systems as it allows increasing the organic load (up to 10 kg COD/m^3/day) compared to conventional treatment systems, without loss of removal efficiency.

Moving bed reactors are characterized by high removal efficiency of carbon and nitrogen, a high concentration of fixed biomass (up to 20–40 g/L), a low HRT, and reduced sludge production because of the increase in the sludge age (Lazarova and Manem 1994).

The basic idea behind the moving bed reactor is to have a continuously operating, non-cloggable biofilm reactor with low headloss and high specific biofilm surface area. This is achieved by having the biofilm (or biomass) grow on small carrier elements that move along with the water in the reactor. The movement is caused by aeration in the reactor (Figure 4.3). Once the organic mater is degraded in the reactor, effluent of the moving bed must pass through a settler in order to separate the suspended biomass, as in a classical-activated sludge system.

There exist several types of supports mainly made of synthetic materials. For example, in a case the elements are made of polyethylene (density 0.95 g/cm^3) and shaped like small cylinders (about 10 mm in diameter and in height) and longitudinal fins on the outside and inside (Figure 4.4). In order to keep them in the reactor, a sieve (light opening about 7 mm) is placed at the outlet. The agitation is so arranged that the carrier elements are constantly being moved away from the sieve.

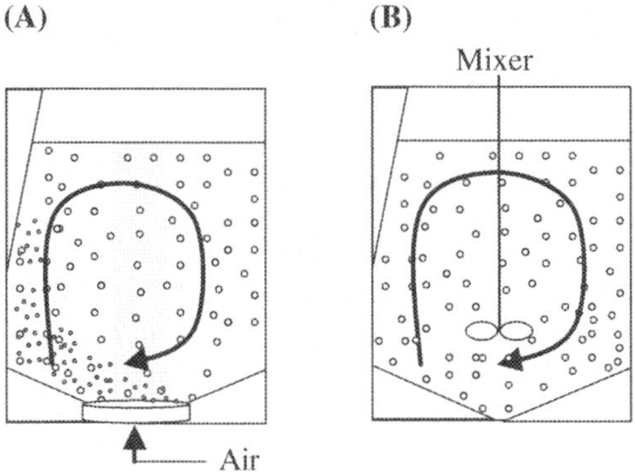

Figure 4.3 Schematic representation of moving bed systems. Carrier media mixed with air (A) and with a mixer (B).

Figure 4.4 Packing material colonized by 4-chlorophenol degrading bacteria in a polyethylene support (Moreno-Andrade *et al.* 2004).

The type and degree of filling of carrier elements in the reactor may be decided for each case, giving a lot of flexibility in the specific biofilm area. A maximum a filling of about 70% (volumetric filling of carrier material in empty reactor) is used, corresponding to a specific, potential growth area of biofilm of about $500\,m^2/m^3$. As the growth is much less on the outside of the cylindrical carrier than the sheltered inside area, the maximum effective specific growth area is expected to be about $350\,m^2/m^3$ at 70% filling.

Other material that has been used is polyurethane cubes. Filling rates vary from 20% to 37%. They provide a large specific surface area for bacterial growth ($1000–5000\,m^2/m^3$). However, only 20–40% of this area is really used due to clogging of the pores. This material requires continuous purge of biomass. The main problem with these processes is to keep porous cubes moving and to prevent their flotation.

There is no need for backwashing or recycling of biomass and the headloss through the reactor is insignificant. The capacity of a reactor of a given volume may be altered by simply changing the degree of filling. Access to fouled or broken diffusers can be provided by pumping carrier elements to other reactors, temporarily increasing their degree of filling, or by pumping the carriers to an empty tank.

Moving bed reactors have been tested in applications for both industrial and municipal wastewater treatment (Ødegaard *et al.* 1994). Table 4.5 presents some industrial applications of this system.

Table 4.5 Industrial applications of moving bed reactors.

Type of wastewater	Characteristics	References
Potato chips industry	Carrier area 250 m^2/m^3; filling ratio 50%; organic load 2–4 kg COD/m^3/day; COD influent: 1.5–2.4 g/L; removal >95%; SS effluent <28 mg/L; total $P < 0.37$ mg/L.	Ødegaard *et al.* (1994)
Dairy wastewater	Carrier area 268 m^2/m^3; filling ratio 54%; influent total COD 3.3 g/L; removal 85% at 12 kg COD/m^3/day. Full-scale plant.	Rusten *et al.* (1992)
Pulping whitewater	Thermophilic aerobic treatment (55°C). Carrier area 268 m^2/m^3; filling ratio 54%; soluble COD: 2.2 g/L; soluble organic load 2.5–3.5 kg COD/m^3/day; removal 60–65%; HRT 13–22 h: lab-scale reactor	Jahren *et al.* (2002)

4.2.3.2 Fluidized bed reactors

In FB reactors the solid particles can be kept in suspension by the power input due to liquid flow (the classical fluidization) or by the gas flow. The latter is applied in airlift bioreactors, in which the bubble movement causes a liquid flow in the riser, which suspends the solids. Airlift reactors have been used for secondary and tertiary municipal wastewater treatment (Nicolella *et al.* 2000).

The liquid to be treated is pumped through a bed of small media at a sufficient velocity to cause fluidization (Figure 4.5). In the fluidized state the media provide a large specific surface for attached biological growth and allow biomass concentrations in the range of 10–40 kg/m^3 to develop. Aeration is done by recirculating the liquid from the reactor to an oxygenator where air, or possibly oxygen, is bubbled (Cooper 1981). To overcome problems related to high recirculation rates needed when there is a high oxygen demand in the reactor, the reactor might be aerated directly (three-phase FB reactor) (Schügerl 1997).

Design considerations In FB and airlift reactors, there is a considerable difference between the local-energy dissipation rates at the bottom and top. In the reactor top, biofilm quickly grows. With increasing retention time of the carrier in the reactor, the biofilm thickness increases and film detachment occurs, because the microorganisms in the deep regions die away and lyse due to the lack of oxygen. As a consequence of stratification and size distribution, biodegradation rate, biofilm composition, and biofilm-specific activity change along the height of the bed. The extent of axial liquid mixing in beds of light particles is significantly less than that in beds of heavy particles, and therefore the particle density effect on axial liquid mixing is important.

The HRT should be shorter than the inverse of the maximum growth rate of the suspended bacteria. At longer retention times, only a low amount of attached

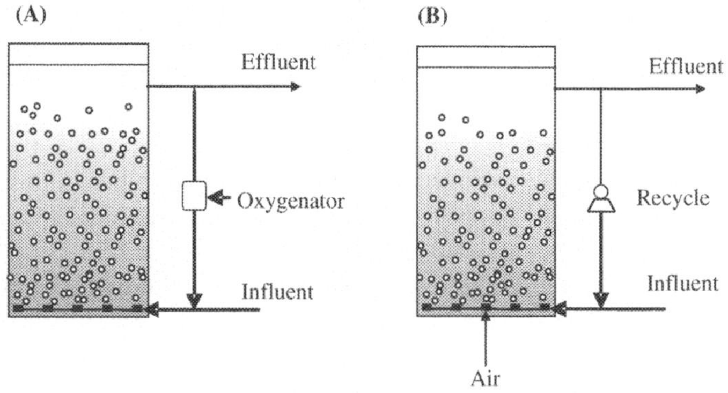

Figure 4.5 FB reactor. Oxygen or air can be supplied by an oxygenator (A) or directly at the bottom of the reactor (B).

biomass remains on the carrier material as patchy biofilms (Tijhuis *et al.* 1994). The biomass concentration attains a maximum at a suitable retention time of the carrier in the reactor. Therefore, a continuous feeding and drain of the carrier is necessary. In the bottom, the attachment rate is slow and the abrasion may be important.

For a reliable operation of biofilm reactors, high biomass concentrations, present in stable granular biofilms, are essential. For this reason, the carriers should be completely covered with biofilm. Biofilm development is the difference between biofilm growth and attachment on one hand, and detachment processes on the other hand. Biofilm development is influenced by various processes, including adsorption and desorption of microorganisms to and from the solid surface, attachment of microorganisms to the surface, biofilm growth, and biofilm detachment. At steady state, the balance between biofilm growth and detachment determines the physical structure of the biofilm, and thereby the settling and fluidization characteristics. It has been found that the biofilm thickness is inversely proportional to the specific degradation rate. A moderate surface loading and high detachment forces yield smooth and strong biofilms (Huang and Wu 1996). High surface loading and low detachment forces lead to low amount of biomass because of the formation of weaker biofilm (Tijhuis *et al.* 1996).

An accurate knowledge of the hydrodynamic aspects of a FB bioreactor is critical in order to assure its reliable design and operation. In these systems, biofilm growth alters particle size, apparent density, shape and roughness, and all these factors have a strong influence on the hydrodynamic behavior of the reactors. Successful design and operation of particle type biofilm reactors rely on information on settling and fluidization characteristics, such as FB height as a function of liquid velocity. Information on FB height is important because it establishes the solids residence time and the specific biofilm surface area in the biologically active zone (Nicolella *et al.* 2000).

FB reactors can be classified as three phase, for example, gas–liquid–solid (GLS) reactors. The solid phase is constituted by biofilm particles suspended in the bulk liquid. The gas phase is constituted by air (or oxygen). In aerobic systems, oxygen is transferred from the air bubbles through the bulk liquid to the biofilm particles, where it is consumed by biochemical reactions. Gas–liquid and liquid solid mass transfer coefficients are, together with reaction kinetic parameters, important design parameters of FB reactors.

Liquid mixing in FB reactors affects the interphase mass transfer, the reactant concentration distribution, and ultimately the reactant conversions. Information on axial liquid mixing is therefore crucial to reactor design and optimization. Axial liquid mixing in liquid–solid and GLS/FB is often analyzed using the axial dispersion model. In these systems, a distribution of solid size and/or density results in a distribution of terminal settling velocities leading to a classification within the FB. This classification is caused by a sorting effect where particles with higher settling velocity are found at the bottom of the bed.

Applications The FB technology is typically most useful for treatment of streams contaminated with organic or inorganic compounds (e.g. ammonia) requiring long solids residence time conditions (longer than 15 days) for biological oxidation and low (less than 100 mg/L) concentrations of SS. Typical operational conditions are: *liquid velocity*: 10–30 m/h (activated carbon and sand); *organic load*: 10 kg COD/m^3/day; height to diameter ratio from 2 to 5 (Sutton *et al.* 1999; Nicoella *et al.* 2000).

4.2.4 Membrane bioreactors

4.2.4.1 Description of the process

The term MBR is applied to the system developed for the wastewater treatment that combines a biological stage and a membrane module bioreactor. Combining membrane technology with biological reactors for the treatment of wastewaters has led to the development of three generic membrane bioreactors: for separation and retention of solids (Figure 4.6); for bubble-less aeration within the bioreactor (Figure 4.7A); and for extraction of priority organic pollutants from industrial wastewaters (Figure 4.7B).

Membranes when coupled to biological processes are most often used as a replacement for sedimentation, that is, for separation of biomass (bacteria and viruses). Membranes can also be coupled with bioprocesses for wastewater treatment in two other ways. Firstly, they can be used for the mass transfer of gases, usually oxygen for aerobic processes. Secondly, membranes can be used for the controlled transfer of nutrients into a bioreactor or the extraction of pollutants from wastewaters, which are untreatable by conventional biological processes. The target pollutants are then removed in a reactor with the correct environmental conditions for biological treatment (Livingston 1994). Stephenson *et al.* (2000) presented a more detailed description and applications of the different membrane bioreactors used in wastewater treatment.

There are two types of configurations for the MBR when the membrane array is used as a replacement for a decanter: the membranes can be placed either outside (Figure 4.6A) or inside (Figure 4.6B) the bioreactor. For the external configuration, the mixed liquor is filtered under pressure in a specific membrane module, whereas

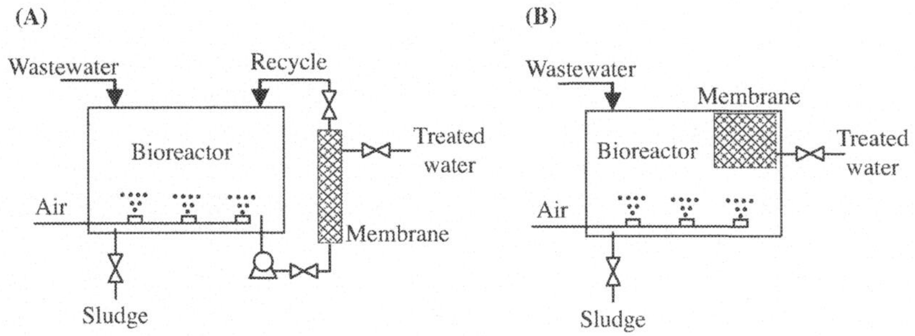

Figure 4.6 Membrane bioreactors for separation and retention of solids. Membranes outside (A) or inside the bioreactor (B).

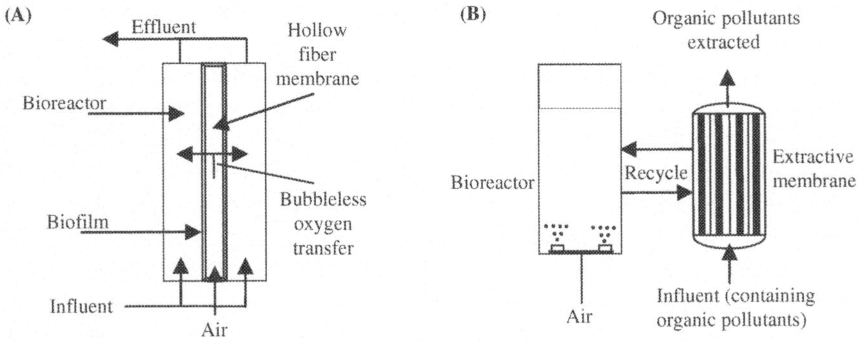

Figure 4.7 Membrane bioreactors for bubble-less aeration within the bioreactor (A) and for extraction of priority organic pollutants from industrial wastewaters (B).

for the submerged configuration, the filtration is carried out in the aeration basin by suction removal of the effluent. In the external system, the permeate flux generally varies between 50 and 120 L/h/m^2 and the transmembrane pressure (TMP) is in the range of 1 to 4 bar. In the submerged configuration, the permeate flux varies from 15 to 50 L/h/m^2 and the TMP is about 0.5 bar. The submerged configuration appears to be more economical based on energy consumption for two main reasons: no recycle pump is needed since aeration generates a tangential liquid flow in the vicinity of the membranes, and the operating conditions are much milder than in an external MBR system because of the lower values of TMP and tangential velocities. Generally, hollow fiber membranes are used in submerged MBR.

To avoid membrane fouling, it is often necessary to carry out tangential filtration in an external MBR system, especially when concentrated effluents and/or concentrated biomass are encountered. A higher-energy cost can be justified. In such cases, it is necessary to control the shear stress on the biological flocs. The shear stress is greater in the external membrane modules because of the high recycle flow ratio. Generally, tubular membranes are used in external MBR systems.

4.2.4.2 Advantages

When compared to traditional-activated sludge systems, the MBR offers many attractive advantages: (1) The traditional secondary clarifier is replaced by a membrane module. This module is more compact and the quality of rejected water is independent of the variations of sludge settling velocity. (2) The MBR allows the biomass concentrations to be higher than for traditional treatment plants (from 20 to 30 g/L).

The membrane bioreactor is characterized by a complete retention of the biomass inside the bioreactor because of the use of membrane separation, which controls and increases the MCRT independently from the HRT. High SRTs increase the sludge concentration and the applied organic load, thereby increasing the pollutant degradation. As the settling problem of the sludge is avoided, biological degradation can be more complete, resulting in higher treatment efficiencies. Nevertheless, increasing the biomass concentration generates a reduction in the oxygen mass transfer rate depending on the type of wastewater and reactor used.

The higher performances of the MBR against the traditional activated sludge process not only are explained by the high concentration of biomass, but also because of its composition and quality. It has been found that in the MBR process the flocs are smaller and more active with a higher volatile fraction in the mixed liquor and a greater diversity of species especially in terms of free-swimming bacteria than in the activated sludge system. Enzyme activity was also seen to be higher in the MBR and this was attributed to washout in the activated sludge systems (Cicek *et al.* 1999).

Other advantages of this system are: (1) the volume of the aeration tank can be also reduced since a higher concentration of biomass can be stored in the bioreactor. (2) the production of sludge, the disposal of which is often difficult, is decreased by a factor of 2 to 3, resulting in a reduction of the overall operating costs (Gander *et al.* 2000). (3) the membrane bioreactor is perfectly integrated in the industrial process because the wastewater can directly be treated *in situ*, allowing water reuse and concomitant reduction of the manufacturing costs linked to water consumption.

Despite these potential advantages, the implementation of MBRs for wastewater treatment has been scarce due to high membrane costs and large-energy inputs for membrane operation (Visvanathan *et al.* 2000). A considerable amount of research to date, therefore, has focused on membrane operation, specifically on understanding membrane fouling to reduce MBR operational costs (Defrance *et al.* 2000).

Yoon *et al.* (2004) presented a methodology to obtain the most economical operational condition of MBR. To minimize operational costs, aeration and sludge treatment costs were estimated for various operational conditions. It was found that sludge treatment cost turned out to overwhelm the aeration cost over the reasonable operational conditions. Therefore, sludge minimization was considered to be a key for the economical operation of MBR. Economically optimum MCRT and target MLSS turned out to be 16 h and 11,000 mg/L, respectively.

4.2.4.3 *Membrane characteristics*

The preferred membrane materials for MBR are invariably polymeric on the simple basis of cost. Geometries employed range from flat plate/plate and frame to tubular or hollow fiber. The choice of configuration is profoundly influenced by the MBR process configuration, namely by whether the membrane element is placed within the bioreactor or external to it (Stephenson *et al.* 2000). Generally, tubular membranes are used in external MBR. The common materials are cellulose acetate; polysulphone, polypropylene, PTFE, polyamide (0.04–0.1 μm of pore diameter); ceramic materials (0.1–0.2 μm), zirconium dioxide (0.05 μm), and alumina (0.2 μm).

4.2.4.4 *Membrane fouling*

Fouling is the general term given to the process by which a variety of species present in the water increase the membrane resistance by adsorbing or depositing onto its surface, adsorbing onto the pore surfaces within the bulk membrane material (pore restriction) or by complete pore-blocking.

It has been demonstrated that there exists a critical flux, below which the membrane fouling can be neglected and thus membrane cleaning is not required (Field *et al.* 1995b). It is important, therefore, to choose an adequate initial permeate flux or TMP. Conventional techniques for limiting membrane fouling are as follows: (1) reduction of membrane fouling by aeration in the vicinity of membranes by filtration below the critical flux, by the addition of coagulants, by high-frequency back-pulsing, or by utilizing a high recycle velocity; (2) removal of the fouling material after formation by chemical washing (backwashing or back-pulsing). Membrane fouling is influenced by the membrane chemical nature, but also, by the membrane operational parameters (Ramesh-Babu and Gaikar 2001). For example, the use of hollow fiber microfiltration membranes introduces TMP gradients, which have an impact on flux rates. The magnitude of the flux depends on the design of the hollow fiber (length, internal diameter, permeability) and on the properties of the cake.

4.2.4.5 *MBR for industrial wastewater treatment*

The MBR technology has been successfully applied to a wide range of industrial wastes. Applications include oily wastes, food wastes, tannery effluents, gray waters (shampoo, oil, and soap), textile and dye effluents, and landfill leachates. The exact operating conditions of an MBR are usually waste specific. Reported organic loading rates range between 0.25 and 16 kg COD/m^3/day with corresponding removal efficiencies of 90–99.8%.

Loading rates are higher than municipal applications owing to the high strength of wastes. Feed concentrations of 68 g COD/L for a brewery effluent, 42 g COD/L for a food processing waste effluent and 29 g COD/L for an oily waste, are typical examples of the treated wastewaters. Similar performances were achieved with fruit juice, cotton mill, and tannery effluent at respective loading rates of 5.9, 0.25 and 3.5 kg COD/m^3/day. MCRT used in MBR varies from 6 to 300 day. HRT are generally much greater (reported in days) in industrial applications than those applied for municipal wastewaters. Depending on the type of wastewater HRT could vary from 24 to 240 h with a typical value around 48 to 72 h (Stephenson *et al.* 2000; Marrot *et al.* 2004).

4.3 ANAEROBIC SYSTEMS

The anaerobic systems are well-established units for the biodegradation of organic matter from wastewater. The implementation and successful applications of the anaerobic systems was mainly due to the development of high-rate reactors. One of the main characteristics of the high-rate systems is the uncoupling of solid and HRT, resulting in high retention of active biomass. In these reactors, the bacterial adhesion resulting in biofilms is the major mechanism that enables a high retention of active biomass. The biofilm formation can be due to bacterial immobilization on either an inert carrier or by self-immobilization. The most significant examples of reactors using either stationary or mobile inert carrier materials are the fluidized bed and fixed film reactors (FB and FF, respectively), whereas the upflow anaerobic sludge and expanded granular sludge bed reactors (UASB and EGSB, respectively) are the main examples of reactors using self-immobilization of biomass-forming granules.

For a full review about biofilm formation see Hulshoff Pol *et al.* (2004) and Lens *et al.* (2003). This chapter section presents the most relevant high-rate reactors with potential application for the anaerobic treatment of industrial wastewater.

4.3.1 UASB reactor

The application of anaerobic systems as the main biological step in wastewater treatment systems was scarce until the development of the UASB reactor by Lettinga and his group in the 1970s in the former Wageningen Agricultural University in The Netherlands (Lettinga *et al.* 1980). With more than 850 full-scale installations currently in operation worldwide treating a wide range of industrial wastewaters and sewage (Kleerebezem and Macarie 2003), the UASB reactor is by far the most popular high-rate reactor. The UASB reactor is based on the following concepts (Lettinga *et al.* 1984):

(a) Anaerobic granular sludge has or acquires good sedimentation properties. Therefore, mechanical mixing is not required.
(b) The required contact between the granular sludge and the wastewater is granted by the feeding of the wastewater at the bottom of the reactor and by the biogas production.
(c) The wash-out of biomass can largely be prevented using a three phase GLS separator at the top of the reactor.

Consequently, the main characteristics of the UASB reactor are the granular sludge, the influent distribution system, and the three-phase GLS separator as shown in Figure 4.8; the wastewater is evenly distributed over the bottom section of the reactor and flows upward through the granular sludge bed, where biochemical reactions take place converting the organic pollutants to biogas (a mixture of

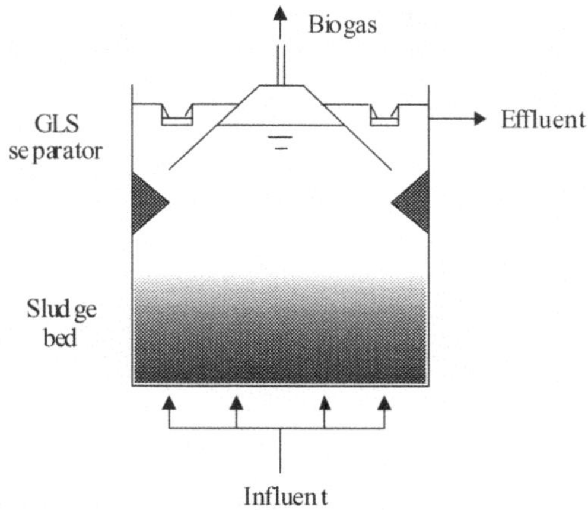

Figure 4.8 Schematic diagram of an UASB reactor.

methane and carbon dioxide). The biogas, together with the upflow velocity, provides adequate mixing of the sludge–wastewater mixture avoiding channeling in the sludge bed. The biogas is collected in the GLS separator. The sludge–water mixture flows to the settler section in the top of the reactor where biomass is allowed to settle and return to the sludge bed (Kleerebezem and Macarie 2003).

The success of the UASB reactor relies on the establishment of a dense sludge bed, in the order of 30–50 kg VSS/m^3, that allows the application of high volumetric COD loadings compared to other anaerobic systems. The sludge bed is composed of a high concentration of granules of diameters ranging from 0.1 to 5 mm, depending upon the wastewater treated. Reported typical settling velocities for granules are in the range between 20 and over 50 m/h and SVI values less than 20 mL/g (Schmidt and Ahring 1996). Specific methanogenic activity (SMA) values of granular sludge are in the range between 0.5 and 2 g COD-CH_4/g VSS-day. However, SMA values up to 7.1 g COD-CH_4/g VSS-day have been reported for thermophilic granules (Schmidt and Ahring 1996).

4.3.1.1 UASB reactor design

Generally, the design rules of biological reactors are all based on high removal efficiency of degradable organic matter. Consequently, if the substrate composition and strength of a wastewater are known, a basic design of a high-rate anaerobic system can be established. According to Lettinga and Hulshoff Pol (1991), the main design criteria of UASB reactors are, among others: applicable organic load, upflow velocity, three-phase separator, and influent distribution system.

Applicable organic load The UASB reactors are generally designed based on the organic volumetric load (OVL) (kg COD/m^3-day) that is defined as follows:

$$\mathrm{OVL} = \frac{Q\,S_0}{V} \tag{4.4}$$

Table 4.6 Suggested OVL in granular sludge UASB reactors at 30°C in order to obtain a COD removal efficiency between 85% and 95%.

Wastewater strength (g COD/L)	Insoluble COD (%)	OVL (kg COD/m^3-day)	
		Little TSS removal	High TSS removal
1–2	10–30	8–12	2–4
	30–60	8–14	2–4
	60–100	NA	NA
2–6	10–30	12–18	3–5
	30–60	12–24	2–6
	60–100	NA	2–6
6–9	10–30	15–20	4–6
	30–60	15–24	3–7
	60–100	NA	3–8
9–18	10–30	15–24	4–6
	30–60	NA	3–7
	60–100	NA	3–7

NA: not applicable; TSS: total suspended solids.
Adapted from Lettinga and Hulshoff Pol (1991).

where Q: influent flow rate (m^3/day), S_0: influent COD (kg COD/m^3), and V: volume of reactor (m^3). From Equation (4.4) the volume of the reactor, V, can be obtained:

$$V = \frac{Q\,S_0}{\text{OVL}} \tag{4.5}$$

For most industrial wastewaters, the OVL (based on degradable COD) is the critical factor for the reactor volume. Its value depends on the quantity and quality of the granular sludge; the nature, type, and concentration of the pollutants; the temperature; the required treatment efficiency and the desired safety regarding peak loads. Tables 4.6 and 4.7 show suggested OVL for various strengths of wastewater and a range of temperature.

One critical point for the operation of the process is the presence of SS in the wastewater as it could have a detrimental effect on the performance of the UASB reactor. Consequently, in many cases the best solution is to remove the SS in advance thus avoiding they would become adsorbed onto or absorbed into the granular sludge.

Upflow velocity Liquid upflow velocity (V_{UP}) is an important design parameter. A high V_{UP} increases the turbulence in the system enhancing the contact between the granular sludge and the incoming wastewater, avoiding the appearance of concentration gradients inside the system. However, an excessive V_{UP} could cause washout of valuable biomass producing a reduction of the COD removal efficiency. Typical V_{UP} values are in the range of 0.5–2 m/h for UASB reactors (de Man *et al.* 1988).

In the same way, the reactor height (H) has important implications for the performance and costs of the UASB reactor. V_{UP} is directly related to H and as it should

Table 4.7 Suggested OVL in granular sludge UASB reactors as a function of temperature treating mainly dissolved COD.

Temperature (°C)	OVL (kg COD/m^3-day)[a]
40	15–25
30	10–15
20	5–10
15	2–5
10	1–3

[a]Conservative figures.
Adapted from Lettinga *et al.* (1983).

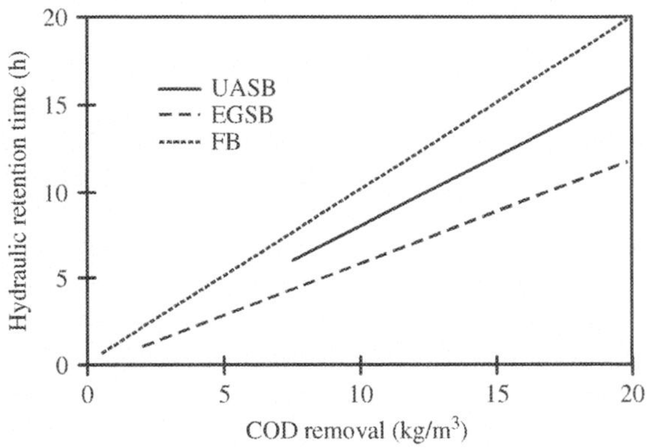

Figure 4.9 HRT required as a function of COD concentration to be removed for three types of high-rate anaerobic reactors. Lower limits of the curves are set by the hydraulic operational limits (Adapted from Kleerebezem and Macarie 2003).

not exceed a certain value to avoid biomass washout, H is also limited. Suggested H values for UASB reactors are between 4 and 6 m (van Haandel and Lettinga 1994). The relationship between V_{UP} and H for a UASB reactor is as follows:

$$V_{UP} = \frac{Q}{A} = \frac{V}{\text{HRT}\,A} = \frac{H}{\text{HRT}} \tag{4.6}$$

where A: reactor cross-section area (m^2), HRT: hydraulic retention time (h). In general, UASB reactors are not adequate for the treatment of low-strength soluble wastewaters due to their limited hydraulic capacity as shown in Figure 4.9.

Three-phase separator The GLS separator is an essential part of the UASB reactor. According to Lettinga *et al.* (1984), the main objectives of the GLS separator are as follows:

(a) Collect the biogas, thus avoiding turbulence in the settling zone.
(b) Separation of the biomass from the wastewater enabling it to slide back into the sludge bed and preventing biomass washout.

(c) Restricting excessive expansion of the sludge bed.
(d) Improve effluent solids removal.

The design of the GLS separator is in principle of quite simple construction; however, there are some special designs that have been introduced to the market by some engineering firms (Frankin 2001). Some general guidelines for the design of the GLS separator are (Lettinga and Hulshoff Pol 1991; van Haandel and Lettinga 1994):

(a) The inclination of the GLS separator plates should be in the range between 45° and 60°.
(b) The surface area in the openings between the gas collectors should be at least one-third of the cross-sectional reactor surface area.
(c) The volume of the GLS separator may vary from 15% to 20% of the total volume reactor.
(d) The height of the GLS separator should be between 1.5 and 2 m for 5–7 m reactor heights.
(e) A gas–liquid interface under the GLS separator should be created; either with a submerged separator or with a separator placed above the water surface.

Influent distribution system The main goal of the distribution system is to obtain a homogeneous distribution of the influent in order to avoid channeling of the wastewater through the granular sludge bed and formation of dead zones. This point is of particular importance mainly when the biogas production rates are lower than 0.5 m^3/m^3-day, as could be the case when treating very low-strength wastewaters. The area covered by one inlet point will be based on both the OVL and the biomass type (flocculent or granular) present in the reactor. For the case of granular sludge, 0.5–2 and 2–4 m^2 are the maximum areas per inlet point to obtain a satisfactory treatment for OVL of 1–4 kg COD/m^3-day and above 4 kg COD/m^3-day, respectively (Lettinga and Hulshoff Pol 1991).

Other important characteristics of UASB reactors are the shape and construction materials. The reactors can have a cylindrical or rectangular form and due to corrosion problems, concrete coated with plastic material is used for reactor walls whereas for all other specific devices non-corrosive materials such as polyvinyl chloride (PVC) and plastic covered hardwood are used (van Haandel and Lettinga 1994). In cases where the reactor volume exceeds 400 m^3, it is suggested to construct a modular reactor because of a number of advantages that exist over a one-compartment reactor (Lettinga and Hulshoff Pol 1991).

4.3.2 EGSB reactor

The UASB reactor represented a remarkable progress for the environmental technology and mainly for the anaerobic processes. Nevertheless, some modifications were suggested in order to expand its field of applications resulting in the EGSB reactor (de Man *et al.* 1988). The features of both reactors are similar; however, in the EGSB the granular sludge bed is expanded due to the application of V_{UP} higher than those imposed in UASB reactors (van der Last and Lettinga 1992). A schematic diagram of the EGSB reactor is shown in Figure 4.10.

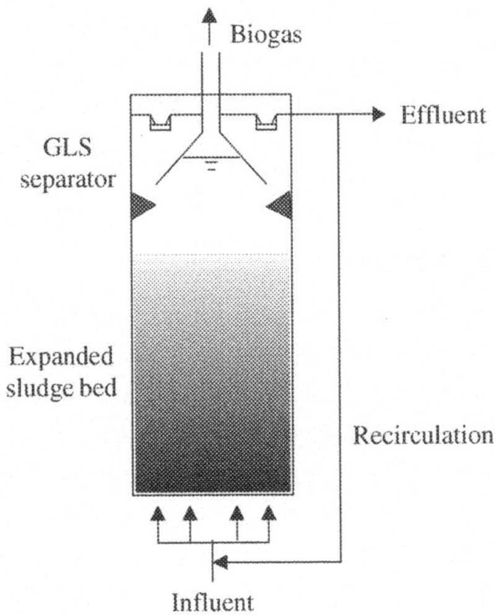

Figure 4.10 Schematic diagram of an EGSB reactor.

High V_{UP}, exceeding 5–6 m/h, is achieved by applying high liquid recirculation rates (Kato *et al.* 1994). Additionally, EGSB reactors are tall reactors with a limited diameter (high height/diameter ratio) and a relatively small footprint. As a result of the V_{UP} applied mainly granular sludge will be retained in the EGSB reactor. In principle, the high hydraulic and gas (m^3/m^2-h) loads applied to the EGSB reactor will improve the granular sludge–wastewater contact in two ways (Kato 1994):

(a) Expanding the sludge bed allowing the even distribution of the wastewater by preventing dead zones and short circuiting.
(b) The turbulence enables convective transport of substrates from the bulk into the biofilm increasing the total rate of substrate transport beyond that of diffusion alone.

However, a recent study indicated that a direct relationship between V_{UP} and substrate consumption could not be found. Instead, it was demonstrated that the anaerobic biofilms play a more relevant role in fully expanded EGSB reactors. Apparently, the characteristics of granular sludge (size and inner structure) are the main factors responsible of the internal mass transport limitations of the anaerobic sludge (González-Gil *et al.* 2001).

Due to the characteristics of the EGSB reactor (high V_{UP} and recirculation ratios), the system can be applied for the treatment of low-strength wastewaters (Kato *et al.* 1994), and for the treatment of wastewaters from the chemical and petrochemical industries where high recycle rates may decrease the potential toxicity of such streams (Razo-Flores *et al.* 1999; Macarie 2000). It has been proposed that the lowest feasible COD influent concentration that can be treated in an EGSB reactor is 13 mg/L at OVL of 5 kg COD/m^3-day (Kato 1994). On the other hand, OVL up to 40 kg COD/m^3-day can be applied in EGSB reactors (Seghezzo *et al.* 1998).

The design of the EGSB reactors is similar to the one described for UASB reactors. Consequently, the information presented in Tables 4.6 and 4.7 is also applicable for EGSB reactors. As shown in Figure 4.9, the EGSB reactor is not hydraulically limited when treating strongly diluted wastewater; however, it must be clear that this system is not adequate for the removal of SS.

4.3.3 Anaerobic FB reactor

Before the mid-1970s, these systems were found only at laboratory or pilot level. In the USA, the first industrial-scale application to wastewater treatment was made in a soft drinking bottling plant. Since then, aerobic and anaerobic FB reactors have been used extensively at commercial scale for municipal wastewater nitrification, denitrification, ammonia removal, beer industry, black waters, and the treatment of wastes that contain complex organic compounds from metallurgical, chemical, and petrochemical plants (Heijnen *et al.* 1989; Sutton and Mishra 1991; Nicolella *et al.* 2000).

Sand, porous glass beads, pumice stone, and basalt are some support media used, with a size range of 0.2–0.8 mm, besides lighter supports with apparent densities around 1100 and 1200 kg/m^3, designed specifically to reduce fluidization rates. Activated supports to improve pollutant removal processes, such as ionic interchange resins and granular-activated carbon (GAC), also have been used. The less appropriate probably has been GAC and sand, due to its high density (Speece 1996); therefore there is a need to design lighter media.

Flow regimes and hydrodynamics of FB reactors These systems are made up of two to three-phase reactors (liquid–solid, gas–solid, and liquid–gas–solid), where the support particles expand with a liquid and/or gas flow, in such a way that particles are always suspended and have no fixed position, moving easily in the surroundings where each one remains localized in a small volume into the bed (Henze and Harremoës 1983; Kunii and Levenspiel 1991). Gas is used to increase turbulence and mixing but also to diminish bed density, so fluidization liquid velocities used in a three-phase reactor are lower (Fan *et al.* 1982; Muroyama and Fan 1985).

Traditional fluidization (Figure 4.11) uses heavier than water support media, typically sand, and an up-flow liquid current is used to expand the bed, therefore high superficial liquid velocities and gas recirculation will be necessary to reach a soft fluidization regime (Muroyama and Fan 1985).

Contrary, during inverse fluidization, a down-flow liquid current is used as the continuous phase, since support media used are lighter with apparent densities varying from 75 to 930 kg/m^3, like polystyrene, polypropylene, or polyethylene. Support particles are suspended or floating in the liquid phase and the liquid level control is made in the upper part of the reactor (Figure 4.12).

In this case, lower operating flow velocities used as a consequence of the support flotation capacity and low density, may also reduce the high-cost associated to pumping equipment in the case of industrial reactors (Fan *et al.* 1982; Karamanev and Nikolov 1992; Krishnaia *et al.* 1993).

During the biological treatment of wastewaters, the biogas produced in an inverse FB reactor tends to raise countercurrent to the liquid flow direction and

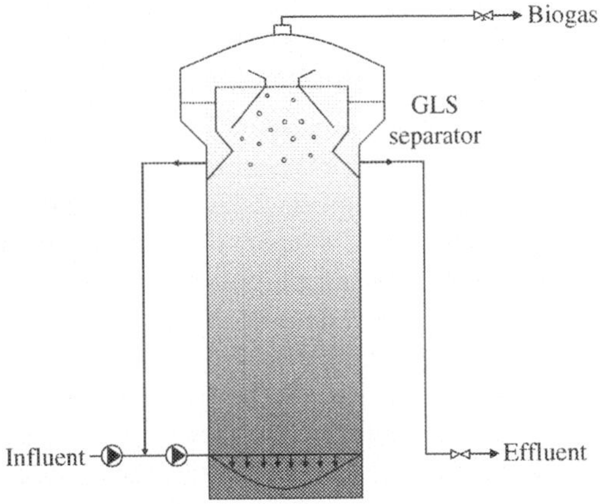

Figure 4.11 Schematic diagram of an industrial-scale up-flow FB.

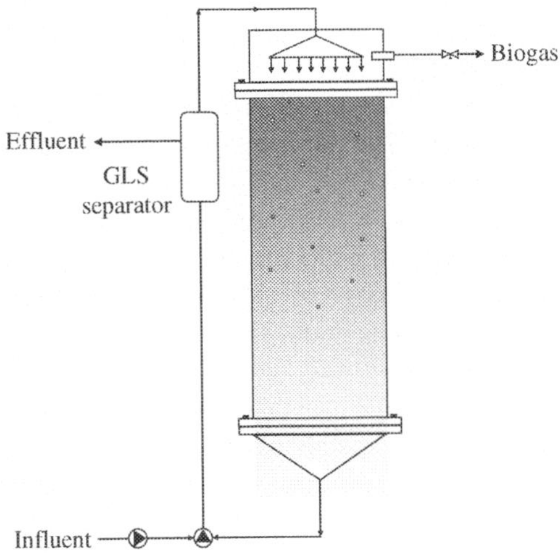

Figure 4.12 Schematic diagram of an experimental set-up of an inverse FB.

small gas bubbles will remain for larger time in the bed, forming a dispersion with the liquid phase, diminishing effectively in this way the bed density. Problems related to gas hold up may arise in this kind of fluidization (Fan *et al.* 1982; Karamanev and Nikolov 1992; Krishnaia *et al.* 1993).

The inverse FB has been tested only at laboratory level and most of the papers reported dealt with biofilm formation, treatment performance with winery, municipal, and synthetic wastewater and hydrodynamic studies (Fan *et al.* 1982; Meraz *et al.* 1997; García-Calderón *et al.* 1998; Buffière *et al.* 2000; Castilla *et al.* 2000).

Classical dimensions for FB or height/diameter (H/D) ratios, may vary between 2 and 20, to avoid dragging of support particles from the fluidization zone. The reactor length includes a disengagement zone, consisting normally of a widening in column diameter, which allows the support particles to diminish its terminal velocities and go back to the bed.

Bed expansion is controlled by fluid vertical velocity and is considered to be 30 to 50% of the stagnant bed, but shall reach 100%, that will only be attained with high recirculation rates, using short HRT and hence, liquid superficial velocities reported vary from 10 to 30 m/h in up-flow beds. The magnitude of the necessary expansion velocities is a function of support particles sedimentation velocities, which are from 150 to 300 m/h for supports heavier than water (Heijnen *et al.* 1989; Sutton and Mishra 1991; Nicolella *et al.* 2000).

In the case of the lighter support media used in the inverse FB reactor, information about support particles terminal velocity is scarce and may be around 27 m/h for a spherical shaped mineral support with a density of 690 kg/m^3 (García-Calderón *et al.* 1998). Due to the flotation capacity of support particles used in this configuration, the down flow liquid velocity will be the necessary to exert the appropriate force contrary to the flotation force, to drag the particles down the bottom of the reactor. Although expansion velocities reported are found from 2 to12 m/h, the minimum velocities will be applied when gas velocity and hold up is the highest, as the case of the inverse turbulent FB (García-Calderón *et al.* 1998; Castilla *et al.* 2000).

The operational restriction for this prototype due to the down-flow pattern during high-strength wastewater treatment is biomass accumulation onto the support that changes bioparticles density, the hydrodynamic regime, and gas hold up. When the designed loading rate is surpassed, biogas and biomass accumulation increases, leading to an excessive bed expansion that carries out reactive support from the bed (Meraz *et al.* 1997; García-Calderón *et al.* 1998; Buffière *et al.* 1999).

Fluidization parameters such as liquid velocity (u) and bed voidage (ε) or bed expansion attained by liquid flow rate and gas hold up, are determinant in the design and operation of FB reactors, whether up-flow or down-flow regime is used. Such variables related to support sphericity (ϕ_s) and density of liquid or gas flows ($\rho_{l,g}$), can be described with some accuracy through the empirical Equation (4.7),

$$\frac{1.75}{\varepsilon_{mf}^3\phi_s}\mathrm{Re}_{p,mf}^2 + \frac{150(1-\varepsilon_{mf})}{\varepsilon_{mf}^3\phi_s^2}\mathrm{Re}_{p,mf} = \mathrm{Ar} \tag{4.7}$$

where the Reynolds number, $\mathrm{Re}_{p,mf}$, that considers the particles diameters (d_p) at the onset of fluidization or minimum fluidization conditions (mf), is defined as,

$$\mathrm{Re}_{p,mf} = \frac{d_p u_{mf}\rho_{l,g}}{\mu} \tag{4.8}$$

and the Archimedes number, Ar, also at the onset of fluidization, is defined as,

$$\mathrm{Ar} = \frac{d_p^3\rho_{l,g}(\rho_s - \rho_{l,g})g}{\mu^2} \tag{4.9}$$

where g, is acceleration of gravity and μ, is dispersion viscosity.

The equation states that when the onset of fluidization occurs, a balance between drag forces for upward or downward moving fluids is equivalent to the support particles weight or flotation force (Muroyama and Fan 1985).

Gas hold-up or the residence time of gas in the liquid phase, directly influences bed voidage (and consequently design volume) as much as mass transfer between gas and liquid phase. The gas fraction to be present in the bed, e_g, has to be taken also into account for FB design and numerous correlations have been proposed to relate gas fraction with bed expansion.

The most important design parameter in a FB is bed height, which will depend directly on bed expansion. Also, bed expansion along with bed height and liquid velocity measurements allows biofilm thickness and biomass concentration estimations (Nicollela *et al.* 1995; 1997).

Start up and operation Inoculation with high-activity sludge and volatile solids content is recommended during start up of these reactors. Inoculation is often made with 30% to 60% sludge in volume and reactors are closed for several days or weeks, to allow microorganisms to be in contact with the support media the largest time possible. Another strategy is to inoculate continuously the reactor with anaerobic digesters supernatant together with the wastewater to be treated, apparently start up times diminish, but for industrial-scale reactors is not practical due to the supernatant volumes required (Heijnen *et al.* 1989).

Loading rate strategies used during start up are those known as the maximum efficiency profile, where a loading rate less than 1 kg/m^3-day and a mass loading rate of 0.1 kg COD/kg SS-day are applied at HRT of 1 day. Loading rate is doubled when maximum removal efficiency has been attained. The second one is the maximum loading rate profile, where high loading rates are applied, considering the highest mass loading rate that the reactor tolerates, with an HRT of 1 day. Loading rate will be increased although removal level is still low; for example, around 1500–2000 mg/L volatile fatty acids in the effluent. In this case, addition of alkali is advisable to control acidification due to overcharge, so this start-up profile is expensive.

The very small particle size provides high specific surfaces, ranging from 2000 to 8000 m^2/m^3, and allows the accumulation of large amounts of biomass during the immobilization stage at the start-up. During operation biomass accumulates, reaching around 10–40 kg/m^3 and may attain 90 kg/m^3, as in the case of the anaerobic systems. Specific activity of the biomass may vary from 0.7 to 2.0 kg COD/kg biomass-day in up-flow FB reactor and from 0.8 to 4.4 kg COD/kg biomass-day, in the inverse FB reactor.

The thickness of biofilms may vary from 60 to 200 μm for particle supports between 0.3 and 0.5 mm, conferring a removal capacity of 40 kg/m^3-day. Biofilms are compact when HRT are between 4 and 40 h and will be thicker at higher feeding organic loading rates. Due to this and to H/D ratios used, treatment capacities higher than other configurations may be achieved, from 20 to 100 kg COD/m^3-day in very compact equipment (Henze and Harremoës 1983; Heijnen *et al.* 1989; Ehlinger 1994; Holst *et al.* 1997).

A comparison between the up-flow and the inverse FB reactor is presented in Table 4.8, where some aspects related to reactor performance are shown.

Table 4.8 Experimental data reported for the treatment of municipal wastewater (MWW) and a synthetic one containing glucose and acetate (SWW) in an inverse FB reactor compared with those reported for up-flow FB reactors by Heijnen *et al.* (1989)[a].

Type of FB reactor	Up-flow	Down-flow	
		SWW	MWW
Media support	GAC, sand	Polyethylene	Polyethylene
Mean diameter (mm)	0.2–0.8	0.4	0.4
Superficial velocity (m/h)	5–35	1.6–7.3	7.6–10.9
Biomass concentration (kg VS[b]/m^3 reactor)	40–150	13.8	1.1
Specific activity (kg $COD_{removed}$/kg VS-day)	0.7–2.0	0.8–4.4	2.6–4.2

[a] Taken from Castilla *et al.* (2000); [b]VS: immobilized biomass.
MWW characteristics: 266 ± 99 mg COD_t/L; 59 ± 35 mg VSS/L; pH of 8.2 ± 0.2.
SWW characteristics: 0.2–2.0 g COD_t/L; pH of 7.0 ± 0.2.

Biofilm accumulation in the FB reactor In general, biofilms may be layered structures where the highest growth rate microorganisms will be found at the outside, whilst slower growing organisms will be found inside protected from the fluid shear stress. In the particular case of anaerobic bioreactors, biofilms will be dominated by methanogens due to the preferential acidogenic population in the peripheral layers that will detach easily (Nicolella *et al.* 2000).

Biofilm thickness can be controlled through high superficial gas or liquid velocities applied during expansion that may produce enough shear stress and also dynamic friction or attrition between particles to keep biofilm thin and compact. Nevertheless, eventually biofilm increases along with particles density. In up-flow reactors with excessive biofilm, compactness is lost and mean diameter of particles increases, particles density reduces and as a consequence, a higher bed expansion is attained, dragging out reactive support from the bed.

Many devices have been used to eliminate excessive biofilm that increase costs, such as pumps, gravitational screens with bars and wedges, aspersion washing systems or hydraulic separators coupled to external cyclones. Also, for up-flow FB reactors, turbulence at the top may be used, causing bed segregation due to the particles that have thin or no biofilm at all (Sutton and Mishra 1991; Nicolella *et al.* 2000).

FB advantages and disadvantages The main advantage that the FB reactors represent is small volume and area requirement due to high volumetric efficiency associated to the development of high biomass concentration (Sutton and Mishra 1991). Other advantages related to conventional suspended biomass reactors are:

(1) shear stress due to fluidization by friction with liquid or dynamic friction between particles, that renders compact and thin biofilms with high depuration capacity;
(2) improved liquid mixing and turbulence around particles that facilitate diffusion of the substrate into the biofilm;

(3) instant toxic high concentrations in the influent will have minimal effects on the biofilm due to the mixing regime;
(4) wash-out of fine solid inert particles;
(5) no SS retention and no blockage problems will be present (Henze and Harremoës 1983; Heijnen *et al.* 1989; Nicolella *et al.* 2000).

The lack of SS retention is related to the fluidization regime, therefore, removal devices to attain higher treatment quality will be necessary (Ehlinger 1994; Holst *et al.* 1997).

Gas produced by biofilms that is not easily released may diminish the support particles' density, so devices to separate liquid from support particles and gas such as deflectors are necessary (Sutton and Mishra 1991; Holst *et al.* 1997).

Mechanical devices used for FB that increase the investment costs, are liquid diffusers that must provide uniform fluidization to keep a uniform flow distribution inside the system. Other devices required are nozzles, to prevent or avoid blocking due to sedimentable particulate matter or support particles, flow controllers, pressure drop sensors, liquid level sensors, GLS separators (Fan *et al.* 1982; Henze and Harremoës 1983; Heijnen *et al.* 1989; Kunii and Levenspiel 1991; Sutton and Mishra 1991; Ehlinger 1994; Holst *et al.* 1997).

4.3.4 Staged anaerobic systems

With the aim of improving performance and stability of anaerobic digestion, several configurations that involve multiple staged or phased reactors have been investigated. One of the main approaches for staging two digesters consists in the physical separation of the non-methanogenic (hydrolysis/acid production) phase and the methanogenic phase. In this type of configuration (acid/gas staged digestion) the aim is to optimize the conditions for the different microbial groups present in each reactor. The other staging approach takes advantage of the benefits of thermophilic and mesophilic digestion. In this approach (temperature-phased anaerobic digestion or TPAD), a thermophilic reactor is combined with a mesophilic reactor for obtaining the benefits of both while avoiding the problems associated with both systems when operated independently (Han *et al.* 1997). The two possible configurations (mesophilic/thermophilic or thermophilic/mesophilic) have been investigated. A less utilized approach is the use, in series, of two or more thermophilic digesters (staged thermophilic digestion) or two mesophilic digesters (two-staged mesophilic digestion) for special applications. Table 4.9 summarizes basic information of the staged anaerobic systems mentioned here. In the next sections a more detailed description of the two main staged processes, for example, acid/gas staged digestion and TPAD is provided.

4.3.4.1 Acid/gas staged digestion

Two distinct groups of microorganisms, non-methanogenic bacteria (hydrolytic/ acidogenic bacteria) and methanogenic archaea, are involved in anaerobic digestion. Each of these groups has different characteristics regarding their physiology, nutritional requirements, growth characteristics, metabolic characteristics, environmental

Table 4.9 Options for staged anaerobic systems.

Staged system	Temperature	SRT (d)	Comments
Acid/gas			
Acid stage (pH = 6)	M	1–3	Two-stage has greater VS reduction than single-stage
Gas stage (neutral pH)	M or T	>10	Better foaming control It can meet Class A sludge requirements if a thermophilic stage is included
Acid/gas			
Acid stage (pH = 6)	T	1–2	
Gas stage (neutral pH)	M or T	>10	
TPAD			
First stage	T	3–5	Greater capability for absorbing shock loading Two-stage has greater VSS destruction than single stage
Second stage	M	7–15	Destruction of odorous compounds (fatty acids) Improved stability of digestion operation It can meet Class A sludge requirements
TPAD			
First stage	M	>7–10	
Second stage	T	>5	
Two-staged mesophilic			
First stage	M	7–10	Biosolids are less odorous and easier to dewater
Second stage	M	variable	
Staged thermophilic			
First stage: large digester	T	17–22	Several smaller digesters are used to reduce pathogen short circuiting to achieve Class A sludge requirements
Second stage: smaller digesters	T	Two each	

M: mesophilic; T: thermophilic.
Adapted from Metcalf and Eddy (2003).

optima, and sensitivity to environmental stress. The acid/gas staged digestion permits selection and enrichment of different bacteria in each digester by independently controlling the digester operating conditions. Therefore, by operating a first-stage digester at short HRT and low pH it is possible to optimize acid-forming bacterial growth and consequently promoting hydrolysis of the organic feed into simpler components that are further metabolized into fermentation products, including volatile acids. Then, the fermentation products produced in the first stage (acid stage) can be fed to a second-stage digester in which a methanogenic population is promoted by operation at neutral pH and optimal HRT. In this second stage, the fermentation products are used for methane production (gas stage). Most of the carbon dioxide is produced during the acid stage. Therefore, significantly higher methane than carbon dioxide concentration is produced during the gas stage. Complete separation had not been achieved in most cases and some methane production has been observed in the acid stage.

The acid/gas staged digestion has the following advantages (Fannin and Biljetina 1987):

(a) Ability to maintain appropriate densities of acid and methane formers.
(b) Maximization of conversion rates through independent control of temperature, pH, redox potential, biomass recycle, retention time, and other parameters for each stage.
(c) Enrichment of the product gas with methane.
(d) Greater stability with respect to feedstock loading, pH, and toxic shock loads.
(e) Higher solids reduction at lower retention time and, therefore, reduced total volume per unit of processed influent.
(f) Reduced gas clean up costs where required, due to the increased methane content of the product gas.

The main disadvantage is a higher capital investment for the additional equipment, controls, and labor requirements.

Different individual digester concepts for each stage have been investigated. The selected digester concept depends upon the type and solid concentrations of the feeds (Fannin and Biljetina 1987). Soluble feeds are generally highly biodegradable and contain low concentrations of particulate-associated solids, therefore, requiring short HRTs and SRTs for conversion to methane. The completely stirred tank reactor (CSTR) has been widely used for the digestion of low- to high-strength soluble feeds, such as sewage sludge. For soluble feeds requiring minimal hydrolysis shorter HRTs can be achieved using combinations of reactors such as anaerobic filters or UASB reactors. Low to medium solids feeds generally consist of two fractions, one that is highly soluble and biodegradable and another one formed by particulate matter that is more resistant to digestion. For this type of feeds it is required a reactor that achieves longer SRTs and MCRTs in comparison to HRTs in order to degrade the particulate fraction.

Anaerobic SBR (ASBRs) have been used for the hydrolysis/acidogenesis of the particulate matter fraction. Another approach that has been investigated is a liquefaction process consisting of a reactor operated as a batch-fed percolation system to promote hydrolysis of the particulate matter. Early application of this approach was used for the digestion of wheat straw and grass by Colleran (1982). The percolation system was initially fed with dry feed (>25% solids concentration) and then water is percolated through the reactor (optimum liquefaction occurred at 10–15% solids concentration). Then the percolating liquid containing dissolved organics was treated in a high-rate digester such as a UASB reactor. Percolating systems have been also used for treatment of high solids feeds.

Table 4.10 Examples of applications of acid/gas staged digestion (Laffite-Trouque and Forster 2000; Andersson and Björnsson 2002; Held *et al.* 2002; Bouallagui *et al.* 2004).

Waste(s)	Digestion concept	HRT	OLR	COD or VS removal	CH_4 yield
Activated sludge and syrups	*Acid stage*: CSTR, 55°C	4 h	n.r.	n.r.	n.r.
	Gas stage: CSTR, 35°C	8–15 days	0.42–0.63 kg VS/ m^3-day	41–52% VS_{deg}	0.12–0.34 m^3/ kg VS_{in}
Liquid fraction of organic waste from households	*Acid stage*: CSTR, 40°C, pH 7	24 days	9.8 kg COD/ m^3-day	67%[a] COD removal	0.52 m^3/ kg VS_{deg}
	Gas stage: UFAF, 40°C, pH 6.8	6.2 days	12.2 kg COD/ m^3-day	37.5% COD removal	0.31 m^3/ kg VS_{deg}
Beet tops, ley crops, straw	*Acid stage*: n.r.	n.r.	n.r.	n.r.	n.r.
	Gas stage: PBR, 35–37°C	24 h	2.4–25 g COD/L-day	50–73% tCOD removal	0.19–0.42 m^3/kg COD
Fruit and vegetable waste	*Acid stage*: ASBR, 35°C, pH 5.5	3 days	3.7–10.1 g COD/L-day	35–45% tCOD removal	n.r.
	Gas stage: ASBR, pH 7.3–7.5	10 days	0.72–1.65 g COD/L-day	67.9–92.7% tCOD removal	0.36–0.45 m^3/ kg COD_{in}

UFAF: up flow anaerobic filter; PBR: packed bed reactor.
[a]Complete separation was not achieved in the acid stage and CH_4 formation was observed;
tCOD: total COD; COD_{in}: COD influent; VS_{deg}: VS degraded; VS_{in}: VS influent; n.r.: not reported.

Acid/gas staged digestion has been used for a variety of applications. Table 4.10 summarizes operational and performance parameters of some examples of acid/gas staged digestion applications.

4.3.4.2 Temperature-phased anaerobic digestion

Single-stage mesophilic digestion has been the standard treatment for the stabilization of municipal wastewater sludge and for the treatment of cattle and swine manure. However, the mesophilic process has the following drawbacks: (a) it requires relatively long HRT because of the high content of poorly biodegradable organic matter present in the manure and in the waste activated sludge; (b) it is not so efficient in the reduction of volatile solids (15–30% for cattle waste); and (c) it is not so efficient in the inactivation of pathogens. Alternatively, thermophilic processes have shorter HRT by taking advantage of the higher metabolic rate of thermophilic microorganisms. Additionally, inactivation of pathogens is more efficient at thermophilic temperatures. However, the effluent quality from thermophilic digestion is poor and the residual sludge is more difficult to dewater (Kim *et al.* 2002). Also, thermophilic digestion is more sensitive to changes in operational conditions (e.g. temperature and organic loading) as well as to influent sludge characteristics such as the presence of toxic substances (van Lier 1996; Kim *et al.* 2002).

In order to alleviate the drawbacks of the mesophilic and thermophilic digestion processes, the TPAD process was developed at Iowa State University by Dague and collaborators (Kaiser and Dague 1994; Han *et al.* 1997). It combines thermophilic digestion (usually 55°C) as a first stage and mesophilic digestion (usually 35°C) as a second stage. The combined system uses the thermophilic stage to provide higher volatile solids (VS) reductions at shorter HRT, and the mesophilic stage as a polishing step of the drawbacks of the thermophilic stage, that is, removing the volatile fatty acids accumulated during the thermophilic stage and improving thickening properties of the thermophilic sludge.

The TPAD process has the following advantages (Han *et al.* 1997; Han and Dague 1997):

(a) Ability to achieve higher bioconversion and methane production rates than those achieved by mesophilic systems.
(b) Ability to treat higher solids and organic loadings at relatively shorter HRT.
(c) TPAD systems size would be less than one half the size of conventional single-stage systems.
(d) Better ability to deactivate pathogenic organisms and consequently to produce Class A biosolids, the highest quality of stabilized sludge for surface reuse regulated by the US Environmental Protection Agency.
(e) Capability for absorbing shock loadings like other two-stage processes.

The main disadvantages are a higher capital investment and higher operational costs due to higher-energy requirements. Table 4.11 summarizes operational and performance parameters of examples of TPAD applications.

4.4 ANAEROBIC–AEROBIC SYSTEMS

4.4.1 Aerobic step as a post-treatment

The numerous favorable characteristics of anaerobic processes such as: low cost, operational simplicity, and low solids production have contributed to highlighting the anaerobic systems for the treatment of domestic sewage, especially through UASB reactors. Despite these advantages, the UASB reactors have difficulties in producing effluents that can comply with the environmental standards. Therefore, a post-treatment step is of great importance as a manner of adapting the treated effluent to the environmental discharge standards. In this case the main objective of the post-treatment is to complement the organic matter removal, as well as to promote the removal of components that are barely affected by the anaerobic

Table 4.11 Examples of applications of TPAD (Dugba and Zhang 1999; Sung and Shanta 2003; Song *et al.* 2004).

Waste	Digestion concept	HRT	OLR	VS removal	CH_4 production
Dairy wastewater	*First stage*: thermophilic 55°C, ASBR *Second stage*: mesophilic 35°C, ASBR	3 days (s)[a] 0.6 days (t)[b] 2.4 days (m)[c]	2–4[d] g VS/L-day	43.8–44.1[e] % VS_{deg}	0.41–0.82[f] L/L-day
Dairy cattle manure	*First stage*: thermophilic 58°C, CSTR *Second stage*: mesophilic 38°C, CSTR	14 days (s) 4 days (t) 10 days (m)	1.87–5.82[d] g VS/L-day	39.7–41.5[e] % VS_{deg}	0.21–0.22[f] L/g VS_{fed}
Sewage sludge	*First stage*: thermophilic 55°C, CSTR *Second stage*: mesophilic 35°C, CSTR	30 days (s) 10 days (t) 20 days (m)	2.90[d] g VS/L-day	50.7–58.8[e] % VS_{deg}	0.42–0.47[f] L/g VS_{deg}

[a]s: HRT for complete system; [b]t: HRT for thermophilic stage; [c]m: HRT for mesophilic stage; [d]OLR for thermophilic stage; [e]VS removed for complete system; [f]CH_4 production for complete system; VS_{deg}: VS degraded; VS_{fed}: VS in fed.

treatment (nutrients and pathogens). Balancing the advantages and disadvantages of both systems, recent research has indicated the benefits in combining the processes, with the anaerobic stage being followed by the aerobic stage.

The advantages of the anaerobic–aerobic process are (van Haandel and Marais 1999; von Sperling *et al.* 2001): (a) lower-energy consumption; (b) lower chemical consumption for dewatering; (c) less sludge to be disposed; (d) less equipment required; and (e) higher operational simplicity.

Several technological alternatives have been applied for the anaerobic–aerobic combination. For example in the alternative of UASB-activated sludge, an important feature of the system is the return of the aerobic excess sludge to the anaerobic reactor, where the solids undergo stabilization, considerably simplifying the plant flowsheet and the sludge processing. This approach of returning the activated sludge wastage to the UASB reactor has been tested experimentally by Souza and Foresti (1996) and showed to give good results. Von Sperling *et al.* (2001) investigated the performance of a pilot-scale UASB-activated sludge plant operated for a period of 261 days. The plant received actual municipal wastewater. The plant showed good COD removal, with efficiencies ranging from 69 to 84% for the UASB reactor, from 43 to 56% for the activated sludge system only and from 85 to 93% for the overall system. The final effluent SS concentration was very low, with averages ranging from 13 to 18 mg/L. Other alternatives have been done coupling UASB reactors to submerged biofilters (Gonçalves *et al.* 1998) or UASB to trickling filter systems (Chernicharo and Nascimento 2001).

4.4.2 Anaerobic step to enhance the mineralization of recalcitrant compounds

Many toxic compounds present in industrial wastewaters are difficult to remove by using only anaerobic or aerobic environments. Electrophilic aromatic pollutants with multiple chloro, nitro, and azo groups have proven to be persistent to biodegradation by aerobic bacteria. These compounds are readily reduced by anaerobic consortia to lower chlorinated aromatics or aromatic amines but are not mineralized further. The reduction increases the susceptibility of the aromatic molecule for oxygenolytic attack. Sequencing anaerobic and aerobic treatment steps provide enhanced mineralization of many electrophilic aromatic pollutants. The combined activity of anaerobic and aerobic bacteria can also be obtained in a single treatment step if the bacteria are immobilized in particulate matrices (e.g. biofilm, soil aggregate, etc.). The anaerobic microniches established inside the biofilms can be applied to the reduction of electron withdrawing functional groups in order to prepare recalcitrant aromatic compounds for further mineralization in the aerobic outer layer of the biofilm (Zitomer and Speece 1993; Field *et al.* 1995). Aside from mineralization, polyhydroxylated and chlorinated phenols as well as nitroaromatics and aromatic amines are susceptible to polymerization in aerobic environments. Consequently, an alternative approach for bioremediation systems can be directed toward incorporating these aromatic pollutants into detoxified humic-like substances (Field *et al.* 1995a).

The mineralization of some recalcitrant pollutants has been possible by using the sequential anaerobic–aerobic treatments. Usually, the combined treatments are carried out in two separate reactors connected in series. An alternative is the use of one single reactor in which the anaerobic and aerobic phases are operated in a sequence (Shaw *et al.* 2002; Buitrón *et al.* 2003b). This procedure may offer the advantage of a permanent exchange of metabolite between the aerobic and the anaerobic microorganisms, which favor the establishment of trophic chains.

Another advantage of the anaerobic–aerobic treatment in front of the aerobic process is when volatile organics are present in the wastewater. In this case the volatile compounds are degraded in the anaerobic stage avoiding their volatilization. The alternating anaerobic–aerobic phases can be conducted by using the SBR. In this process the selection and enrichment of desired microbial populations could be accomplished easily by alternating anaerobic–aerobic phases through the control of the aeration policy.

Azo dyes are organic compounds difficult to biodegrade by either anaerobic or aerobic environments alone due to their high stability to microbial attack. Generally, bacterial azo dye biodegradation proceeds in two stages. The first stage involves a reductive cleavage of the azo linkages, resulting in the formation of – generally colorless but potentially hazardous – aromatic amines. The second stage involves degradation of the aromatic amines. Azo dye reduction usually requires anaerobic conditions, whereas bacterial biodegradation of aromatic amines is an almost exclusively aerobic process. A wastewater treatment process in which anaerobic and aerobic conditions are combined is therefore the most logical concept for removing azo dyes from wastewater. Cruz and Buitrón (2001) demonstrated that the mineralization of the azo dye disperse blue 79 could be accomplished with an anaerobic–aerobic system. In the anaerobic stage 98% of biotransformation of the azo dye was obtained. The total amines produced in the anaerobic biofilter were effectively biodegraded in the aerobic biofilter.

van der Zee and Villaverde (2005) present a review concerning the combined anaerobic–aerobic treatment of azo dyes. These authors concluded that, taken the reported results as a whole, the complete reduction of azo dyes at a reasonable time scale it is not a problem, provided that the prerequisites for azo dye reduction are met, that is, anaerobic conditions (absence or limitation of competing electron acceptors like oxygen, nitrate, and nitrite), availability of an electron donor and the absence of toxicity toward the anaerobic biomass.

REFERENCES

Andersson, J. and Björnsson, L. (2002) Evaluation of straw as a biofilm carrier in the methanogenic stage of two-stage anaerobic digestion of crop residues. *Bioresource Technol.* **85**, 51–56.

Betancur, M.J., Moreno, J. and Buitrón, G. (2004) Event-driven control for treating toxicants in aerobic sequencing batch bioreactors. In: *Proceedings of the 9th International Symposium on Computer Applications in Biotechnology (CAB9)*, CDROM file 1074, March 28–31, Nancy, France.

Bouallagui, H., Torrijos, M., Godon, J.J., Moletta, R., Ben Cheikh, R., Touhami, Y., Delgenes, J.P. and Hamdi, M. (2004) Two-phases anaerobic digestion of fruit and vegetable wastes: bioreactors performance. *Biochem. Eng. J.* **21**, 193–197.

Buitrón, G. and Ortiz, J. (1998) Biodegradation of phenolic compounds with a sequencing batch reactor, *52th Industrial Waste Conference*, Purdue University, Ed. Ann Arbor Press, Chelsea Michigan, Section 7, Chapter 27, 263–269.

Buitrón, G., Betancur, M., Moreno, G. and Moreno, J. (2003a) Oxidation-reduction potential as control variable for the anaerobic stage during the anaerobic–aerobic *p*-nitrophenol degradation. *Biotechnol. Progress* **19**, 1822–1927.

Buitrón, G., Melgoza, R M. and Jiménez, L. (2003b) Pharmaceutical wastewater treatment using an anaerobic/aerobic sequencing batch biofilter. *J Environ. Sci. Health. Part A Toxic/Hazard Subst. Environ. Eng.* **38**(10), 2077–2088.

Buitrón, G., Schoeb, M-E., Moreno-Andrade, I. and Moreno, J. (2005) Comparison of two control strategies for an automated sequencing batch reactor under extreme peaks of 4-chlorophenol. *Wat. Res.* **39**(6), 1015–1024.

Castilla, P., Meraz, M., Monroy, O. and Noyola, A. (2000) Anaerobic treatment of low concentration wastewater in an inverse fluidized bed reactor. *Wat. Sci. Technol.* **41**(4–5), 245–251.

Chernicharo, C.A.L. and Nascimento, M.C.P. (2001) Feasibility of a pilot-scale UASB/trickling filter system for domestic sewage treatment. *Wat. Sci. Technol.* **44**(4), 221–228.

Chiesa, S.C. and Irvine, R.L. (1985) Growth and control of filamentous microbes in activated sludge: an integrated hypothesis. *Wat. Res.* **19**, 471–479.

Chiesa, S.C., Irvine, R.L. and Manning Jr, J.F. (1985) Feat/famine growth environments and activated sludge population selection. *Biotechnol. Bioeng.* **27**, 562–568.

Cicek, N., Franco, J.P., Suidan, M.T. and Urbain, V. (1998) Using a membrane bioreactor to reclaim wastewater. *J. Amer. Wat. Works Assoc.* **90**(11), 105–113.

Cohen, Y. (2001) Biofiltration – the treatment of fluids by microorganisms immobilized into the filter bedding material: a review. *Bioresource Technol.* **77**, 257–274.

Colleran, E. (1982) The application of the anaerobic filter design to biogas production from solid and liquid agricultural wastes. In: *Symposium Papers, Energy from Biomass and Wastes VI* (ed. D.L. Klass). Institute of Gas Technology, Chicago, p. 43.

Cooper, P.F. (1981) The use of biological fluidised beds for the treatment of domestic and industrial wastewaters. *Chem. Eng.* **371**, 373–376.

Cruz, A. and Buitrón, G. (2001) Biodegradation of disperse blue 79 using sequenced anaerobic/aerobic biofilters. *Wat. Sci. Technol.* **44**(4), 159–166.

de Man, A.W.A., van der Last, A.R.M. and Lettinga, G. (1988) The use of EGSB and UASB anaerobic systems for low strength soluble and complex wastewaters at temperatures ranging from 8 to 30°C. In: *Proceedings of the 5th International Symposium on Anaerobic Digestion* (eds. E.R. Hall and P.N. Hobson). Pergamon, Oxford, pp. 197–209.

Defrance, L., Jaffrin, M.Y., Gupta, B., Paullier, P. and Geaugey, V. (2000) Contribution of various constituents of activated sludge to membrane bioreactor fouling. *Bioresource Technol.* **73**, 105–112.

Dugba, P.N. and Zhang, R. (1999) Treatment of dairy wastewater with two-stage anaerobic sequencing batch reactor systems – thermophilic versus mesophilic operations. *Bioresource Technol.* **68**, 225–233.

Eckenfelder Jr, W.W. and Musterman, J.L. (1995) *Activated Sludge Treatment of Industrial Wastewater*. Technomic Publishing Company, Lancaster Pennsylvania, pp. 281.

Ehlinger, F. (1994) Anaerobic biological fluidized beds: operating experiences in France, *Proceedings of the 7th International Symposium on Anaerobic Digestion*, pp. 315–323.

EPA (1999) Sequencing batch reactors. Wastewater technology fact sheet. US Environmental Protection Agency, Office of Water, Washington, DC, EPA 932-F-99-073, September.

Fan, L-S., Muroyama, K. and Chern, S-H. (1982) Hydrodynamic characteristics of inverse fluidization in liquid–solid and gas–liquid–solid systems. *Chem. Eng. J.* **24**, 143–150.

Fannin, K.F. and Biljetina, R. (1987) Reactor designs. In: *Anaerobic Digestion of Biomass* (eds. D.P. Chynoweth, and R. Isaacson). Elsevier Applied Science, London and New York, pp. 141–169.

Field, J.A., Stams, A.J.M., Kato, M. and Schraa, G. (1995a) Enhanced biodegradation of aromatic pollutants in cocultures of anaerobic and aerobic bacterial consortia. *Antonie van Leeuwenhoek Int. J. Gen. Mol. Microbiol.* **67**, 47–77.

Field, R.W., Wu, D., Howell, J.A. and Gupta, B.B. (1995b) Critical flux concept for microfiltration fouling. *J. Membrane Sci.* **100**, 259–272.

Frankin, R.J. (2001) Full scale experiences with anaerobic treatment of industrial wastewaters. *Wat. Sci. Technol.* **44**(8), 1–6.

Gander, M.A., Jeferson, B. and Judd, S.J. (2000) Membrane bioreactors for use in small wastewater treatment plants: membrane materials and effluent quality. *Wat. Sci. Technol.* **41**(1), 205–211.

García-Calderón, D., Buffière, P., Moletta, R. and Elmaleh, S. (1998) Influence of biomass accumulation on bed expansion characteristics of a down-flow anaerobic fluidized bed. *Biotech. Bioeng.* **57**(2), 137–144.

Gonçalves, R.F., de Araújo, V.L. and Chernicharo, C.A.L. (1988) Association of a UASB reactor and a submerged aerated biofilter for domestic sewage treatment. *Wat. Sci. Technol.* **38**(8–9), 189–195.

González-Gil, G., Kleerebezem, R., Seghezzo, L. and Lettinga, G. (2001) Kinetics and mass transfer phenomena in anaerobic granular sludge. *Biotechnol. Bioeng.* **73**, 125–134.

González-Martínez, S. and Wilderer, P.A. (1991) Phosphate removal in a biofilm reactor. *Wat. Sci. Technol.* **23**(7–9), 1405–1415.

Han, Y. and Dague, R.R. (1997) Laboratory studies on the temperature-phased anaerobic digestion of domestic wastewater sludge. *Wat. Env. Res.* **69**(6), 1139–1143.

Han, Y., Sung, S. and Dague, R.R. (1997) Temperature-phased anaerobic digestion of wastewater sludge. *Wat. Sci. Technol.* **36**(6–7), 367–374.

Heijnen, J.J., Mulder, A., Enger, W. and Hoeks, F. (1989) Review on the application of anaerobic fluidized bed reactors in waste-water treatment. *Chem. Eng. J.* **41**, B37–B50.

Held, C., Wellacher, M., Karl-Heinz, R. and Gübitz, G.M. (2002) Two-stage anaerobic fermentation of organic waste in CSTR and UFAF-reactors. *Bioresource Technol.* **81**, 19–24.

Henze, M. and Harremöes, P. (1983) Anaerobic treatment of wastewater in fixed film reactors – a literature review. *Wat. Sci. Technol.* **15**, 1–101.

Holst, T.C., Truc, A. and Pujol, R. (1997) Anaerobic fluidized beds: ten years of industrial experience. *Proceedings of the 8th International Conference on Anaerobic Digestion*, Vol. 2, pp. 142–149.

Huang, J.-S. and Wu, C.-S. (1996) Specific energy dissipation rate for fluidized bed bioreactors. *Biotechnol. Bioeng.* **50**, 643–654.

Hudson, N., Doyle, J., Lant, P., Roach, N., de Bruyn, B. and Staib, C. (2001) Sequencing batch reactor technology: the key to a BP refinery (Bulwer Island) upgrading environmental protection system – a low cost lagoon based reto-fit. *Wat. Sci. Technol.* **43**(3), 339–346.

Hulshoff Pol, L.W., de Castro Lopes, S.I., Lettinga, G. and Lens, P.N.L. (2004) Anaerobic sludge granulation. *Wat. Res.* **38**, 1376–1389.

Irvine, R.L., Wilderer, P.A. and Flemming, H. (1997) Controlled unsteady state processes and technologies – an overview. *Wat. Sci. Technol.* **35**(1), 1–10.

Jahren, S.J., Rintala, J.A. and Ødegaard, H. (2002) Aerobic moving bed biofilm reactor treating thermomechanical pulping whitewater under thermophilic conditions. *Wat. Res.* **36**, 1067–1075.

Kaiser, S.K. and Dague, R.R. (1994) The temperature-phased anaerobic biofilter process. *Wat. Sci. Technol.* **29**(9), 213–223.

Karamanev, D.G. and Nikolov, L.N. (1996) Application of inverse fluidization in wastewater treatment: from laboratory to full-scale bioreactors. *Environ. Progress* **15**(3), 194–196.

Kato, M.T. (1994) *The Anaerobic Treatment of Low Strength Soluble Wastewaters*. PhD Thesis, Wageningen Agricultural University, Wageningen, The Netherlands.

Kato, M.T., Field, J.A., Versteeg, P. and Lettinga, G. (1994) Feasibility of expanded granular sludge bed reactors for the anaerobic treatment of low strength soluble wastewaters. *Biotechnol. Bioeng.* **44**, 469–479.

Keller, J., Watts, S., Battye-Smith, W. and Chong, R. (2001) Full-scale demonstration of biological nutrient removal in a single tank SBR process. *Wat. Sci. Technol.* **43**(3), 355–362.

Ketchum Jr, L.H. (1997) Design and physical features of sequencing batch reactors. *Wat. Sci. Technol.* **35**(1), 11–18.

Kim, M., Ahn, Y.H. and Speece, R.E. (2002) Comparative process stability and efficiency of anaerobic digestion; mesophilic vs. thermophilic. *Wat. Res.* **36**, 4369–4385.

Kleerebezem, R. and Macarie, H. (2003) Treating industrial wastewater: anaerobic digestion comes of age. *Chem. Eng.* April, 56–64.

Kolb, F.R. and Wilderer, EA. (1995) Activated carbon membrane biofilm reactor for the degradation of volatile organic pollutants. *Wat. Sci. Technol.* **31**(1), 205–213.

Krishnaia, K., Guru, S. and Sekar, V. (1993) Hydrodynamic studies on inverse gas–liquid–solid fluidization. *Chem. Eng. J.* **51**, 109–112.

Kunii, D. and Levenspiel, O. (1991) *Fluidization Engineering. Series in Chemical Engineering*, 2nd edn. Butterworth-Heinemann, USA.

Lafitte-Trouqué, S. and Forster, C.F. (2000) Dual anaerobic co-digestion of sewage sludge and confectionery waste. *Bioresource Technol.* **71**, 77–82.

Lazarova, V. and Manem, J. (1994) Advances in biofilm aerobic reactors ensuring effective biofilm activity control. *Wat. Sci. Technol.* **29**(10–11), 319–327.

Lens, P.N.L., O'Flaherty, V., Moran, A.P., Stoodley, P. and Mahony, T. (2003) *Biofilms in Medicine, Industry and Environmental Biotechnology: Characteristics, Analysis and Control. Integrated Environmental Technology Series*, IWA Publishing, London.

Lettinga, G. and Hulshoff Pol, L.W. (1991) UASB-process design for various types of wastewaters. *Wat. Sci. Technol.* **24**(8), 87–107

Lettinga, G., van Velsen, A.F.M., Hobma, S.W., de Zeeuw, W. and Klapwijk, A. (1980) Use of upflow sludge blanket (USB) reactor concept for biological wastewater treatment, specially for anaerobic treatment. *Biotechnol. Bioeng.* **22**, 699–734.

Lettinga, G., Hobma, S.W., Hulshoff Pol, L.W., de Zeeuw, W.J., de Jong, P., Grin, P. and Roersma, R.E. (1983) Design, operation and economy of anaerobic treatment. *Wat. Sci. Technol.* **15**(8–9), 177–195.

Lettinga, G., Hulshoff Pol, L.W., Koster, I.W., Wiegant, W.M., de Zeeuw, W.J., Rinzema, A., Grin, P.C., Roersma, R.E. and Hobma, S.W. (1984) High-rate anaerobic wastewater treatment using the UASB reactor under a wide range of temperature conditions. *Biotechnol. Gen. Eng. Rev.* **2**, 253–284.

Livingston, A.G. (1994) Extractive membrane bioreactors: a new process technology for detoxifying chemical industry wastewaters. *J. Chem. Technol. Biotechnol.* **60**, 117–124.

Macarie, H. (2000) Overview of the application of anaerobic treatment to chemical and petrochemical wastewaters. *Wat. Sci. Technol.* **42**(5–6), 201–213.

Mace, S. and Mata-Alvarez, J. (2002) Utilization of SBR technology for wastewater treatment: an overview. *Ind. Eng. Chem. Res.* **41**, 5539–5553.

Marrot, B., Barrios Martínez, A., Moulin, P. and Roche, N. (2004) Industrial wastewater treatment in a membrane bioreactor: a review. *Environ. Prog.* **23**(1), 59–68.

Melgoza, R.M, Cruz, A. and Buitrón, G. (2004) Anaerobic/aerobic treatment of colorants present in textile effluents. *Wat. Sci. Technol.* **50**(2), 149–155.

Meraz, M., Monroy, M. and Noyola, A. (1997) Start-up and operation of an inverse fluidized bed. *Proceedings of the 8th International Conference on Anaerobic Digestion*, Vol. 3, pp. 248–251.

Metcalf and Eddy, Inc. (2003) *Wastewater Engineering Treatment and Reuse*, 4th edn. Mc Graw Hill, New York, pp. 1819.

Moreno-Andrade, I., Martinez, D. and Buitrón, G. (2004) Effect of starvation and shock loads of 4-chlorophenol on the performance of a moving bed SBR, *Proceeding of the 3rd IWA Specialised Conference on Sequencing Batch Reactor Technology-SBR3*, February 22–26, Noosa, Australia.

Muroyama, K. and Fan, L.S. (1985) Fundamentals of gas–liquid–solid fluidization. *AICHE J.* **31**, 1–34.

Nicolella, C., van Loosdrecht, M.C.M. and Heijnen, J.J. (2000) Wastewater treatment with particulate biofilm reactor. *J. Biotech.* **80**, 1–33.

Ødegaard, H., Rusten, B. and Westrum, T. (1994) A new moving bed biofilm reactor–applications and results. *Wat. Sci. Technol.* **29**(10–11), 157–165.

Qasim, S.R. (1999) *Wastewater Treatment Plants. Planning, Design, and Operation*, 2nd edn. Technomic Pub. Co., Lancaster Pennsylvania, pp. 1107.

Ramesh-Badu, P. and Gaikar, V.G (2001) Membrane characteristics as determinant in fouling of UF membranes. *Separ. Purif. Technol.* **24**, 23–34.

Razo-Flores, E., Smulders, P., Prenafeta-Boldú, F., Lettinga, G. and Field, J.A. (1999) Treatment of anthranilic acid in an anaerobic expanded granular sludge bed reactor at low concentrations. *Wat. Sci. Technol.* **40**(8),187–194.

Rother, E. and Cornel, P. (2004) Optimising design, operation and energy consumption of biological aerated filters (BAF) for nitrogen removal of municipal wastewater. *Wat. Sci. Technol.* **50**(6), 131–139.

Rusten, B., Ødegaard, H. and Lundar, A. (1992) Treatment of dairy wastewater in a novel moving bed biofilm reactor. *Wat. Sci. Technol.* **26**(3–4), 702–711.

Schmidt, J.E. and Ahring, B.K. (1996) Granular sludge formation in upflow anaerobic sludge blanket (UASB) reactors. *Biotechnol. Bioeng.* **49**, 229–246.

Schügerl, K. (1997) Three-phase-biofluidization – application of three-phase fluidization in the biotechnology – a review. *Chem. Eng. Sci.* **52**(21/22), 3661–3668.

Seghezzo, L., Zeeman, G., van Lier, J., Hamelers, H.V.M. and Lettinga, G. (1998) A review: the anaerobic treatment of sewage in UASB and EGSB reactors. *Bioresource Technol.* **65**, 175–190.

Shaw, C.B., Carliell, C.M. and Wheatley, A.D. (2002) Anaerobic/aerobic treatment of coloured textile effluents using sequencing batch reactors. *Wat. Res.* **36**, 1993–2001.

Song, Y.C., Kwon, S.J. and Woo, J.H. (2004) Mesophilic and thermophilic temperature co-phase anaerobic digestion compared with single-stage mesophilic and themophilic digestion of sewage sludge. *Wat. Res.* **38**, 1653–1662.

Souza, J.T. and Foresti, E. (1996) Domestic sewage treatment in an upflow anaerobic sludge blanket – sequencing batch reactor system. *Wat. Sci. Tech.* **33**(3), 73–84.

Speece, R.E. (ed.) (1996) *Anaerobic Biotechnology for Industrial Wastewaters*. Ed. Archae Press, TN, USA.

Stephenson, T., Judd, S., Jefferson, B. and Brindle, K. (2000) *Membrane Bioreactors for Wastewater Treatment*. IWA Publishing, United Kingdom, pp. 179.

Sung, S. and Santha, H. (2003) Performance of temperature-phased anaerobic digestion (TPAD) system treating dairy cattle wastes. *Wat. Res.* **37**, 1628–1636.

Sutton, P.M. and Mishra, P.N. (1991) Biological fluidized beds for water and wastewater treatment: a state-of-the-art review. *Environ. Tech.* **3**(8), 52–56.

Sutton, P.M., Hurvid, J. and Hoeksema, M. (1999) Biological fluidized bed treatment of wastewater from byproduct coking operations: full-scale history. *Water Environ. Res.* **71**, 5–9.

Tijhuis, L., Loosdrecht van, M.C.M. and Heijnen, J.J. (1994) Formation and growth of heterotrophic aerobic biofilms on small suspended particles in airlift reactors. *Biotechnol. Bioeng.* **44**, 595–608.

Tijhuis, L., Hijman, B., Loosdrecht van, M.C.M. and Heijnen, J.J. (1996) Influence of detachment, substrate loading and reactor scale on the formation of biofilms in airlift reactors. *Appl. Microbiol. Biotechnol.* **45**, 7–17.

Tilche, A., Bortone, G., Malaspina, F., Piccinini, S. and Stante, L. (2001) Biological nutrient renoval in a full-scale SBR treating piggery wastewater: results and modelling. *Wat. Sci. Technol.* **43**(3), 363–371.

Torrijos, M. and Moletta, R. (1997) Winery wastewater depollution by a sequencing batch reactor. *Wat. Sci. Technol.* **35**(1), 249–257.

Vallés-Morales, M.J., Mendoza-Roca, J.A., Bes-Piá, A. and Iborra-Clar, A. (2004) Nitrogen removal from sludge water with SBR process: start-up of full-scale plant in the municipal wastewater treatment plant in Ingolstadt, Germany. *Wat. Sci. Technol.* **50**(10), 51–58.

van der Last, A.R.M. and Lettinga, G. (1992) Anaerobic treatment of domestic sewage under moderate climatic (Dutch) conditions using upflow reactors at increased superficial velocities. *Wat. Sci. Technol.* **25**(7), 167–178.

van der Zee, F.P. and Villaverde, S. (2005) Combined anaerobic–aerobic treatment of azo dyes – a short review of bioreactor studies. *Wat. Res.* **39**, 1425–1440.

van Haandel, A.C. and Lettinga, G. (1994) *Anaerobic Sewage Treatment: A Practical Guide for Regions with Hot Climate*. John Wiley and Sons, Chichester.

van Haandel, A. and Marias, G. (1999) *O Comportamento do sistema de Lodo Ativado*. Teoria e Aplicações para proyectos e operação. Universidade Federal de Paraíba, Campina Grande Brazil, pp. 488.

van Lier, J.B. (1996) Limitation of thermophilic anaerobic wastewater treatment and the consequences for process design. *Antonie van Leeuwenhoek* **69**, 1–14.

Vargas, A., Soto, G., Moreno, J. and Buitrón, G. (2000) Observed based time-optimal control of an aerobic SBR for chemical and petrochemical wastewater treatment. *Wat. Sci. Technol.* **42**(5–6), 163–170.

Vesilind, P.A. (ed.) (2003) *Wastewater Treatment Plant Design*. Water Environment Federation and IWA Publising. pp. 7–1, 7–22.

Visvanathan, C., Ben Aim, R. and Parameshwaran, K. (2000) Membrane separation bioreactors for wastewater treatment. *Crit. Rev. Environ. Sci. Technol.* **30**, 1–48.

von Sperling, M., Freire, V.H. and Chernicharo, C.A.L. (2001) Performance evaluation of a UASB-activated sludge system treating municipal wastewater. *Wat. Sci. Technol.* **43**(11), 323–328.

Wanner, J. (1992) Comparison of biocenoses from continuous and sequencing batch reactor. *Wat. Sci. Technol.* **25**, 239–242.

Wilderer, P.A., Irvine, R.L., Goronszy, M.C., Artan, N., Demoulin, G., Keller, J., Morgenroth, E., Nyhuis, G., Tanaka, K. and Torrijos, M. (2001) *Sequencing Batch Reactor Technology*, Scientific and Technical Report No. 10, IWA Publishing, London, pp. 76.

Yoon, S-H., Kim, H.-S. and Yeom, I.-K. (2004) The optimum operational condition of membrane bioreactor (MBR): cost estimation of aeration and sludge treatment. *Wat. Res.* **38**, 37–46.

Zilverentant, A.G. (1997) Pilot-testing, desing and full-scale experience of a sequencing batch reactor system for the treatment of the potentially toxic wastewater from a road and rail car cleaning site. *Wat. Sci. Technol.* **35**(1), 259–267.

Zitomer, D.H. and Speece, R.E. (1993) Sequential environments for enhanced biotransformation of aqueous contaminants. *Environ. Sci. Technol.* **27**(2), 227–244.

Zwerger, W., Beler-Baykal, B. and Wilderer, P.A. (2000) Application of granular zeolite as support material for biofilms and as temporary sink for ammonium. In: *Proeeedings of the 2nd Imernational Symposium on Sequencing Batch Reactor Teehnology.* IWA Publishing, London, **1**, pp. 92–101.

5

Application of biological treatment systems for nitrogen-rich wastewaters

N. Bernet and M. Spérandio

5.1 INTRODUCTION

Many effluents contain nitrogen in much higher concentrations than municipal wastewaters. In industrial wastewaters, nitrogen can be in the form of organic nitrogen (urea, proteins, etc.) like in food industry effluents and animal wastewaters, in the form of ammonia like in landfill leachate, sludge liquor, wastewater after anaerobic pre-treatment and effluent from different industries such as fertilizer, chemical, metallurgical and mining industries. In more rare cases, industrial wastewaters may contain nitrate (fertilizer industry, nuclear industry). Table 5.1 gives a few examples of industrial nitrogen-rich wastewaters.

Organic nitrogen, when present, is rapidly converted to ammonia (via hydrolysis and ammonification processes) and the treatment processes generally consider ammonium-nitrogen as the form to be removed.

The most widely used process to remove ammonia is biological nitrification–denitrification. This process has been developed for the treatment of municipal

 Advanced Biological Treatment Processes for Industrial Wastewaters edited by F.J. Cervantes, S.G. Pavlostathis and A.C. van Haandel. ISBN: 1843391147. Published by IWA Publishing, London, UK.

Table 5.1 Characteristics of industrial nitrogen-rich wastewaters (all parameters in mg/L except for pH).

	Leachate from municipal solid waste	Anaerobic digester supernatants		Pectin producing industry
Total N	0–3320	1605	–	1600
$N\text{-}NH_4^+$	0–1250	1150	600–700	500
$N\text{-}NO_x$	0–11.26	<1	<10	900
COD	0–195,000	1184	800–1300	8100
SS	140,900	56	<400	800
HCO_3^-	0–1272	5100	–	–
pH	1.5–9.5	8.1–8.4	8.2–8.4	7.6–8
TP	–	12	15	–
BOD_5	480–72,500	230	–	–
Origin and reference	El-Fadel *et al.* 2002	Rotterdam, van Dongen *et al.* 2001	Munich, Arnold *et al.* 2000	Denmark, Pedersen *et al.* 2003

N: nitrogen; COD: chemical oxygen demand; SS: suspended solids; TP: total phosphorus; BOD_5: biochemical oxygen demand.

wastewater generally containing 30–80 mg N/L. Adaptation of conventional nitrification–denitrification to the treatment of effluents containing up to several grams of nitrogen per liter generates high construction investment and operation costs (high oxygen demand for nitrification, Chemical oxygen demand (COD) supply for denitrification). Moreover, nitrification is very sensitive to the presence of toxic compounds such as heavy metals or organic compounds and different methods have been developed to asses the toxicity of industrial wastewaters upon nitrification (Juliastuti *et al.* 2003; Eilersen *et al.* 2004; Ren 2004). Novel cost effective alternatives have been proposed recently and developed in the few past years: nitrogen removal over nitrite, Anaerobic ammonium oxidation (Anammox), simultaneous nitrification–denitrification (SND). These concepts have been designed using either suspended biomass systems (activated sludge (AS) process including sequencing batch reactor, chemostat) or biofilm reactors (fixed-bed, mobile-bed).

These new concepts are of particular interest in the case of industrial effluents with a low COD/N ratio, for which addition of external carbon sources is necessary to cover the demand for organic substrate in denitrification.

The characteristics of the wastewater to be treated will be decisive on the choice of the strategy used. From the main parameters, in addition of the COD/N ratio, alkalinity, pH and temperature can be cited.

5.2 SUSPENDED BIOMASS PROCESSES

5.2.1 AS process

5.2.1.1 Configuration and design criteria

AS process is commonly designed and operated to achieve nitrification and denitrification. This can be obtained in a single sludge system (mixed autotrophic and heterotrophic bacteria) or in two separated sludge systems (Figure 5.1).

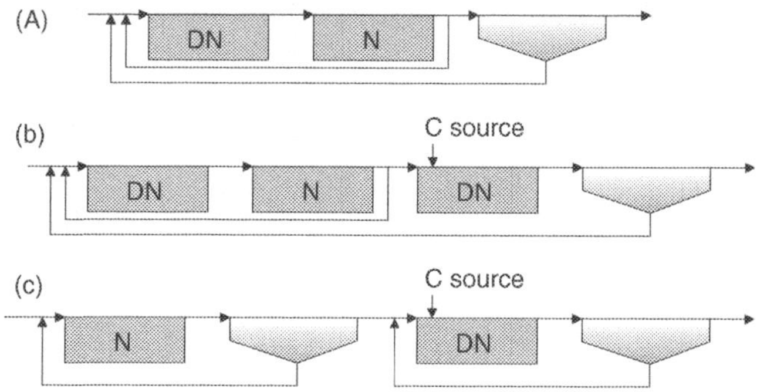

Figure 5.1 Suspended growth processes for nitrification/denitrification (N/DN). (A) MLE system, (B) pre and post-denitrification system, and (C) separated sludge system.

Performing nitrification and denitrification with a unique suspended biomass is a commonly used strategy. This needs to alternate aerobic and anoxic conditions, in multi-reactor systems or in single-reactor systems by switching on/off aerators. On the other hand, separated sludge systems allow to combine either two AS systems, or an AS system with an attached biomass reactor, like a tertiary nitrification step (e.g. trickling filters, biofilters, fluidized bed, Biofilm airlift suspension). This second option is useful to upgrade existing plant when surface constraints are high (van Benthum *et al.* 1998a). More recently, suspended sludge system has been also upgraded by addition of fluidized carriers, which can support nitrifiers, leading to hybrid suspended and attached-growth reactor (van Benthum *et al.* 1997; Ochoa *et al.* 2002; Pambrun *et al.* 2004).

Basically, selection of a biological nutrient removal (BNR) process configuration depends on the wastewater characteristics (nitrogen forms, biodegradability of organic matter, COD/N ratio of wastewaters), the effluent requirement and the volume constraints.

When the influent organic matter is valuable for denitrification and if COD/N ratio is higher or closed to the carbon needs for denitrification (e.g. 3.7 g COD/g $N\text{-}NO_3^-$ for nitrate denitrification or 2.3 g COD/g $N\text{-}NO_2^-$ for nitrite denitrification based on methanol stoichiometry), the pre-denitrification principle is applied (see Figure 5.1). In the so-called MLE process (Modified Ludzack–Ettinger), nitrate produced in the aerobic nitrification tank are recirculated through an internal loop to the anoxic zone. The recirculation flow rate is calculated to reach a desired nitrate concentration in effluent depending on the reject constraints (commonly designed between 200% and 500% of the influent flow rate). Pre-denitrification configuration could be chosen when nitrate or nitrite are a major nitrogen source in wastewater. In that case, internal recirculation loop could be unnecessary. A limit of the MLE process is that a defined fraction of the nitrate produced in the aerobic tank are not denitrified and still present in the rejected water.

Nitrate and nitrite removal can also be achieved in a post-denitrification zone (Figure 5.1). This can be adapted if the carbon source of wastewater is unfavorable for denitrification or if stringent nitrogen constraints are imposed on rejected water. If no secondary carbon is added in the post-anoxic zone, denitrification

Table 5.2 Heterotrophic denitrification rate with different carbon sources.

Carbon source	Denitrification rates (20°C) (mg N-NO_3^-/(g VSS) h)
Easily biodegradable (VFA, simple carbohydrates, methanol, ethanol)	5–10
Slowly biodegradable	2–4
Endogenous activity	0.2–1

rate will be low based on anoxic endogenous respiration of heterotrophic bacteria. For this reason, post-denitrification is classically performed with the help of a readily biodegradable substrate (Table 5.2).

The principal design criteria for nitrogen removal in As process are the sludge retention time (SRT) and the aerobic fraction of biomass. Both are chosen in order to maintain aerobic autotrophic bacteria responsible for nitrification, considering the minimal annual temperature of the process. The aerobic SRT is directly linked to the observed growth rate ($\hat{\mu}_a$) and the decay rate (b_a) by the following expression:

$$SRT_{\text{aerobic}} = \frac{1}{\hat{\mu}_a - b_a} \quad \text{with} \quad \hat{\mu}_a = \hat{\mu}_{a,20^\circ C}\,\theta^{(T-20)}$$

Therefore, the minimal aerobic SRT is temperature dependent, and the chosen process SRT is generally at least two or three times higher than the minimal value (Metcalf and Eddy 1991). Obviously, the maximal growth rate of a given sludge can be reduced by the presence of inhibiting compounds and this aspect should be evaluated for each industrial wastewater.

In comparison with the suspended sludge process treating domestic wastewater, a sludge treating industrial nitrogen-rich effluent can contain a very high concentration of autotrophic bacteria. Like the autotrophic to heterotrophic biomass ratio, the nitrification rate mainly depends on the wastewater COD/N ratio, the aerobic sludge age and the temperature. Consequently, for wastewater with a very low COD/N ratio, strictly aerobic, the possible applied nitrogen load and the maximal specific nitrification rate can reach very high value, up to 2 kg N-NH_4^+/m-day and 0.5–1 g N-NH_4^+/(g SS)-day, respectively at 30°C (Table 5.3).

In the case of unfavorable influent organic matter for denitrification (low COD/N ratio, or low biodegradability), as the organic carbon for denitrification is mainly supplied by an external source, it can be proposed to design a separated-biomass system, with two successive AS processes in series (Figure 5.1C). The first process is designed for nitrification (and carbon removal if necessary), and the second one is designed for post-denitrification (and removal of slowly biodegradable substances). The major advantage of this configuration is to maintain an autotrophic-rich biomass in the first process and to make possible the separated optimization of nitrification and denitrification stage. However, the major disadvantage is a greater number of unit processes required than for combined system.

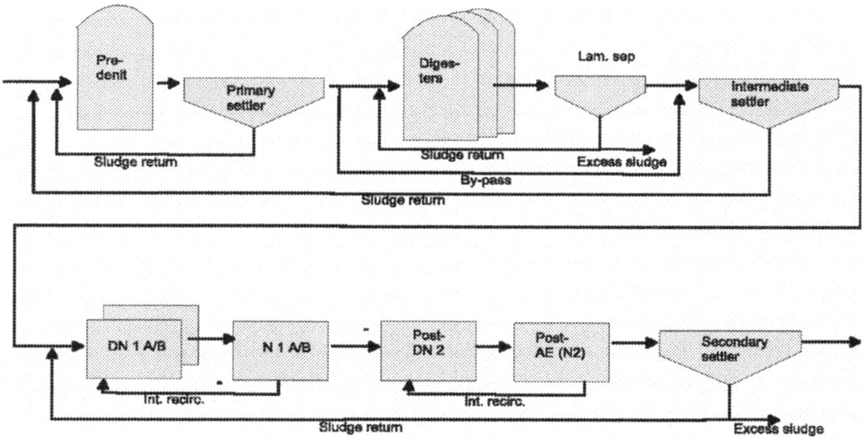

Figure 5.2 Schematic flow sheet of CPKelco WWTP (Pedersen *et al.* 2003).

5.2.1.2 *Case study of CPKelco, pectin producing factory*

The Wastewater Treatment Plan (WWTP) of the texturising producer CPKelco is one of the largest industrial plants in Denmark, treating an average flow of 3400 m^3/day, and described in details by Pedersen *et al.* (2003). Wastewater characteristics are summarized in Table 5.3. The total nitrogen load is composed mainly of oxidized form (800–1000 mg N-NO_x/L) and ammonia (500 mg N/L). The schematic diagram of the plant is shown in Figure 5.2. Organic matter, principally made of easily biodegradable carbohydrates, is first used in a pre-denitrifying AS process for removed NO_x of the influent. Then, a fraction of the flow is treated in three anaerobic digesters followed by lamella settler, whereas another fraction is by-passing to an intermediate settler. All the wastewater from this settler is led to a multi-basin AS process, with four successive anoxic and aerobic zones (pre-DN/N/post-DN/N).

Initial pre-denitrifying reactor receives a NO_x load of about 2 kg N-NO_x/m^3-day at 40°C. Nitrifying capacity of the last As process is summarized in Table 5.3. The total volume of denitrifying zones (DN1 and DN2) of the last AS process, working at 38–40°C, are designed for treating a nitrate or nitrite load of about 0.75 kg N-NO_x/m^3-day, hence a mean specific denitrifying rate of about 6 mg N-NO_x/(g SS)-h. Due to the easily biodegradable organic matter of wastewater, and with the help of an external carbon complement when necessary, complete denitrification was obtained. The main reported problem was partial loss of biomass and consequently loss of nitrification activity (due to probable decrease of SRT). This problem was due to filamentous bacteria, which are known to develop on sugars, and it was solved by using alcohol instead of carbohydrate as a secondary carbon source for denitrification. At the end, good effluent quality was reached. The total nitrogen concentration was lower than 45 mg/L (97% removed), COD was stabilized at about 500 mg/L (94% removed) and SS decreased close to 50 mg/L (94% removed).

Table 5.3 Nitrification performance of suspended sludge processes treating nitrogen-rich effluents.

Reference	Process	Effluent type	COD/ N ratio	Applied load (kg N/ m^3-day)	Nitrification rate (g N/ (g SS)-day)	T. (°C)
Pedersen *et al.* 2003	AS N/DN	Pectin Industry, Carbohydrate, ammonia	5.4	0.32	0.1–0.3	39
Lai *et al.* 2004	SBR, N/DN	Sludge digester supernatant	1.5–1.9	1.05–1.2	–	20
Pambrun *et al.* 2004	SBR, N	VFA, ammonium	0.5–1	1.5–2.0	0.5	30
Arnold *et al.* 2000	SBR, N	Sludge digester supernatant	1.5–2	0.6–0.8	0.11–0.14	32
Fux *et al.* 2003	SBR, N/DN	Sludge digester supernatant	0.9	1.4	0.16	35
Hellinga *et al.* 1998	Chemostat, N/DN	Sludge digester supernatant	1.1	1	1	35

5.2.2 Sequencing batch reactor

Use of sequencing batch reactor for treating nitrogen-rich wastewater has been largely reported in the last decade (Arnold *et al.* 2000; Fux *et al.* 2003; Lai *et al.* 2004). This process can be applied for nitrification, nitrification/denitrification (N/DN) or only denitrification. The major advantage of this process is the design simplicity, due to the absence of a separate settler and the associated cost. This process generally allows reaching low ammonia concentration in the effluent, high nitrification and denitrification rates and is known to produce sludge with very good settling properties.

Concerning design parameters, when sequential batch reactor (SBR) is applied to nitrogen-rich wastewater, a major aspect to take into account is the possibility of inhibition due to batch feeding mode and consequent high pollutant concentration maintained in the reactor during a given time after feeding.

Most SBR processes are operated in step-feed mode with a given number of feeding period, and successive anoxic and aerobic periods, followed by the settling and the withdrawing periods (Figure 5.3). Sludge is generally wasted before settling period. Total cycle time is variable, commonly about 5–10 h, depending on the hydraulic retention time (HRT), the concentration of influent and the settling capacities. Adaptation of the number of feeding sequences allows limiting the possible inhibition by, for example, ammonia during nitrification. Table 5.3 shows different examples of performance obtained in SBR processes with the associated operating conditions.

5.2.3 Nitrite build-up in suspended biomass process

5.2.3.1 Factors influencing nitrite accumulation in nitrification

A possible way of N/DN optimization is to carry out a partial nitrification, consisting in stopping the oxidation of ammonia at the stage of nitrite, and then converting

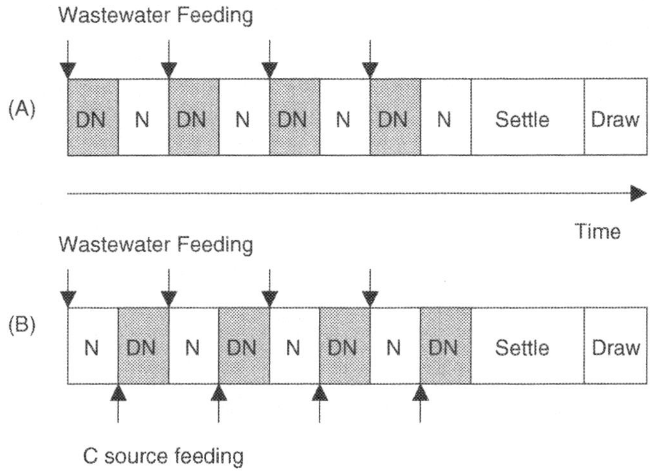

Figure 5.3 Example of step-feed options for sequencing batch reactor time sequence. (A) For wastewater with organic matter and, (B) for wastewater with only nitrogen pollution and with external source for denitrification.

nitrite to gaseous nitrogen by heterotrophic denitrification. By this way, 25% of the consumed oxygen is saved, as well as 40% of the carbon required for denitrification (which becomes closed to 2 g COD/g $N_{removed}$ with methanol) (Voets *et al.* 1975; Turk and Mavinic 1986). This would be of particular interest when removing nitrogen from wastewaters with an unfavorable COD/N ratio.

In suspended biomass, effective inhibition of nitrite oxidation depends on the concentration of NH_4^+/NH_3, NO_2^-/HNO_2 and/or pH, oxygen, temperature. Chemostat reactor and SBR are the most commonly proposed processes. In the first one, nitrite-oxidizing bacteria (NOB) can be washed out by selecting optimal HRT and working at temperature of 30–40°C. In the second one, NH_3 inhibition, and/or oxygen limitation, helped by temperature, can be the major mechanisms used for avoiding the growth of NOB.

Inhibition of nitrite and nitrate oxidizers by free ammonia NH_3 and HNO_2 have been widely studied. Due to possible acclimatization, inhibition concentration level can significantly vary from a process to another (Anthonisen *et al.* 1976; Abeling and Seyfried 1992). Overall it is now obvious that NOB are more sensitive to inhibition than ammonia-oxidizing bacteria (AOB). For example free ammonia (NH_3) inhibits NOB at concentration of about 0.5–2 mg N-NH_3/L, whereas AOB inhibition starts at concentrations higher than 10 mg N-NH_3/L.

Figure 5.4 shows the influence of total ammonia nitrogen (TAN) concentration and pH on the observed maximal growth rate of NOB at 30°C. For example, in a bioreactor in which concentration is around 300 mg/L, and for a pH of 8, the observed growth rate is around 0.1/day, and then NOB can be washed out if sludge age is maintained equal or lower than 10 days.

Temperature is also a parameter which largely influences nitrite accumulation potential as the nitrite-oxidizer maximal growth rate increase more slowly than

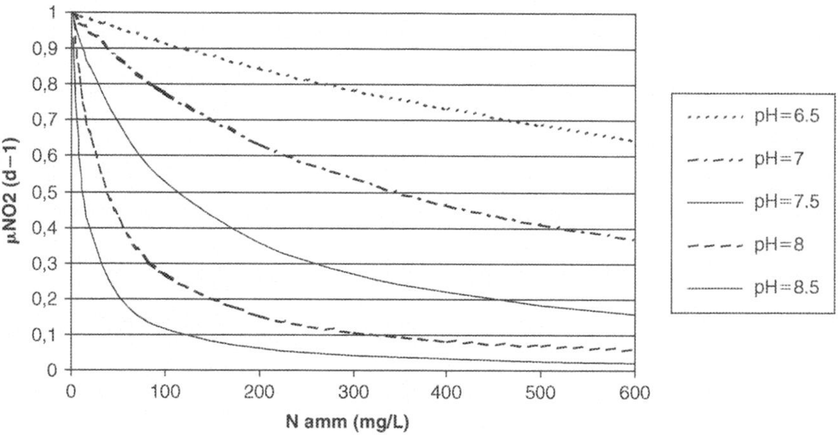

Figure 5.4 Theoretical evolution of observed growth rate of NOB at 30°C as a function of total ammonium nitrogen and pH.

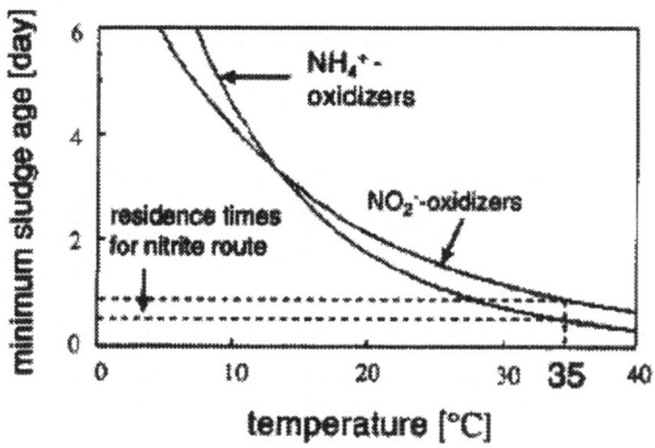

Figure 5.5 Influence of temperature on the minimum sludge age for maintaining ammonium and nitrite oxidizers in nitrifying reactors (Hellinga *et al.* 1998).

ammonium oxidizers with temperature (Hellinga *et al.* 1998). Consequently, when the temperature is around 30–35°C, it is possible to select a sludge age which allows to maintain AOB and to wash out NOB (Figure 5.5). It is the basis of the nitritation application in chemostat process without sludge retention.

Low dissolved oxygen (DO) concentration, around 0.2–0.5 mg/L, is another possible condition for limiting NOB growth (von Münch *et al.* 1996; Zeng *et al.* 2003). This solution leads to SND in the same reactor and offers the potential to save the costs for a second anoxic-aerated tank. However, lower nitrification rates are consequently obtained, due to partial limitation of AOB, and nitrous oxide (N_2O) production is a possible problem (Zeng *et al.* 2003). It is suggested that nitrite are produced and directly consumed inside the microbial aggregates and hence this process would be particularly adapted to granular and biofilm systems (see Section 5.2).

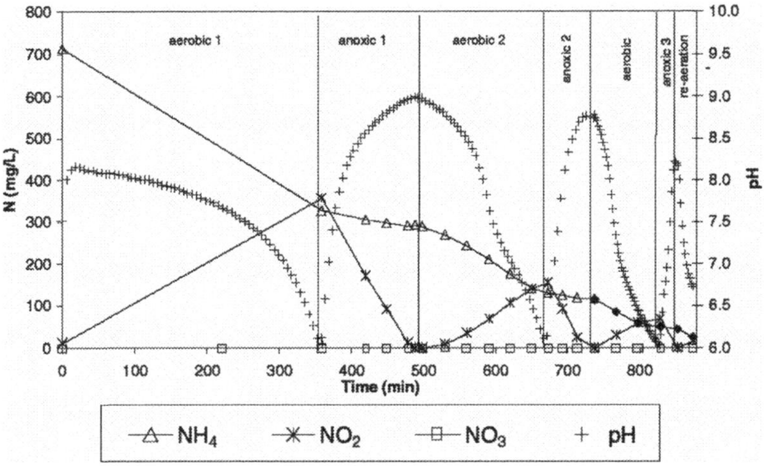

Figure 5.6 Typical cycle of SBR process treating effluent from sludge anaerobic digester at 30°C (Lai *et al.* 2004).

5.2.3.2 SBR application for nitritation–denitritation ("nitrite route")

Applications of SBR for nitritation–denitritation of effluents from sludge anaerobic digestion (supernatant of dewatering process) have been widely reported (Lai *et al.* 2004). On Figure 5.6, a typical SBR cycle is shown with three successive aerobic and anoxic phases.

Effluent was fed to the reactor only once at the beginning of the cycle. Hence, the ammonia nitrogen concentration reaches 700 mg/L and inhibits NOB (30°C, pH 8.2; see Figure 5.4). As bicarbonates are limited in digester sludge effluent, with a classical molar HCO_3^-/N ratio of 1, nitrification of only 50% of total ammonia is possible at each aerobic phase. Anoxic period allows denitrification (with ethanol as carbon source in the presented case) and then produces sufficient alkalinity to nitrify 50% more ammonia during the following aerobic phase. After three successive cycles, nitrification of more than 88% of ammonia was obtained. This sequence was tested in reactor at 20°C, and after a stabilization period, complete nitrogen removal was obtained with a HRT of about 1 day (Figure 5.7).

Due to the production of nitrite as the only nitrification product, the COD/N ratio requirement for complete denitrification was only 1.5–1.9, compared to 3–4, in the case of nitrogen removal via nitrate. Stable nitrite build-up in SBR at 20°C is less evident than at high temperature. Unstable pH values during a cycle and alternatively changes from NH_3 to HNO_2 inhibition is a probable solution for long-term limitation of NOB growth at low temperature (Lai *et al.* 2004).

5.2.3.3 Chemostat application for nitritation–denitritation ("nitrite route")

A single stirred bioreactor with no biomass retention can be proposed for treating nitrogen-rich wastewaters. The advantage of this system (patent SHARON®

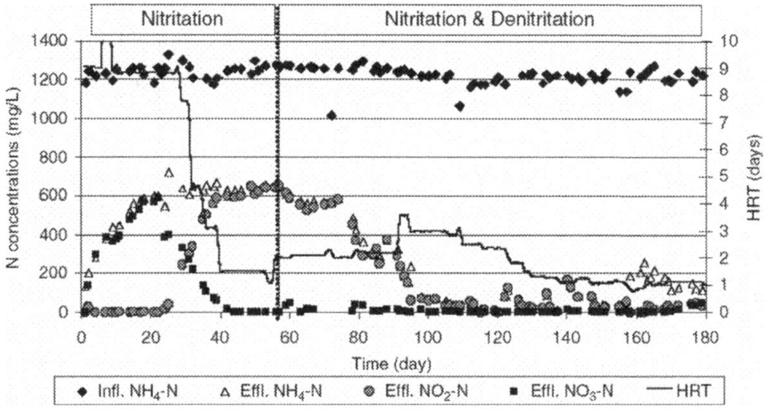

Figure 5.7 Performance of SBR process treating effluent from sludge anaerobic digester at 20°C (Lai *et al.* 2004).

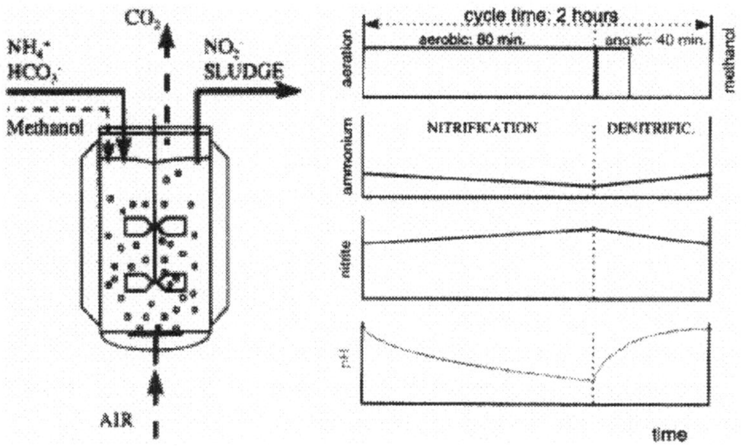

Figure 5.8 Description of SHARON® process for nitritation–denitritation (Hellinga *et al.* 1998).

process, Hellinga *et al.* 1998; van Dongen *et al.* 2001) is that SRT is controlled by HRT. Thus, at 35°C, it is possible to maintain only AOB if a HRT of about 1 or 2 days is stabilized in the system (Hellinga *et al.* 1998; Fux *et al.* 2002; 2003). Denitrification can be achieved by switching off aeration during a given time with simultaneous addition of a carbon source (Figure 5.8). Due to the continuous feeding, ammonia and nitrite are accumulated, respectively during anoxic and aerobic period. Therefore, the maximal performance depends on the inlet concentration, dilution rate, cycle time and influent alkalinity.

5.2.3.4 Design comparison of SBR and chemostat processes

As the chemostat reactor is operated at a given HRT, the volume demand of this process only depends on wastewater flow. As a comparison, in SBR process, the

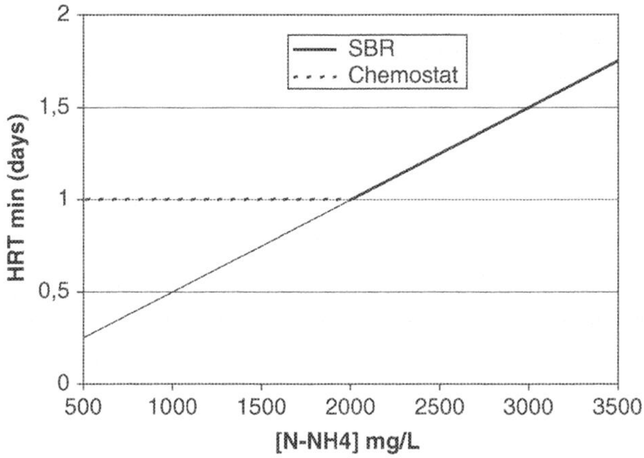

Figure 5.9 Comparison of HRT in chemostat and SBR processes for nitritation at 30°C.

maximal volumetric load depends mainly on oxygen maximal transfer capacities. For example, at 30°C, considering that maximal reported oxygen transfer rate is close to 200 g/m^3-h (which corresponds to a K_L value of about 1000/d), it is possible to treat a maximal nitrogen load of about 2.5 kg N/m^3-day. Consequently, the minimal aerobic HRT of SBR process increases with nitrogen concentration of wastewater (Figure 5.9). From this, it comes that the SBR process can be more compact than chemostat process if the wastewater concentration is lower than 1500 mg/L, both systems need similar volume when concentration is around 2000 mg/L, and chemostat process will be difficult to apply for higher concentration due to high oxygen demand. If denitrification is included, HRT needed are fairly twice higher but the comparison of processes is the same. Fux *et al.* (2003) confirmed that chemostat volume is twice bigger than SBR volume when treating a wastewater with a concentration of 1000 mg N/L, leading to higher investment costs (0.51/kg N for SBR, €0.77/kg N for chemostat).

5.2.4 Optimal control of alternated systems

Based on basic on-line sensor like DO, pH, redox potential, or more sophisticated one like oxygen uptake rate (OUR), nitrification and denitrification phases can be controlled in order to maximize the removed nitrogen rate of the process.

Concerning pH, this parameter gives a clear indication of nitrification when decreasing and indicates denitrification when increasing. It is then possible to switch from aerobic to anoxic period when a minimal threshold value is reached in the case of bicarbonate limited wastewater like digester supernatant, as it is shown in Figure 5.10 (Lai *et al.* 2004).

When wastewater alkalinity is high, pH change range is lower and the interpretation of signal depends on CO_2 stripping and specific strategies should be applied, based on derivatives or "valley" detection. DO and redox potential profiles also show rapid drops and bending points when ammonia and nitrate are depleted (Figure 5.10), and then are commonly used for optimal control of nitrification and denitrification (Wareham *et al.* 1993; Plisson Sauné *et al.* 1996;

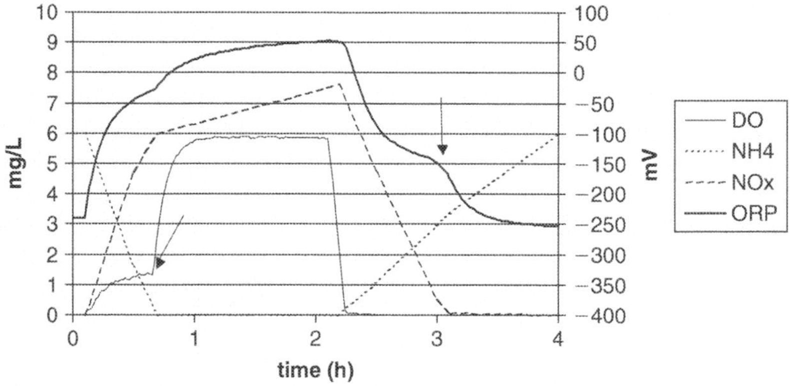

Figure 5.10 Detection of end of nitrification and denitrification by DO and oxidation–reduction potential (ORP). SBR process treating leachate at 25°C.

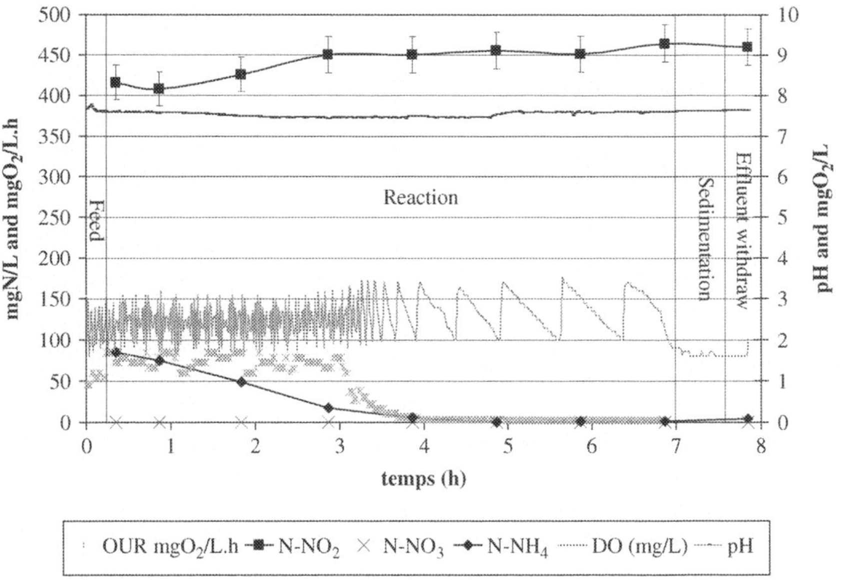

Figure 5.11 Example of nitritation cycle followed by OUR measurement (Pambrun *et al.* 2004).

Mauret *et al.* 2001). Adaptation of aerobic and anoxic period time allows to guaranty maximal N removal despite instability of nitrification and denitrification rate as well as wastewater characteristics.

Moreover, if nitrite build-up is performed in SBR process by means of NH_3 inhibition, the detection of ammonia depletion is crucial in order to avoid too long aerobic period which allows the nitrite oxidizers to develop. For this application, on line estimation of OUR by frequent switch of the aeration is very helpful (Pambrun *et al.* 2004; Third *et al.* 2004) as a rapid drop of activity is observed at the end of nitrification (Figure 5.11).

5.3 BIOFILM PROCESSES

As autotrophic organisms are strongly involved in biological nitrogen removal, including nitrifying bacteria (ammonia and nitrite oxidizers) and anaerobic ammonium oxidizers, the use of biofilm processes is of particular interest. Anammox was discovered in a biofilm process, a denitrifying fluidized-bed reactor (Mulder *et al.* 1995).

Biofilm processes benefit of the presence of gradients, either in the reactor in the case of plug-flow systems like some biofilters, or in the biofilm itself, especially oxygen gradients, from the outer layer to the inner part of the biofilm.

5.3.1 Nitrite build-up in nitrifying biofilm

In biofilm processes, unlike in the SHARON system, HRT and sludge age or SRT are decoupled. This allows higher nitrite production rates. However, nitrite oxidizers are not so easily removed as in a chemostat.

Free ammonia (NH_3) is the most commonly reported method to inhibit nitrite oxidation (Section 5.2.3.1.). Nitrogen removal via-nitrite was studied for the treatment of the effluent of a digested potato starch effluent and compared with conventional nitrification–denitrification (Abeling and Seyfried 1992). The nitrification tank was filled with a plastic support. pH and free ammonia inhibition were used to control nitrite accumulation. Concentrations of 1–5 mg NH_3/L were shown to inhibit nitratation, but not nitritation. The carbon needs for denitrification were 60% in comparison with denitrification via nitrate.

Free ammonia was shown to be the major parameter leading to nitrite build-up in a submerged biofilter, when compared with other parameters such as pH, temperature, alkalinity, ammonium load (Fdz-Polanco *et al.* 1994; 1995; 1996; Villaverde *et al.* 1997a, b). However, this inhibition is also dependent on the concentration of nitrifying bacteria and on the threshold concentration of free ammonia causing inhibition (Rols *et al.* 1994; Villaverde *et al.* 2000). To achieve free ammonia inhibition in upflow submerged biofilm reactors, Ceçen (1996) showed that free ammonia inhibition could be enhanced by operating at low DO concentration, high bulk ammonium and pH values.

High temperatures that have been used to selectively wash out NOB in the SHARON process can be used in biofilm systems, combined with other factors like oxygen limitation and/or free ammonia inhibition, to favor nitrite build-up (Bougard *et al.* 2005).

Finally, low DO concentrations have been shown to favor nitrite build-up, probably because of a difference of affinity constants for oxygen between AOB and NOB. In an airlift reactor, Garrido *et al.* (1997) achieved a complete conversion of ammonia, with 50% nitrite accumulation, at a DO concentration between 1 and 2 mg/L. Bernet *et al.* (2001) observed a complete conversion of ammonia at a DO concentration of 0.5 mg/L with a conversion of 80% to nitrite. A stable nitrite accumulation has been recently obtained in a granule-type airlift with DO control. After 90 days of operation at low DO concentration of less than 1.0 mg/L, a nitrite conversion rate of 2.6 g NO_2-N/L-day could be achieved (Tokutomi 2004).

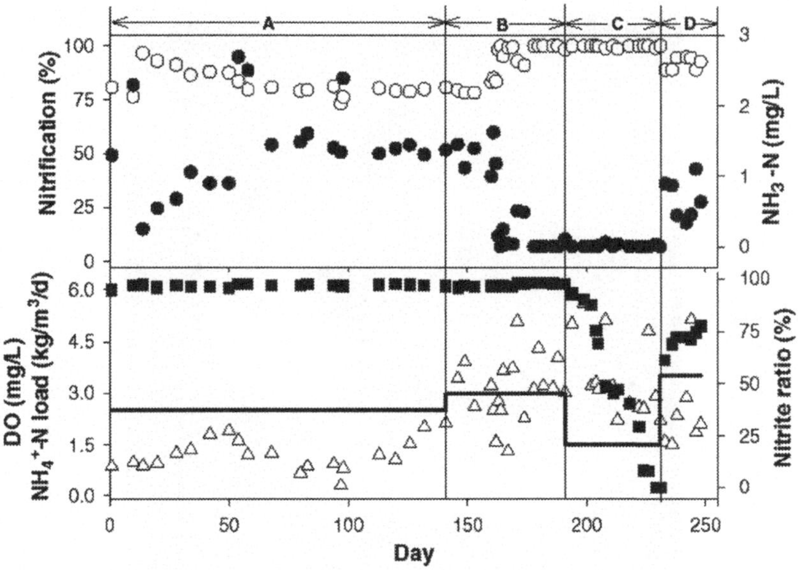

Figure 5.12 Profiles of free ammonia (NH_3-N, ●), DO (△), Nitrification efficiency (○), NH_4-N load (—) and Nitrite ratio (■) as function of time in a biofilm airlift reactor (Kim *et al.* 2003).

The best results can be obtained when different factors are combined. Thus, Kim *et al.* (2003) obtained nitrite build-up in a biofilm airlift reactor combining free ammonia inhibition and low DO concentration. The reactor was operated at a loading rate of 1.5–3.5 kg N/m^3-day (Figure 5.12).

During period C, the loading rate was decreased, causing a decrease of free ammonia and a recovery of nitrite oxidation in 40 days. This slow recovery was attributed to oxygen supply limitation to the inner layer of NOB in the biofilm. Fluorescence *in situ* hybridization analysis of cryosectioned nitrite accumulating nitrifying biofilm showed that the β-subclass of *Proteobacteria*, where ammonia oxidizers belong, was distributed outside of the biofilm whereas the α-subclass of *Proteobacteria*, where nitrite oxidizers belong, was found mainly in the inner part of the biofilm (Figure 5.13). The spatial distribution showed that nitrite oxidizers are more susceptible to oxygen limiting conditions than ammonia oxidizers.

However, Fux *et al.* (2004a) reported on difficulties in maintaining long-term partial nitritation of ammonium-rich sludge digester liquids in a lab-scale moving-bed. The maximum nitrite production rate amounted to 2.7 g NO_2^--N/m^2-d (3 g O_2/m^3-day, 30.5°C), thus doubling the dilution rate compared to CSTR operation with suspended biomass for a supernatant with 700 g NH_4^+-N/m^3. Whenever, the available alkalinity was fully consumed, an optimal amount of nitrite was produced. However, a significant amount of nitrate was produced after 11 months of operation, making the effluent unsuitable for anaerobic ammonium oxidation. Due to the SRT is relatively long in biofilm systems, slow growth of nitrite oxidizers occurs. None of the selection criteria applied – a high ammonium loading

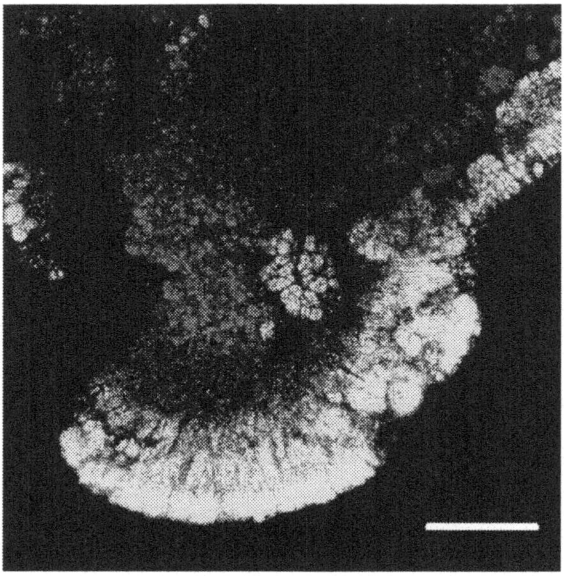

Figure 5.13 Simultaneous *in situ* hybridization of nitrite accumulating biofilm from the reactor with CY-3 labeled probe Alf 1b and fluorescein isothiocyanate (FITC) labeled probe Bet 42a. Cells of α-subclass of *Proteobacteria* are in the darker inner region; cells of β-subclass of *Proteobacteria* are in the brighter outer region. Bar = 50 μm (Kim *et al.* 2003).

rate, high free ammonia or low oxygen concentration – led to selective suppression of nitrite oxidation. The authors recommend a completely mixed stirred tank reactor (CSTR) or SBR with suspended biomass for full-scale operation.

Nitrite produced in a nitritation process can be removed either by conventional denitrification, with savings of organic carbon supply of 40% compared with nitrate denitrification and, as a consequence, a lower sludge production, or by autotrophic nitrite reduction.

5.3.2 Nitritation–denitritation in biofilms

Nitrification and denitrification can be integrated in one high-rate and compact biofilm reactor (van Loosdrecht *et al.* 2000). The wastewater to be treated has to be recycled between the aerobic and anaerobic compartments of a single reactor. Indeed, nitrification–denitrification in a continuously fed single aerobic compartment is not possible because of the stratification of the biofilm, because of their lower growth rate and yield, nitrifying bacteria tend to grow inside the biofilm, whereas heterotrophic denitrifying bacteria form a layer at the outside of the biofilm, therefore, limiting nitrifying activity by lack of oxygen.

Combining nitrification and denitrification in two compartments of the same reactor or in a sequencing batch reactor helps to out-compete nitrite oxidizers from the system. The nitrite is removed by denitrification at high rate, thus out-competing nitrite oxidizers. This strategy has been tested, coupling a biofilm airlift suspension reactor for nitrification with a denitrifying chemostat (van Benthum *et al.* 1998a), or coupling both reactions in an intermittently aerated biofilm airlift reactor (van Benthum *et al.* 1998b). Low COD/N ratios of 3.0 and 2.6 kg COD/kg N were obtained, respectively in accordance with the "nitrite route".

5.3.3 Nitritation-Anammox in biofilms

The recently discovered Anammox process offers the opportunity to develop fully autotrophic ammonium removal systems (Jetten *et al.* 2002). Aerobic and anaerobic ammonium oxidation can be combined to remove ammonium as molecular nitrogen. Both processes can be coupled in a single oxygen-limited step like in CANON (completely autotrophic nitrogen removal over nitrite), OLAND (oxygen limited autotrophic nitrification–denitrification), and the "aerobic deammonification" process. Another possibility is to combine Anammox with partial nitrification in two different reactors like in the SHARON–Anammox process.

5.3.3.1 SHARON–Anammox

Nitrogen removal from a municipal sludge digestion effluent has been carried out by combination of a partial nitrification process (SHARON) and Anammox (van Dongen *et al.* 2001).

Anammox process has been carried out in a granular sludge SBR process fed with the effluent from a SHARON reactor (Table 5.4).

In a lab-scale gas-lift reactor inoculated with granular biomass from an Anammox SBR, a N-removal rate as high as 8.9 kg N/m^3-day has been reported (Sliekers *et al.* 2003). A recent study showed that flotation problems can occur with Anammox granular sludge, in SBR and in gas-lift reactor, when the loading rate exceeds the maximum specific Anammox activity of the sludge (Dapena-Mora *et al.* 2004a).

In 2002, the first full-scale Anammox reactor was started at Dokhaven WWTP Rotterdam (The Netherlands). The technology applied is a gas-lift loop type granular reactor. This plant has a capacity to treat 500 kg N/day.

Anammox was tested in fixed-bed reactors (Fux *et al.* 2004b). High specific nitrogen elimination rates of 3.5 kg N/m^3-day but after a start-up periods of more

Table 5.4 Conversion in a granular sludge SBR Anammox reactor fed with a nitrified effluent from a SHARON reactor (van Dongen *et al.* 2001).

Parameter	Unit	Steady-state operation
Test period	Day	10
Influent NH_4-N	kg/m^3	0.55 ± 0.10
Influent NO_2-N	kg/m^3	0.60 ± 0.10
NH_4-N conversion	kg/m^3-day	0.35 ± 0.08
NO_2-N conversion	kg/m^3-day	0.36 ± 0.01
Effluent NO_2-N	kg/m^3	0
Volumetric conversion	kg N_{tot}/m^3-day	0.75 ± 0.20
Sludge conversion	kg N_{tot}/(kg SS-day)	0.18 ± 0.03

than 1 year. Up-flow anaerobic sludge blanket reactors (UASB) were also applied to Anammox (Schmidt *et al.* 2004) achieving 99% ammonia removal at a loading rate of 0.14 kg N/m^3-day.

Fux and Siegrist (2004) compared nitrogen removal from sludge digester liquid by conventional nitrification–denitrification and partial nitritation-Anammox, from environmental and economical points of view. It appears that if similar performances of 85–90% N removal can be achieved with both systems, the second option is more sustainable (Figure 5.14): no greenhouse gas production (CO_2, N_2O), no organic carbon needed and, therefore, a very low sludge production.

Mulder (2003) comparing different nitrogen removal systems, including conventional nitrification–denitrification, nitrification–denitrification via nitrite and autotrophic N-removal, using six indicators of sustainability (sludge production, energy consumption, resource recovery, area requirement and N_2O emission) reached the same conclusion.

As it is detailed in Table 5.6, partial nitritation Anammox costs are lower, mainly because no organic carbon source is needed and, therefore, less excess sludge is produced. Oxygen demand is lower due to partial ammonium oxidation to nitrite. These calculations are only valid for big WWTPs and the problem of the start-up of the Anammox unit, together with the lack of experience of the process at full scale, have to be considered.

The main practical drawback of this autotrophic nitrogen removal is the extremely low growth rate of the Anammox organisms (doubling time of 11 days). A few studies report on efficient enrichments of Anammox biomass from municipal WWTP sludge (Toh *et al.* 2002; Dapena-Mora *et al.* 2004b; Zheng *et al.* 2004) or UASB granules (Thuan *et al.* 2004).

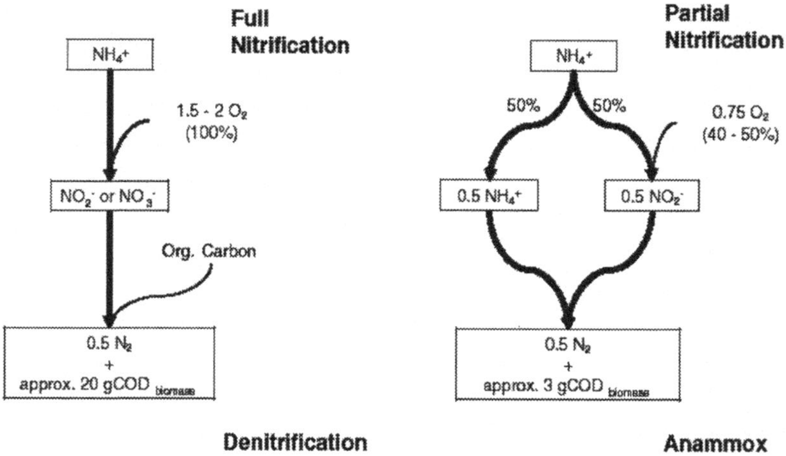

Figure 5.14 Comparison of oxygen and organic carbon consumption of classical N/DN (left) and partial nitritation/Anammox (right). More sludge is produced by the classical process (yield of heterotrophic denitrification: 0.3 g COD-biomass/g COD_{dosed}). The acid consumption for the Anammox process is neglected (Fux and Siegrist 2004).

Pynaert *et al.* (2004) propose a procedure for start-up of autotrophic nitrogen removal rotating biological contactor (RBC) by sequential addition of aerobic nitrifying sludge and, once a well nitrifying biofilm has developed, of an anaerobic granular sludge as a source of Anammox organisms. After 250 days of start-up, N elimination was about 1.8 g N/L-day.

5.3.3.2 CANON

The idea of the CANON system is to combine partial nitrification to nitrite and Anammox in a single reactor (Third *et al.* 2001). This concept was studied in SBR but the gas–liquid mass transfer of oxygen was limiting, leading to relatively low N-removal rates. However, the system was shown to be very stable during the period of test (3 months) (Sliekers *et al.* 2002).

Application of CANON in a gas-lift reactor led to a N-conversions rate of 1.5 kg N/m^3-day, what is 20 times higher compared to the removal rates previously obtained (Sliekers *et al.* 2003). Anammox has the advantage that these organisms can be located in the same biofilm as the aerobic nitrifiers (Jetten *et al.* 2002), which is not the case for heterotrophic denitrification (see Section 5.3.2).

A mathematical model combining nitrification and Anammox in a biofilm reactor (CANON) was developed (Hao *et al.* 2002). Simulations were conducted to test the sensitivity of different parameters to the performance of the system. It was concluded that the one step process is probably not optimal if high ammonium removal efficiency is required. In this case, a two-step process like SHARON–Anammox will be more efficient.

5.3.3.3 "Deammonification" and OLAND in RBC

Nitrogen losses have been observed in full-scale nitrifying RBC treating landfill leachate at Mechernich in Germany (Baumgarten and Seyfried 1996) and at Kölliken in Switzerland (Siegrist *et al.* 1998) under low DO conditions. These important losses, between 40% and 70% of the influent load (Hippen *et al.* 2001), cannot be explained by heterotrophic denitrification due to the low COD removal: $COD_{removed}/N_{removed}$ is 1.3 at Meternich (Helmer *et al.* 1999).

The term "aerobic deammonification" was given by Hippen *et al.* (1997) to the direct biological conversion of ammonia to molecular nitrogen in (micro) aerobic conditions. Then, because there are both aerobic conditions in the outer biofilm and anoxic conditions at the core of the biofilm, the term "aerobic/anoxic deammonification" was adopted by Helmer *et al.* (1999).

Due to the potential interest of this process for the treatment of nitrogen rich wastewater, research was developed to understand the nitrogen conversion processes in the biofilm. It was demonstrated that nitritation was performed by ammonia oxidizers in the outer part of the biofilm, in combination with Anammox in the deeper layer by microorganisms similar to those responsible for the Anammox process (Helmer-Madhok *et al.* 2002; Egli *et al.* 2003). The Anammox organisms identified in the plants of Mechernich and Kölliken were both close to *Candidatus Kuenenia stuttgartiensis* (Helmer-Madhok *et al.* 2002; Egli *et al.* 2003).

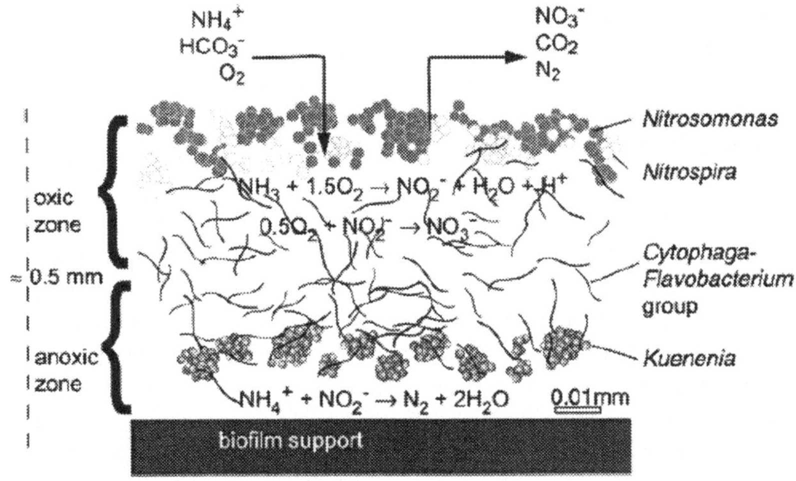

Figure 5.15 Schematic drawing of the community structure of the biofilms in the Kölliken BC and of the main nitrogen conversion reactions (Egli *et al.* 2003).

A scheme of the microbial organization of the biofilm, proposed by Egli *et al.* (2003), is presented in Figure 5.15.

Pynaert *et al.* (2003) propose a similar hypothesis for a highly loaded lab-scale RBC inoculated with an active concentrated nitrifying culture and fed with a synthetic wastewater under oxygen limitation. The autotrophic N removal under DO limitation was termed OLAND by Kuai and Verstraete (1998). The RBC removed $89 \pm 5\%$ of the influent N at the highest surface load of approximately 8.3 g N/m^2-day, with N_2 as the main product. The microbial community was dominated by *Nitrosomonas*-like species for the aerobic ammonia oxidizers and close relatives to *Kuenenia stuttgartiensis* for the probable anaerobic ammonia oxidizing community. The authors state that a part of the denitrification activity could be due to nitrifiers and to heterotrophs.

5.3.4 Case study: Leachate purification plant at Mechernich (Germany)

This is the first application report of aerobic/anoxic deammonification, widely described and studied by the Institute for Water Quality and Waste Management at the University of Hanover (Germany) (Baumgarten and Seyfried 1996; Hippen *et al.* 1997; Helmer and Kunst 1998; Helmer *et al.* 1999; Helmer *et al.* 2001; Hippen *et al.* 2001; Helmer-Madhok *et al.* 2002).

The plant has been running completely since 1994. The initial biological step consisted of two steps (Baumgarten and Seyfried 1996):

- Nitrification in a biological contactor plant with micro-sieves placed before and after it. The biological contactor is divided into four lines with three biological contactor cylinders in series.
- Denitrification in an AS system with following intermediate reaction and a settling tank.

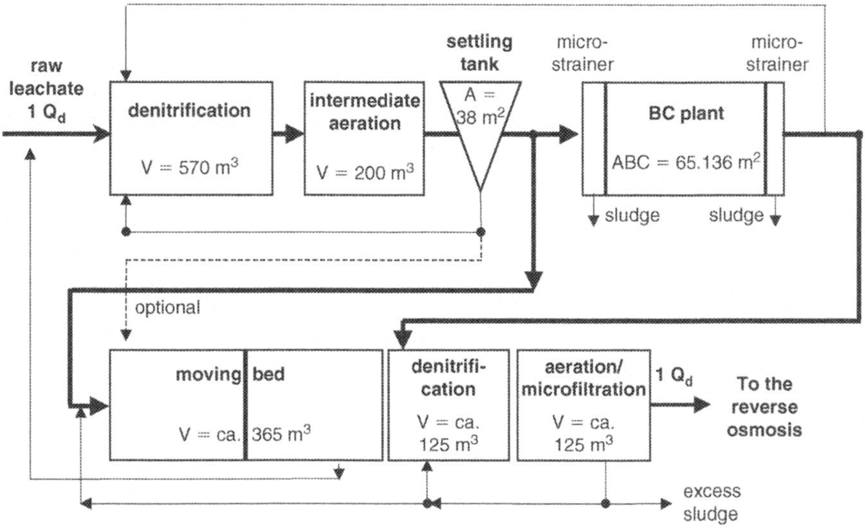

Figure 5.16 Flow sheet of the pre-treatment plant in Mechernich/Germany (Hippen *et al.* 2001).

Table 5.5 Operation parameters and characteristics of leachate in the influent of the RBC (Hippen *et al.* 2001).

Parameter	Unit	Value (range)
Water quantity[a]	m^3/day	102 (0–182)
Temperature	°C	28 (27–30)[2b]
pH	–	8.3 (7.4–8.7)
DO[c]	mg O_2/L	0.7–1
NH_4-N	mg N/L	209 (32–681)
CSB	mg O_2/L	1645 (442–2900)
TOC	mg C/L	310 (216–412)[d]
$B_{A,NH4\text{-}N}$	g N/m^2-d^2	1.5 (0.5–2.6)
$DB_{A,NH4\text{-}N}$	g N/m^2-d^2	1.5 (0.5–2.6)

[a]Flow without recycle.
[b]Almost constant through usage of heat exchangers.
[c]Values of the cascade with deammonification.
[d]Data of the first half of 1994.

It has been extended by a secondary denitrification unit and a moving bed plant parallel to the existing RBC plant to effect nitrification (Figure 5.16). This last stage was started at the end of 1998. The BC plant was initially designed for a nitrogen loading of 2 g N/m^2-d. The operation parameters and leachate data (influent of the RBC) are given in Table 5.5.

5.4 CONCLUSION: FROM TREATMENT TO RECOVERY

Table 5.6. gives a comparison of new and conventional treatment processes (Schmidt *et al.* 2003). It includes a process called NO_x, which is autotrophic

Table 5.6 Operational characteristics of wastewater treatment processes for nitrogen removal (Schmidt *et al.* 2003).

	Conventional nitrification denitrification	No_x	OLAND	SHARON	Anammox	Canon	Aerobic deammonification
Aerobic ammonia oxidizers	Many	*Nitrosomonas eutropha*	Unknown	*Nitrosomonas eutropha*	absent	*Nitrosomonas eutropha*	Unknown salt tolerant ammonia oxidizer
Aerobic nitrite oxidizers	Many	Absent	Unknown	Absent	Absent	Absent	*Nitrobacter*
Anaerobic ammonia oxidizers	Absent	Absent	Unknown	Absent	*Brocadia anammoxidans*, *Kuenenia stuttgartiensis*	*Brocadia anammoxidans*, *Kuenenia stuttgartiensis*	*Kuenemia stuttgartiensis*
Biofilms or suspension	Biofilms/ suspension	Suspension	Biofilms	Suspension	Biofilms	Biofilms	Biofilms
NH_4^+ loading (kg N/ m^3-d)	2–8	5	0.1	0.5–1.5	10–20	2–3	1–2
N-removal efficiency	95%	95%	85%	90%	90%	90%	60%
Process complexity	Separate oxic and anoxic compartments or periods, methanol dosing	Separate oxic and anoxic compartments, methanol dosing, membrane for sludge retention	Aeration needs to be tuned to ammonia loading	Separate oxic and anoxic compartments or periods, methanol dosing	Preceding partial nitrification needed	Aeration needs to be tuned to ammonia loading	Aeration needs to be tuned to ammonia loading
Application status	Established	Pilot plant	Laboratory	Two full-scale plants	Full scale initiated	Laboratory	Two full-scale plants

nitrification–denitrification by *Nitrosomonas*-like bacteria in the presence of NO_x (NO/NO_2) (Zart and Bock 1998).The process has been tested at pilot-scale for the treatment of wastewater from intensive fish farming and at a municipal WWTP (sludge liquor). One can cite a new anaerobic N–S biological interaction involving simultaneous anaerobic ammonium oxidation and sulfate reduction, ammonium being the electron donor and sulfate the electron acceptor. This process was discovered in a granular activated carbon (GAC) anaerobic fluidized bed reactor treating vinasse from an ethanol distillery of sugar beet molasses (Fdz-Polanco *et al.* 2001). Until now, the microbiological processes and microorganisms involved remain unclear.

In the conventional and new processes presented in the chapter to remove nitrogen, relatively large amounts of resources (energy and chemicals) are required. Removal technologies have to be changed to make wastewater management more affordable and sustainable in term of nutrient management (Wilsenach *et al.* 2002). Especially in the case of nitrogen rich wastewaters, nitrogen should be recovered in a mineral form to be used in agriculture. The most promising options are the production of struvite or Magnesium–Ammonium–Phosphate (MAP) and stripping of ammonia (Janus and van der Roest 1997; Mulder 2003).

Struvite is a white crystalline substance consisting of ammonium, magnesium and phosphate in equal molar concentration : $MgNH_4PO_4.6H_2O$. Struvite forms according to the following reaction (Doyle and Parsons, 2002) :

$$Mg^+ + NH_4^+ + PO_4^{3-} + 6H_2O \rightarrow MgNH_4PO_4 \,.\, 6H_2O$$

This technology has been mainly applied to the treatment of anaerobic digester liquor (Battistoni *et al.* 1997; von Münch and Barr 2001; Yoshino *et al.* 2003). The principal objective was to avoid fouling due to struvite precipitation by forming struvite where it can be controlled, out of the digester.

Ammonium removal by struvite precipitation has also been applied to industrial wastewaters (Tünay *et al.* 1997, Altinbas *et al.* 2002). MAP precipitation in leather tanning industry exhibited a very satisfactory performance (Tünay *et al.* 1997). Altinbas *et al.* (2002) applied MAP precipitation technology to two industrial effluents from an opium alkaloid processing industry and from a baker's yeast industry. Ammonium removals of 61–80% were achieved at a pH value of 9.2 at the stoichiometric ratio ($Mg:NH_4:PO_4 = 1:1:1$), whereas 83% removal was obtained at the same pH value but above the stoichiometric ratio ($Mg:NH_4:PO_4 = 1:1:1.1$).

Air (or steam when available) stripping of ammonia is operated after increasing the pH with NaOH or $Ca(OH)_2$. Then, free ammonia is converted to ammonium sulfate by desorption with sulfuric acid. The ammonium sulfate solution might be used as raw product for fertilizer production.

From an economical point of view, Siegrist (1997) concluded from a compared study that separate treatment of anaerobic supernatant with physical (air stripping) and chemical (MAP precipitation) processes is significantly more expensive than biological nitrification–denitrification. But MAP precipitation could be economic if struvite would be marketable, as a slow-release fertilizer or to be used in fertilizer manufacture, at a price outweighing the chemical costs for Mg dosing and pH adjustment (Doyle and Parsons 2002). The same comment can be done concerning ammonium sulfate in the case of ammonia stripping.

REFERENCES

Abeling, U. and Seyfried, C.F. (1992) Anaerobic–aerobic treatment of high-strength ammonium wastewater-nitrogen removal via nitrite. *Water Sci. Technol.* **26**(5–6), 1007–1015.

Altinbas, M., Ozturk, I. and Aydin, A.F. (2002) Ammonia recovery from high strength agro industry effluents. *Water Sci. Technol.* **45**(12), 189–196.

Anthonisen, A.C., Loehr, R.C., Prakasam, T.B.S. and Srinath, E.G. (1976) Inhibition of nitrification by ammonia and nitrous acid. *J. Water Pollut. Control Fed.* **48**(5), 835–852.

Arnold, E., Bohm, B. and Wilderer, P.A. (2000) Application of activated sludge and biofilm sequencing batch reactor technology to treat reject water from sludge dewatering systems: a comparison. *Water Sci. Technol.* **41**(1), 115.

Battistoni, P., Fava, G., Pavan, P., Musacco, A. and Cecchi, F. (1997) Phosphate removal in anaerobic liquors by struvite crystallization without addition of chemicals: preliminary results. *Water Res.* **31**(11), 2925–2929.

Baumgarten, G. and Seyfried, C.F. (1996) Experiences and new developments in biological pretreatment and physical post-treatment of landfill leachate. *Water Sci. Technol.* **34**(7–8), 445–453.

Bernet, N., Peng, D., Delgenès, J.P. and Moletta, R. (2000) Nitrification at low oxygen concentration in biofilm reactor. *J. Environ. Eng. ASCE*, **127**(3), 266–271.

Bougard, D., Bernet, N., Chèneby, D. and Delgenès, J.P. (2006) Nitrification of a high-strength wastewater in an inverse turbulent bed reactor: effect of temperature on nitrite accumulation. *Proc. Biochem.* **41**(1), 104–113.

Ceçen, F. (1996) Investigation of partial and full nitrification characteristics of fertilizer wastewaters in a submerged biofilm reactor. *Water Sci. Technol.* **34**(11), 77–85.

Dapena-Mora, A., Campos, J.L., Mosquera-Corral, A., Jetten, M.S.M. and Mendez, R. (2004a) Stability of the ANAMMOX process in a gas-lift reactor and a SBR. *J. Biotechnol.* **110**(2), 159–170.

Dapena-Mora, A., Van Hulle, S.W.H., Campos, J.L., Mendez, R., Vanrolleghem, P.A. and Jetten, M. (2004b) Enrichment of Anammox biomass from municipal activated sludge: experimental and modelling results. *J. Chem. Technol. Biotechnol.* **79**(12), 1421–1428.

Doyle, J.D. and Parsons, S.A. (2002) Struvite formation, control and recovery. *Water Res.* **36**(16), 3925–3940.

Egli, K., Bosshard, F., Werlen, C., Lais, P., Siegrist, H., Zehnder, A.J.B. and van der Meer, J.R. (2003) Microbial composition and structure of a rotating biological contactor biofilm treating ammonium-rich wastewater without organic carbon. *Microbial Ecol.* **45**(4), 419–432.

Eilersen, A.M., Arvin, E. and Henze, M. (2004) Monitoring toxicity of industrial wastewater and specific chemicals to a green alga, nitrifying bacteria and an aquatic bacterium. *Wat. Sci. Technol.* **50**(6), 277–283.

El-Fadel, M., Bou-Zeid, E., Chahine, W. and Alayli, B. (2002) Temporal variation of leachate quality from pre-sorted and baled municipal solid waste with high organic and moisture content. *Waste Manage.* **22**(3), 269–282.

Fdz Polanco, F., Villaverde, S. and Garcia, P.A. (1994) Temperature effect on nitrifying bacteria activity in biofilters: activation and free ammonia inhibition. *Water Sci. Technol.* **30**(11), 121–130.

Fdz Polanco, F., Mendez, E. and Villaverde, S. (1995) Study of nitrifying biofilms in submerged biofilters by experimental design methods. *Water Sci. Technol.* **32**(8), 227–233.

Fdz Polanco, F., Villaverde, S. and Garcia, P.A. (1996) Nitrite accumulation in submerged biofilters – combined effects. *Water Sci. Technol.* **34**(3–4), 371–378.

Fdz-Polanco, F., Fdz-Polanco, M., Fernandez, N., Uruena, M.A., Garcia, P.A. and Villaverde, S. (2001) New process for simultaneous removal of nitrogen and sulphur under anaerobic conditions. *Water Res.* **35**(4), 1111–1114.

Fux, C., Boehler, M., Huber, P., Brunner, I. and Siegrist, H. (2002) Biological treatment of ammonium-rich wastewater by partial nitritation and subsequent anaerobic ammonium oxidation (Anammox) in a pilot plant. *J. Biotechnol.* **99**(3), 295–306.

Fux, C., Lange, K., Faessler, A., Huber, P., Grueniger, B. and Siegrist, H. (2003) Nitrogen removal from digester supernatant via nitrite – SBR or SHARON? *Water Sci. Technol.* **48**(8), 9–18.

Fux, C. and Siegrist, H. (2004) Nitrogen removal from sludge digester liquids by nitrification/denitrification or partial nitritation/anammox: environmental and economical considerations. *Water Sci. Technol.* **50**(10), 19–26.

Fux, C., Huang, D., Monti, A. and Siegrist, H. (2004a) Difficulties in maintaining long-term partial nitritation of ammonium-rich sludge digester liquids in a moving-bed biofilm reactor (MBBR). *Water Sci. Technol.* **49**(11–12), 53–60.

Fux, C., Marchesi, V., Brunner, I. and Siegrist, H. (2004b) Anaerobic ammonium oxidation of ammonium-rich waste streams in fixed-bed reactors. *Water Sci. Technol.* **49**(11–12), 77–82.

Garrido, J.M., van Benthum, W.A.J., van Loosdrecht, M.C.M. and Heijnen, J.J. (1997) Influence of dissolved oxygen concentration on nitrite accumulation in a biofilm airlift suspension reactor. *Biotechnol. Bioeng.* **53**(2), 168–178.

Hao, X.D., Heijnen, J.J. and van Loosdrecht, M.C.M. (2002) Sensitivity analysis of a biofilm model describing a one-stage completely autotrophic nitrogen removal (CANON) process. *Biotechnol. Bioeng.* **77**(3), 266–277.

Hellinga, C., Schellen, A.A.J.C., Mulder, J.W. van Loosdrecht, M.C.M. and Heijnen, J.J. (1998) The SHARON process: an innovative method for nitrogen removal from ammonium-rich waste water. *Water Sci. Technol.* **37**(9), 135–142.

Helmer, C. and Kunst, S. (1998) Simultaneous nitrification/denitrification in an aerobic biofilm system. *Water Sci. Technol.* **37**(4–5), 183–187.

Helmer, C., Kunst, S., Juretschko, S., Schmid, M.C., Schleifer, K.H. and Wagner, M. (1999) Nitrogen loss in a nitrifying biofilm system. *Water Sci. Technol.* **39**(7), 13.

Helmer, C., Tromm, C., Hippen, A., Rosenwinkel, K.H., Seyfried, C.F., and Kunst, S. (2001) Single stage biological nitrogen removal by nitritation and anaerobic ammonium oxidation in biofilm systems. *Water Sci. Technol.* **43**(1), 311–320.

Helmer-Madhok, C., Schmid, M., Filipov, E., Gaul, T., Hippen, A., Rosenwinkel, K.H., Seyfried, C.F., Wagner, M. and Kunst, S. (2002) Deammonification in biofilm systems: population structure and function. *Water Sci. Technol.* **46**(1–2), 223–231.

Hippen, A., Rosenwinkel, K.H., Baumgarten, G. and Seyfried, C.F. (1997) Aerobic deammonification: a new experience in the treatment of wastewaters. *Water Sci. Technol.* **35**(10), 111–120.

Hippen, A., Helmer, C., Kunst, S., Rosenwinkel, K.H. and Seyfried, C.F. (2001) Six years' practical experience with aerobic/anoxic deammonification in biofilm systems. *Water Sci. Technol.* **44**(2–3), 39–48.

Janus, H.M. and van der Roest, H.F. (1997) Don't reject the idea of treating reject water. *Water Sci. Technol.* **35**(10), 27–34.

Jetten, M.S.M., Schmid, M., Schmidt, I., Wubben, M., van Dongen, U., Abma, W., Sliekers, O., Revsbech, N.P., Beaumont, H.J.E., Ottosen, L., Volcke, E., Laanbroek, H.J., Campos-Gomez, J.L., Cole, J., van Loosdrecht, M., Mulder, J.W., Fuerst, J., Richardson, D., van de Pas, K., Mendez-Pampin, R., Third, K., Cirpus, I., van Spanning, R., Bollmann, A., Nielsen, L.P., den Camp, H.O., Schulz, C., Gundersen, J., Vanrolleghem, P., Strous, M., Wagner, M. and Kuenen, J.G. (2002) Improved nitrogen removal by application of new nitrogen-cycle bacteria. *Rev. Env. Sci. Biotechnol.* **1**(1), 51–63.

Juliastuti, S.R., Baeyens, J. and Creemers, C. (2003) Inhibition of nitrification by heavy metals and organic compounds: the ISO 9509 test. *Environ. Eng. Sci.* **20**(2), 79–90.

Kim, D.J., Chang, J.S., Lee, D.I., Han, D.W., Yoo, I.K. and Cha, G.C. (2003) Nitrification of high strength ammonia wastewater and nitrite accumulation characteristics. *Water Sci. Technol.* **47**(11), 45–51.

Kuai, L.P. and Verstraete, W. (1998) Ammonium removal by the oxygen-limited autotrophic nitrification–denitrification system. *Appl. Environ. Microbiol.* **64**(11), 4500–4506.

Lai, E., Senkpiel, S., Solley, D. and Keller, J. (2004) Nitrogen removal of high strength wastewater via nitritation/denitritation using a sequencing batch reactor. *Water Sci. Technol.* **50**(10), 27–33.

Mauret, M., Ferrand, F., Boisdon, V., Spérandio, M. and Paul, E. (2001) Process using DO and ORP signals for biological nitrification and denitrification: validation of a food-processing industry wastewater treatment plant on boosting with pure oxygen. *Water Sci. Technol.* **44**(2–3), 163–170.

Metcalf and Eddy (1991) *Wastewater Engineering, Treatment, Disposal, Reuse,* 3rd edn. McGraw-Hill, New-York.

Mulder, A. (2003) The quest for sustainable nitrogen removal technologies. *Water Sci. Technol.* **48**(1), 67–75.

Mulder, A., van de Graaf, A.A., Robertson, L.A. and Kuenen, J.G. (1995) Anaerobic ammonium oxidation discovered in a denitrifying fluidized bed reactor. *FEMS Microbio. Ecol.* **16**(3), 177–183.

Ochoa, J.C., Colprim, J., Palcio, B., Chatellier, P., Paul, E. (2002) Active heterotrophic and autotrophic biomass distribution between fixed and suspended systems in an hybrid biological reactor. *Water Sci. Technol.* **46**(1–2), 397–404.

Pambrun, V., Paul, E. and Spérandio, M. (2004) Treatment of nitrogen and phosphorus in highly concentrated effluent in SBR and SBBR processes. *Water Sci. Technol.* **50**(6), 269–276.

Pedersen, P.D., Jensen, K., Lyngsie, P. and Johansen, N.H. (2003) Nitrogen removal in industrial wastewater by nitration and denitration – 3 years of experience. *Water Sci. Technol.* **47**(11), 181–188.

Plisson-Sauné, S., Capdeville, B., Mauret, M., Deguin, A. and Baptiste, P. (1996) Real-time control of nitrogen removal using three ORP bending-points: signification, control strategy and results. *Water Sci. Technol.* **33**(1), 275–280.

Pynaert, K., Smets, B.F., Wyffels, S., Beheydt, D., Siciliano, S.D. and Verstraete, W. (2003) Characterization of an autotrophic nitrogen-removing biofilm from a highly loaded lab-scale rotating biological contactor. *Appl. Environ. Microbiol.* **69**(6), 3626–3635.

Pynaert, K., Smets, B.F., Beheydt, D. and Verstraete, W. (2004) Start-up of autotrophic nitrogen removal reactors via sequential biocatalyst addition. *Environ. Sci. Technol.* **38**(4), 1228–1235.

Ren, S.J. (2004). Assessing wastewater toxicity to activated sludge: recent research and development. *Environ. Int.* **30**(8), 1151–1164.

Rols, J.L., Mauret, M., Rahmani, H., Nguyen, K.M., Capdeville, B., Cornier, J.C. and Deguin, A. (1994) Population dynamics and nitrite build-up in activated sludge and biofilm processes for nitrogen removal. *Water Sci. Technol.* **29**(7), 43–51.

Schmidt, I., Sliekers, O., Schmid, M., Bock, E., Fuerst, J., Kuenen, J.G., Jetten, M.S.M. and Strous, M. (2003) New concepts of microbial treatment processes for the nitrogen removal in wastewater. *Fems Microbiol. Rev.* **27**(4), 481–492.

Schmidt, J.E., Batstone, D.J. and Angelidaki, I. (2004) Improved nitrogen removal in upflow anaerobic sludge blanket (UASB) reactors by incorporation of Anammox bacteria into the granular sludge. *Water Sci. Technol.* **49**(11–12), 69–76.

Siegrist, H., Reithaar, S., Koch, G. and Lais, P. (1998) Nitrogen loss in a nitrifying rotating contactor treating ammonium-rich wastewater without organic carbon. *Water Sci. Technol.* **38**(8–9), 241–248.

Sliekers, A.O., Derwort, N., Gomez, J.L.C., Strous, M., Kuenen, J.G. and Jetten, M.S.M. (2002) Completely autotrophic nitrogen removal over nitrite in one single reactor. *Water Res.* **36**(10), 2475–2482.

Sliekers, A.O., Third, K.A., Abma, W., Kuenen, J.G. and Jetten, M.S.M. (2003) CANON and Anammox in a gas-lift reactor. *FEMS Microbiol. Lett.* **218**(2), 339–344.

Third, K.A., Sliekers, A.O., Kuenen, J.G. and Jetten, M.S.M. (2001) The CANON system (Completely autotrophic nitrogen-removal over nitrite) under ammonium limitation: interaction and competition between three groups of bacteria. *Syst. Appl. Microbiol.* **24**(4), 588–596.

Third, K.A., Sepramaniam, S., Tonkovic, Z., Newland, M. and Cord-Ruwisch, R. (2004) Optimisation of storage driven denitrification by using on-line specific oxygen uptake rate monitoring during SND in a-SBR. *Water Sci. Technol.* **50**(10), 171–180.

Thuan, T.H., Jahng, D.J., Jung, J.Y., Kim, D.J., Kim, W.K., Park, Y.J., Kim, J.E. and Ahn, D.H. (2004) Anammox bacteria enrichment in upflow anaerobic sludge blanket (UASB) reactor. *Biotechnol. Bioproc. Eng.* **9**(5), 345–351.

Toh, S.K., Webb, R.I. and Ashbolt, N.J. (2002) Enrichment of autotrophic anaerobic ammonium-oxidizing consortia from various wastewaters. *Microb. Ecol.* **43**(1), 154–167.

Tokutomi, T. (2004) Operation of a nitrite-type airlift reactor at low DO concentration. *Water Sci. Technol.* **49**(5–6), 81–88.

Tünay, O., Kabdasli, I., Orhon, D. and Kolcak, S. (1997) Ammonia removal by magnesium ammonium phosphate precipitation in industrial wastewaters. *Water Sci. Technol.* **36**(2–3), 225–228.

Turk, O. and Mavinic, D.S. (1986) Preliminary assessment of a shortcut in nitrogen removal from wastewater. *Can J. Civil Eng.* **13**, 600–605.

Van Benthum, W.A.J., van Loosdrecht, M.C.M. and Heijnen, J.J. (1997). Process Design for nitrogen removal using nitrifying biofilm and denitrifying suspended growth in a biofilm airlift suspension reactor. *Water Sci. Technol.*, **36**(1), 119–128.

van Benthum, W.A.J., Derissen, B.P., van Loosdrecht, M.C.M. and Heijnen, J.J. (1998a) Nitrogen removal using nitrifying biofilm growth and denitrifying suspended growth in a biofilm airlift suspension reactor coupled with a chemostat. *Water Res.* **32**(7), 2009–2018.

van Benthum, W.A.J., Garrido, J.M., Mathijssen, J.P.M., Sunde, J., van Loosdrecht, M.C.M. and Heijnen, J.J. (1998b) Nitrogen removal in intermittently aerated biofilm airlift reactor. *J. Environ. Eng. ASCE* **124**(3), 239–248.

van Dongen, U., Jetten, M.S.M. and van Loosdrecht, M.C.M. (2001) The SHARON–Anammox process for treatment of ammonium rich wastewater. *Water Sci. Technol.* **44**(1), 153–160.

van Loosdrecht, M.C.M., van Benthum, W.A.J. and Heijnen, J.J. (2000) Integration of nitrification and denitrification in biofilm airlift suspension reactors. *Water Sci. Technol.* **41**(4–5), 97–103.

Villaverde, S., Fernandez, M.T., Uruena, M.A. and Fdz-Polanco, F. (1997a) Influence of substrate concentration on the growth and activity of a nitrifying biofilm in a submerged biofilter. *Environ. Technol.* **18**(9), 921–928.

Villaverde, S., Garcia-Encina, P.A. and Fdz-Polanco, F. (1997b) Influence of pH over nitrifying biofilm activity in submerged biofilters. *Water Res.* **31**(5), 1180–1186.

Villaverde, S., Fdz-Polanco, F. and Garcia, P.A. (2000) Nitrifying biofilm acclimation to free ammonia in submerged biofilters. Start-up influence. *Water Res.* **34**(2), 602.

Voets, J.P., Vanstaen, H. and Verstraete, W. (1975) Removal of nitrogen from highly nitrogenous wastewaters. *J. Water Pollut. Control Fed.* **47**(2), 394–397.

von Münch, E., Lant, P. and Keller, J. (1996) Simultaneous nitrification and denitrification in bench-scale sequencing batch reactors. *Water Res.* **30**(2), 277–284.

von Münch, E. and Barr, K. (2001). Controlled struvite crystallisation for removing phosphorus from anaerobic digester sidestreams. *Water Res.* **35**(1), 151–159.

Wareham, D.G., Hall, K.J. and Mavinic, D.S. (1993) Real-time control of aerobic-anoxic sludge digestion ORP. *J. Environ. Eng. ASCE*, **119**(1), 120–136.

Wilsenach, J.A., Maurer, M., Larsen, T.A. and van Loosdrecht, M.C.M. (2003) From waste treatment to integrated resource management. *Wat. Sci. Technol.* **48**(1), 1–9.

Yoshino, M., Yao, M., Tsuno, H. and Somiya, I. (2003). Removal and recovery of phosphate and ammonium as struvite from supernatant in anaerobic digestion. *Water Sci. Technol.* **48**(1), 171–178.

Zart, D. and Bock, E. (1998) High rate of aerobic nitrification and denitrification by Nitrosomonas eutropha grown in a fermentor with complete biomass retention in the presence of gaseous NO_2 or NO. *Arch. Microbiol.* **169**(4), 282–286.

Zeng, R.J., Lemaire, R., Yuan, Z. and Keller, J. (2003) Simultaneous nitrification, denitrification, and phosphorus removal in a lab-scale sequencing batch reactor. *Biotechnol. Bioeng.* **84**(2), 170–178.

Zheng, P., Lin, F.M., Hu, B.L. and Chen, J.S. (2004) Start-up of anaerobic ammonia oxidation bioreactor with nitrifying activated sludge. *J. Environ. Sci. China* **16**(1), 13–16.

6

Application of biological treatment systems for sulfate-rich wastewaters

M. Fernández-Polanco and P.A. García Encina

6.1 INTRODUCTION

Wastewaters containing organic matter and sulfate are generated by many industrial processes that use sulfuric acid (food and fermentation industry) or sulfate-rich feed stocks (sea food-processing industry). Also the use of less oxidized sulfurous compounds in industrial processes (sulfide in tanneries and Kraft pulping, sulfite in sulfite pulping, thiosulfate in processing of photographs or dithionite in pulp bleaching) results in the generation of sulfate-rich wastewaters. Besides these organic wastewaters, some sulfate-rich effluents contain hardly any organic matter. These are generated during the leaching of sulfur-rich wastes (mine spoils, landfills) or during the scrubbing of sulfur containing off-gases.

Biological sulfate removal is a cost-effective alternative for costly, and sometimes complex, physico-chemical sulfate removal methods (Maree *et al.* 1991). Biological sulfate removal consists of two steps with sulfate reduction to sulfide by sulfate reducing bacteria (SRB) as the first step. Channeling of reducing equivalents toward the bacteria is enhanced by the ability of SRB bacteria to compete

 Advanced Biological Treatment Processes for Industrial Wastewaters edited by F.J. Cervantes, S.G. Pavlostathis and A.C. van Haandel. ISBN: 1843391147. Published by IWA Publishing, London, UK.

with other anaerobic bacteria for the available organic substrate and the sensitivity of the other bacteria for sulfide. For wastewaters that contain no or insufficient electron donor and carbon source for a complete sulfate reduction, addition of an appropriate electron donor is required. The selection of the electron donor depends on the cost of substrate required per unit of reduced sulfate and the biodegradability of the substrate.

In the second step, H_2S produced in the first stage is removed from the liquid or enriched stripping gas using physico-chemical or biological techniques (Table 6.5). The relatively high-energy requirements for stripping or the high chemical and disposal costs constitute important draw-backs of physico-chemical sulfide removal methods. Various biological sulfide removal methods exist such as the partial biological oxidation of sulfide to elemental sulfur (S°). Under oxygen-limited, conditions (dissolved oxygen concentrations below 0.1 mg/L), S° is the major end product of the sulfide oxidation, whereas sulfate is mainly formed under sulfide-limiting conditions. Since elemental sulfur is a colloidal solid, it can be eliminated from the wastewater by gravity sedimentation, eventually after flocculent addition.

A major problem for the biological treatment of sulfate-rich wastewaters is the production of H_2S, that reduce the quality and quantity of the biogas produced (Lens *et al.* 1998a). Gaseous and dissolved sulfides cause physical (corrosion, odor, increased effluent chemical oxygen demand (COD)) or biological (toxicity) constraints that may lead to process failure. H_2S is generated by sulfate-reducing bacteria, in both anaerobic and aerobic (anoxic micro-environments) wastewater treatment systems. No practical methods exist to prevent sulfate reduction. Selective inhibition of SRB by molybdate, transition elements, or antibiotics is unsuccessful at full-scale. Selection of a treatment strategy for a sulfate-rich wastewater depends on the aim of the treatment. This can be (1) removal of organic matter, (2) removal of sulfate or (3) removal of both. Theoretically, wastewaters with a COD/sulfate ratio of 0.67 or higher contain enough COD (electron donor) to remove all sulfate by sulfate-reducing bacteria. If the ratio is lower, addition of extra COD, for example, as ethanol or synthesis gas (a mixture of H_2, CO_2 and CO) is required. Complete COD removal in wastewaters with a COD/sulfate ratio of above 0.67 also requires methanogenic COD degradation. Sulfate can be removed from the waste stream by the coupling of a sulfide oxidation step to the sulfate reduction step. Sulfur is recovered from the wastewater in case H_2S is partially oxidized to insoluble elemental sulfur.

6.2 SULFATE-RICH WASTEWATERS

Sulfate concentration in domestic wastewater is relatively low, ranging from 50 to 200 mg/L (Yoda *et al.* 1987), industrial effluents are the major contributors of sulfate-rich wastewater. Depending on whether the sulfate-rich wastewater contains high or low COD, it can be classified as low or high organic strength sulfate wastewater:

- Low organic strength sulfate wastewater (high sulfate concentration with low organic matter) is generated in acid mine drainage (AMD) and other chemical industries where sulfuric acid is used in the processes. Mining wastewater, for

Table 6.1 Typical characteristics of sulfate-rich industrial wastewaters.

Wastewater source	COD (mg/L)	Sulfate (mg S/L)	References
Molasses fermentation	44,800–55,600	2500–3450	Carrondo *et al.* 1983
Sea food processing	10,000–50,000	600–2700	Mendez *et al.* 1995
Potato-starch factory	17,500–18,000	320	Nanninga *et al.* 1986
Tannery	2900–8200	750–1250	Genschow *et al.* 1996
Distillery slops	95,000	6000	Szendry 1983
Edible oil refinery	1000–8200	3100–7400	Anderson *et al.* 1988
Pharmaceutical plant	28,500	14,800	Mohanrao *et al.* 1970
Pulp and paper			
TMP	2000–5000	200–700	Habets and de Vegt
CTMP	7500–10,400	1200–1500	1991
Citric acid	30,000	4500	Svardal *et al.* 1993

TMP: thermomechanical pulping; CTMP: chemo-thermomechanical pulping.

example, contains about 1980 mg/L of sulfate with COD as low as 100 mg/L (Maree and Strydon 1985). The biological sulfate reduction implies the need of an external electron donor. Various organic compounds, wastes and organic wastewater have been used for this purpose like benzoate, butyrate, tannery effluents and micro-algal biomass (Boshoff *et al.* 2004a, b). In addition, hydrogen gas may be used.

- High organic strength sulfate wastewater (high sulfate concentration with high organic matter) is generated by agro-based or food-processing industry and pulp and paper mills. Due to the high biodegradable organic matter content, anaerobic treatment of such water is widely adopted for economic reasons. Some typical organic strength sulfate wastewaters are presented in Table 6.1.

6.3 PROBLEMS ASSOCIATED WITH SULFATE-RICH WASTEWATERS TREATMENT

6.3.1 Sulfide production in anaerobic digestion

The presence of sulfate ion in wastewaters considerably increases the complexity of the biodegradation routes (Widdel 1988). In the presence of sulfate, acidogenic, acetogenic (AB) and methanogenic bacteria compete with SRB for the available substrates (Figure 6.1). The outcome of this competition is important, as it will determine to what extend sulfide and methane, the end products of the anaerobic mineralization processes, will be produced. The main intermediates in the anaerobic mineralization of complex organic matter are hydrogen, acetate, propionate and butyrate (Figure 6.1). Both from thermodynamic (Table 6.2) and kinetic points of view, SRB should out-compete methane producing bacteria (MPB) during growth on these substrates. Thus, sulfate reduction will theoretically become the dominant process in anaerobic digesters treating sulfate-rich wastewaters.

The importance of this competition increases with a decrease in the COD/sulfate ratio of the wastewater. For waste streams with a COD/sulfate ratio of over 0.67,

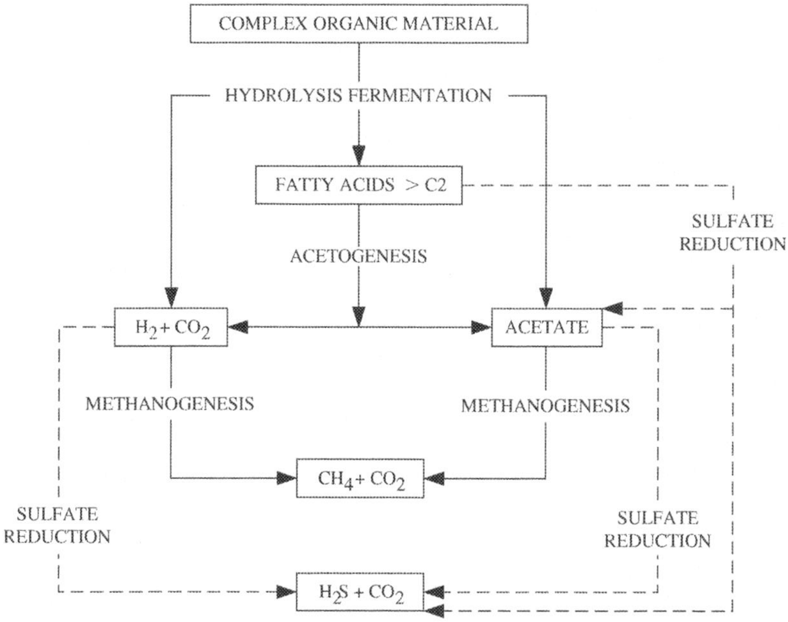

Figure 6.1 Substrate competition between SRB, MPB and AB during anaerobic digestion of organic matter.

Table 6.2 Stoichiometry for the anaerobic degradation of propionate, acetate and hydrogen by SRB and MPB (Data from Thauer *et al.* 1977) ($\Delta G^{\circ\prime}$ at 37°C in kJ/reaction).

Reactions	$\Delta G^{\circ\prime}$
Propionate	
$CH_3CH_2COO^- + 3H_2O \rightarrow CH_3COO^- + HCO_3^- + H^+ + 3H_2$	+76.0
$CH_3CH_2COO^- + 0.75SO_4^{2-} \rightarrow CH_3COO^- + HCO_3^- + 0.75HS^- + 0.25H^+$	−37.7
$CH_3CH_2COO^- + 1.75SO_4^{2-} \rightarrow 3HCO_3^- + 1.75HS^- + 0.5H^+ + 0.25OH^-$	−88.9
Acetate	
$CH_3COO^- + H_2O \rightarrow CH_4 + HCO_3^-$	−31.0
$CH_3COO^- + SO_4^{2-} \rightarrow 2HCO_3^- + HS^-$	−47.6
Hydrogen	
$4H_2 + HCO_3^- + H^+ \rightarrow CH_4 + 3H_2O$	−32.7
$4H_2 + SO_4^{2-} + H^+ \rightarrow HS^- + 4H_2O$	−38.1

there is theoretically enough organic matter (COD) available to completely reduce the sulfate present (Rinzema and Lettinga 1988). A complete removal of the organic matter can only be achieved if, in addition to sulfate reduction, methanogenesis occurs. For COD/sulfate ratios lower than 0.67, the amount of organic matter is insufficient for a complete reduction of the sulfate and extra substrate should be added if removal of sulfate is the objective of the treatment.

Both AB and MPB can, depending on the conditions, successfully compete with SRB for complex substrates (Colleran *et al.* 1995). In anaerobic reactors,

Table 6.3 Factors determining the outcome of the competition between SRB and MPB.

Factor	Reference
Inoculum composition	
Type of seed sludge	McCartney and Oleskiewicz 1991
Bacterial composition	Harada *et al.* 1994; Omil *et al.* 1998
Attachment properties of bacteria	Isa *et al.* 1986a, b
Experimental run time	Harada *et al.* 1994; Omil *et al.* 1998
Inoculation with new bacterial species	Omil *et al.* 1997a
Influent composition	
Type of COD	Polprasert and Haas 1995
Acetate concentration	Yoda *et al.* 1987
Sulfate concentration	Overmeire *et al.* 1994
Sulfide concentration	Omil *et al.* 1996
Operational conditions	
pH	Visser *et al.* 1996
Temperature	Visser *et al.* 1992

hydrogenotrophic SRB (HSRB) out-compete hydrogen utilizing MPB (HMPB) provided sufficient sulfate is present (Rinzema and Lettinga 1988; Visser *et al.* 1993b; Omil *et al.* 1996). This corroborates with the process fundamentals, as HSRB gain more energy from the consumption of molecular hydrogen and have a higher substrate affinity than HMPB, thus decreasing the hydrogen concentration below the threshold value of HMPB (Oude Elferink *et al.* 1994). This explains the rapid inhibition of HMPB when sulfate enters an anaerobic bioreactor.

Also acetate utilizing SRB (ASRB) have a thermodynamic and kinetic advantage over acetate utilizing MPB (AMPB) in their competition for acetate (Widdel 1988). The expected predominance of ASRB over AMPB in excess of sulfate has been confirmed in continuously stirred tank reactors and in the anaerobic contact process (Middleton and Lawrence 1977; Gupta *et al.* 1994). However, the outcome of the competition is less predictable in modern high rate anaerobic reactors with sludge retention based on sludge immobilization. Several studies reported that acetate is completely converted into methane, even in excess of sulfate (Hoeks *et al.* 1984; Mulder 1984), while others report a predominance of ASRB (Omil *et al.* 1996). Recently, two new ASRB were isolated from bioreactors (Oude Elferink *et al.* 1995; 1999). Their growth kinetic properties are only slightly better than those of *Methanosaeta* sp., the most abundant AMPB in bioreactors (Oude Elferink *et al.* 1998). Besides growth kinetics, also many other factors influence the competition between ASRB and AMPB (Table 6.3).

6.3.2 Sulfide toxicity in anaerobic digestion

During anaerobic treatment of sulfate-rich wastewaters, SRB will competitively interact with the other anaerobic bacteria involved in methanogenesis. In the presence of sulfate, propionate and butyrate can be converted by SRB without the participation of methanogens (Harmsen *et al.* 1996). Sulfate reducers can probably combine the properties of syntrophic growth and sulfate reduction (Roest *et al.* 2005). The SRB metabolism resulted in the formation of sulfide, which is the most stable form of sulfur under anaerobic conditions.

Hydrogen sulfide dissociates in water according to the following equations:

$$H_2S \Leftrightarrow H^+ + HS^- \quad K_1 = 1.49 \times 10^{-7}\ (35°C) \quad \text{(Speece 1996)}$$
$$HS^- \Leftrightarrow H^+ + S^{2-} \quad K_2 = 0.8 \times 10^{-17}\ (20°C) \quad \text{(Steudel 2000)}$$

The toxicity of sulfide is regarded as being pH-dependent because only the neutral undissociated hydrogen sulfide (H_2S) molecule can pass through the cell membrane (Speece 1983). In the liquid phase, the total dissolved sulfide is present as the unionized form (H_2S) and as HS^-. As the pK_a value of this acid-base equilibrium is about 7, small pH variations in the pH range 6–8 will significantly affect the free (unionized) H_2S concentration. At neutral pH values, free H_2S accounts to 50% of total dissolved sulfide, whereas at pH 8 it is only around 10% (Figure 6.2).

The sulfide generated during anaerobic treatment will be distributed over the gas phase and the liquid phase according to the following expression:

$$[H_2S]_l = \alpha \cdot [H_2S]_g$$

which $[H_2S]_l$ and $[H_2S]_g$ are, respectively the concentrations of the H_2S in the liquid phase and the gas phase and α is a dimensionless distribution coefficient (Henry constant), whose value is 675 atm (mol fraction)$^{-1}$ at 35°C.

Sulfide can present several problems for the anaerobic digestion process or its implementation. The main manifestations of these problems include:

- the inhibitory effect of H_2S on many bacterial trophic groups involved in anaerobic digestion, thus reducing reactor performance (Koster *et al.* 1986; Hilton and Oleskiewicz 1988; Widdel 1988);
- a reduction of the methane yield, and thus less energy recovery;

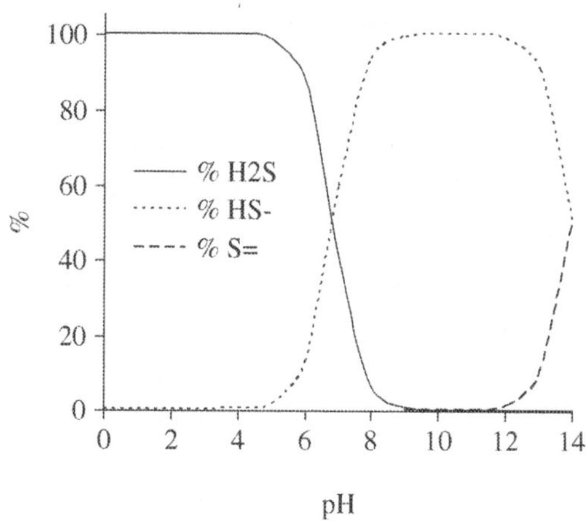

Figure 6.2 Equilibrium for $H_2S/HS^-/S^{2-}$ in aqueous solution as a function of pH.

- malodor, corrosion of piping, pumps, etc., necessity of scrubbing of the biogas and post-treatment of effluents to meet discharge standards (Rinzema 1988);
- accumulation of inert material in the sludge (e.g. metal sulfides);
- bulking problems in the aerobic post-treatment due to the growth of filamentous sulfide oxidizing bacteria such as *Thiothrix* and *Beggiotoa* (Buisman and Lettinga 1990; Ritmann and McCarty 2000);
- reduced COD removal efficiency due to the presence of H_2S in the effluent and oxygen demand on the receiving aquatic media thereby killing aquatic life;
- high toxicity for humans.

The presence of sulfate can also have some beneficial effects during anaerobic treatment of wastewaters:

- Most methanogens lack assimilatory sulfate reductases (Daniels *et al.* 1986) and their sulfur requirements are satisfied by a combination of inorganic sulfide and organic sulfur compounds. Consequently, the production of sulfide by dissimilatory bacterial species during anaerobic digestion may enhance methanogenesis by satisfying the sulfur growth requirements of methanogens.
- The production of sulfide during anaerobic digestion can also be beneficial as the formation of sulfide precipitates can alleviate the toxicity of many heavy metals ions (Lawrence and McCarty 1965; Tursman and Cork 1989). Some of the metallic sulfide, such as arsenic (at low pH), copper, lead, mercury, tin and chromium are practically insoluble, whereas the solubility of other sulfides such as antimony, cadmium, cobalt, iron, nickel and zinc is very low, with metal concentrations below 1.5 mg/L under normal operation conditions in an anaerobic reactor in which sulfate reduction takes place.

The exact mechanism of H_2S toxicity is unclear. One possible mechanism is native protein denaturation through formation of sulfide and disulfide cross-links between polypeptide chains. H_2S may also interfere with coenzymes A and M thorough the formation of sulfide linkages. This may result in interference with the acetyl coenzyme A pathway for CO_2 fixation which is common to both SRB and MPB. H_2S may also affect the internal cell pH.

It is assumed that the inhibitory form of sulfide is the undissociated form H_2S. Thus, one would expect a direct correlation between the H_2S concentration and the extent of inhibition. However, this relation is not always straightforward and other parameters (total sulfide concentration) can correlated better with the observed inhibition. H_2S may become more toxic at higher pH levels, possibly because of the development of strong pH gradients across the cell membrane, which may affect the diffusion properties of the H_2S molecule.

Alternatively, both total sulfide and H_2S may exert an inhibitory effect. The organism, therefore, may have two inhibition thresholds, one for H_2S and one for total sulfide. The levels of undissociated H_2S required for 50% inhibition of the different bacterial groups (IC_{50} value) were found to be much lower than the total sulfide IC_{50} value indicating that H_2S is clearly the most toxic form of sulfide. It is also found that propionate degrading SRB have a much lower threshold for total sulfide than other bacteria studied. By contrast, the level of undissociated H_2S required to cause 50% inhibition of growth is similar to that of other SRB and MPB.

Formation of insoluble heavy metal sulfides can inhibit the SRB activity. The inhibition mechanism that has been postulated has a physical nature. Metal sulfides deposit on the surface of the SRB or are concentrated in the vicinity of the bacterial cells and hinder further metabolism by preventing the contact between the reactants and the cells (Utgikar *et al.* 2002). The effect is external to the cells and non-toxic because SRB retains its ability to reduce sulfate.

6.3.2.1 Effect of sulfide toxicity on MPB

The literature on sulfide toxicity is highly complex and often contradictory. Information on the prevailing pH is often not included, which creates difficulties in determining the role of undissociated H_2S. As regards of the different trophic groups involved in methane production.

The levels reported in the literature for inhibition of MPB are quite varied, with IC_{50} values of 50–125 mg H_2S/L at pH 7–8 for suspended sludge; and 250 and 90 mg H_2S/L at pH values of 6.4–7.2 and 7.8–8.0, respectively for sludge granules (Kroiss and Plahl-Wabnegg 1983; Koster *et al.* 1986; Oleskiewicz *et al.* 1989; McCartney and Oleskiewicz 1993). Oleskiewicz *et al.* (1989) found that the most sensitive trophic group was the propionate oxidizers, with inhibition increasing for electron donors as follows: lactate, butyrate, acetate and propionate.

O'Flaherty *et al.* (1998b) found that syntrophic bacteria were less susceptible to sulfide inhibition than MPB and that their toxicity thresholds were comparable to those of the SRB. However, some evidence has been obtained indicating that inhibition of the syntrophic organisms was irreversible, unlike SRB (O'Flaherty *et al.* 1999). This would leave syntrophic organisms at a competitive disadvantage with SRB in anaerobic sludge.

6.3.2.2 Effect of sulfide toxicity on SRB

Few data are available on the sensitivity of SRB to sulfide toxicity and the published data are quite contradictory. Isa *et al.* (1986a) concluded that SRB growing in a fixed-film reactor were not affected by high levels of sulfide. By contrast, Widdel (1988) reported inhibition of pure cultures of *Desulfotomaculum acetoxidans* at H_2S concentrations of 85 mg/L.

Hilton and Oleskiewicz (1988) found that during the degradation of lactate, inhibition of SRB was directly related to the total sulfide concentration, whereas inhibition of MPB was related to the H_2S concentration.

Visser (1995) found that, for acetate utilizing SRB and MPB, the sensitivity of both groups of bacteria was similar between pH 7.0 and 7.5. However, at higher pH levels, the SRB were found to be considerably less sensitive to sulfide inhibition than MPB.

There is a considerable variation among different groups of SRB with respect to sulfide inhibition. The propionate utilizing SRB were more sensitive to inhibition than other SRB under test. (O'Flaherty *et al.* 1998a) This may explain the commonly reported finding that propionate degradation is the rate-limiting step during anaerobic treatment of sulfate-containing organic wastewaters.

6.3.2.3 *Effect of sludge aggregation*

The IC_{50} values for granular and suspended sludges are very similar at higher pH values, but granular sludge is less inhibited at lower pH values than suspended sludge. Other studies clearly show that sulfide toxicity is mediated at lower concentrations in suspended growth systems than in the attached biofilm of fixed bed systems. Factors such as substrate transport within biofilms/flocs/granules, the site of sulfate reduction and its proximity to the site of methanogenesis, the diffusion of H_2S and dissolved sulfide and pH gradients, etc., clearly play an important role in the ultimate degree of inhibition.

6.3.2.4 *Effect on process operation*

Inhibition decreases the efficiency of reactor performance and can even lead to complete process failure. In general, wastewaters with a COD/sulfate ratio higher than 10 do not pose problems for methanogenic treatment (Rinzema and Lettinga 1988). So far, no models have been developed that allow prediction of the conditions that result in process failure of digesters treating wastewaters with a COD/sulfate ratio higher than 10. The outcome of competitive interactions between SRB and other anaerobic bacteria such as syntrophs and methanogens in digesters sludges, will determine the amount of sulfide in the digester and the risk of process failure. An important consideration in determining the likely outcome of anaerobic treatment of a sulfate-containing wastewater is the origin and microbial composition of the seed sludge: that is, whether it has been acclimatized to sulfate or not.

6.3.3 Other inhibition phenomena

The inhibition phenomena encountered as a result of the presence of sulfur compounds in wastewaters can be divided into three categories: inhibition by sulfide, by sulfite and by cations. Of these, inhibition by sulfide is the most important and will, therefore, be the principal focus of this chapter, but the other inhibition phenomena are also important.

6.3.3.1 *Sulfite inhibition*

Few data are available on sulfite inhibition of anaerobic bacteria. It has been shown in batch experiments that sulfite induces a lag phase in methane production of variable length. The effects of sulfite inhibition appear to be far less severe with adapted sludges, probably owing to the presence of SRB which reduce sulfite to sulfide.

It is generally considered that sulfite inhibition will be insignificant during reactor operation, as SRB populations will develop that eliminate sulfite.

6.3.3.2 *Sodium inhibition*

The effect of sodium on methanogenic digestion has been studied extensively. As for sulfide inhibition thresholds, the literature shows many inconsistencies as indicated by reported values ranging from 6 to 40 g/L for the 50% inhibition of methanogenic bacteria by sodium. These differences can be attributed to the history of the sludge, antagonistic and synergistic effects and the test method used.

The presence of other cations, such as potassium, causes antagonistic or synergistic effects, resulting in a significant change in the sodium sensitivity of anaerobic bacteria. The sulfidogenic activity of granular sludge adapted to sodium levels of 1.5–2 and 5.5–6 g/L, respectively was inhibited at sodium concentrations exceeding 11 g/L and 50% inhibition of the activity was observed at about 15 g/L of sodium for both sludge samples (Visser 1995). Such high sodium concentrations are essential for the growth of many marine SRB, but are inhibitory to freshwater SRB. In practice, the sodium content of a sulfate-rich wastewater is very unlikely to cause process failure (e.g. effect on granulation).

6.3.3.3 Calcium inhibition

Although the calcium ion does not exert direct toxic effects, $CaCO_3$ and/or $Ca_3(PO_4)_2$ precipitates can indirectly upset the reactor performance by scaling. These precipitates are entrapped in the reactor biomass, where they gradually accumulate and ultimately result in a complete loss of the activity of the sludge granules owing to a calcium layer which can completely block substrate transport. Serious scaling of biomass by calcium precipitates may occur at Ca^{2+} concentrations as low as 400 mg/L. Clogging problems can also arise from precipitates in the piping system. Moreover, calcium phosphate precipitation can cause phosphate deficiency and thus limit microbial activity. The effect of high calcium and sulfate concentration can be more important than the H_2S toxicity in pulp and paper mill effluents (Thompson *et al.* 2001).

Scarcely soluble compounds like $CaSO_4$ can be reduced under anaerobic conditions. No substantial differences were noted in the reduction of sodium sulfate or gypsum coming from thermal power plant flue gases desulfurization. (Ghigliazza *et al.* 2000).

6.4 STRATEGIES TO OVERCOME PROBLEMS ASSOCIATED WITH SULFATE-RICH WASTEWATERS TREATMENT

6.4.1 Competition between SRB and MPB

Methods that influence the outcome of the competition between SRB and MPB would be useful to develop fully methanogenic or sulfate reducing sludges, depending on the desired process application. Moreover, they can prevent potential process failures due to sulfide inhibition. Thus far, however, adequate methods to steer the competition between ASRB and AMPB that can be applied at full scale are not available (Van Houten *et al.* 1997; Lens *et al.* 1998a). The best way to steer a reactor toward the predominance of one population over the other, involves the manipulation of the inoculum composition or the environmental conditions, for example, the reactor pH. The MPB/SRB ratio of a sludge can be manipulated by adding pure

Table 6.4 Effect of changes in process conditions on the competition between SRB and MPB: increase of the share of SRB related to the total COD removal (%).

Measure	Increase (%)	Reference
Manipulation of the influent composition		
Increase of acetate concentration	−15	Omil *et al.* 1996
Addition of iron (2 g/L)	0	Isa *et al.* 1986a, b
Addition of transition elements	0	Clancy *et al.* 1992
Availability of the electron donor	NR	Vroblesky *et al.* 1996
Manipulation of the biomass composition		
Addition of *Desulforhadbus amnigenes*	0	Omil *et al,* 1997a
Exposure to oxygen	35	Omil *et al.* 1997a
Manipulation of operational conditions		
Alteration of pH	41	Omil *et al.* 1996
Shock treatment		
Temperature decrease to 15°C	0	Omil *et al.* 1997b
Temperature increase to 65°C	30	Visser *et al.* 1993b
Manipulation of reactor design		
Expanded granular sludge bed (upflow velocity 4–6 m/h)	−30	Omil *et al.* 1996
USSB reactor	10	Lens *et al.* 1998b
Baffled reactor	30	Lens *et al.* 1999

Negative values means an increase in the share of the MPB to the total COD removal. NR: not reported; USSB: staged sludge bed.
Adapted from Huslhoff Pol *et al.* (1998)

cultures of MPB or SRB, or creating unfavorable conditions for the undesired population during short time intervals (Table 6.4). Successful methods that enhance the development of an SRB population after selective inhibition by MPB include shocks of high sulfide concentrations (Omil *et al.* 1996) or high (65°C) temperatures (Visser *et al.* 1993a). However, more research is needed to develop methods that can be used for full-scale applications.

6.4.2 Reducing sulfide concentration in bioreactors

Considering the potential problems related to the occurrence of sulfate reduction in the anaerobic digestion process, a complete suppression of the sulfate reduction and a complete conversion of the organic substrate into methane could be considered as the most optimal option. Therefore, attempts have been made to selectively suppress sulfate reduction by using specific inhibitors (Table 6.5). However, so far, no selective inhibitors for SRB have been found that are suitable for use in full-scale anaerobic reactors. This implies that sulfate reduction cannot be prevented in practice. The main strategies to overcome sulfide toxicity are:

- *Chemical precipitation*. Sulfide precipitates as insoluble metal sulfide with many of the divalent metals such as iron, zinc or copper. Although iron sulfide is more soluble than the other metal sulfides, iron salts (e.g. Fe^{2+} and Fe^{3+}) are widely used because of economic and toxicity considerations. Ferrous sulfide

Table 6.5 Measures to reduce the reactor sulfide concentration, thus allowing the integration of methanogenesis and sulfate reduction.

Dilution of the influent
Non-sulfate-containing process water
Recycle of effluent after a sulfide removal step by:
- sulfide stripping
- sulfide precipitation
- biological sulfide oxidation to elemental sulfur
- chemical sulfide oxidation to elemental sulfur

Decrease of the unionized sulfide concentration
Elevation of the reactor pH
Elevation of the reactor temperature
Precipitation of sulfide
Stripping of the reactor liquid using:
- high mixing degree inside the reactor
- recirculation of biogas after scrubbing
- other stripping gas (e.g. N_2)

Separation of sulfide production and methanogenesis
Two stage anaerobic digestion
Multi-step process

Selective inhibition of SRB
Sulfate analogues (e.g. MO_4^{2-})
Transition elements (e.g. Co, Ni or Zn)
Antibiotics

is essentially insoluble, but continuous precipitation of FeS in the reactor could lead to serious consequences such as the reduction of effective volume or clogging in anaerobic filters.

- *Microbial oxidation of ferrous iron.* H_2S can be indirectly removed by oxidation with ferric iron to generate elemental sulfur in a chemical reactor; the resulting ferrous iron is oxidized by iron-oxidizing bacteria such as *Acidithiobacillus ferrooxidans* and *Leptospirillum ferrooxidans* in a separate biological reactor (Jensen and Webb 1995). The biological step is the rate limiting one, but its performance can be improved using a submerged-membrane bioreactor or an immobilized bioreactor (Park *et al.* 2005).
- *Stripping and precipitation.* H_2S in the biogas is removed in a stripping tower with iron or Fe_2O_3. The Fe^{3+} ions react with the H_2S in the biogas to produce elemental sulfur. A chelating agent is added to the liquid to prevent FeS, $Fe(OH)_2$ or $Fe(OH)_3$ precipitation. The clean biogas is then recycled back to the reactor to strip out the gaseous sulfide from the reactor. The process is operated in continuous. The exhausted Fe_2O_3 can be regenerated by heating in air/O_2. Important disadvantages of the system are the high cost of the chemicals and the control of the exact dosing.
- *Sulfide oxidation with oxygen.* Sulfide can be oxidized with oxygen, biological or chemically (with or without catalyst). In comparison with other methods to control sulfide toxicity, the use of oxygen has many advantages such as the

elimination of expensive chemicals, no treatment of sludge with excess of reagents is necessary, no problems of clogging or precipitation in the reactor appear, improve the organic removal and the process stability, elemental sulfur is the end product (valuable product) and the use of oxygen allows the process automatization to control the oxidant dose.

The biological oxidation of sulfide has been studied in recent years (Buisman and Lettinga 1990; Buisman *et al.* 1990a; Sengül and Müezzinoglu 1991; Janssen *et al.* 1997; Janssen *et al.* 1998). Many of these studies were employed for sulfide elimination rather than to control sulfide toxicity to MPB. Sulfide is oxidized to elemental sulfur under oxygen limiting conditions (<0.1 mg/L) by aerobic sulfide oxidizing bacteria. Acclimatization to low sulfide concentrations enabled the sludge to degrade subsequent high loads which were toxic to non-acclimatized sludge (Burgess and Stuetz 2002)

6.4.3 Process technology

In principle, all common anaerobic bioreactors can be applied for the treatment of sulfate-containing wastewaters, provided proper measures to prevent the occurrence of high H_2S concentrations in the liquid or in the gas phase (Table 6.5), the precipitation of inorganic sulfides (leading to less active biomass), and low mass transfer efficiencies due to the lower biogas generation.

When wastewaters containing organic matter and sulfate are treated in an anaerobic bioreactor, organic matter will be removed both via sulfate reduction and via methanogenesis. In practice, anaerobic treatment always proceeds successfully for wastewaters with COD/sulfate ratios exceeding 10, as for such wastewaters the H_2S concentration in the reactor will never exceed the presumed critical value of 150 mg/L (Rinzema and Lettinga 1988). At COD/sulfate ratios lower than 10, process failures of anaerobic reactors have been reported, whereas in other cases the process proceeds successfully when precautions are taken to prevent sulfide toxicity (Table 6.5) (e.g. determination of the optimal pH to maintain the concentration of unionized H_2S as low as possible). When the COD/sulfate ratio is lower than 1, sulfidogenesis will prevail. In this case, sulfidogenic wastewater treatment may be of commercial interest in the context of sulfur recovery for re-use (Visser 1995), although the addition of organic matter could be necessary.

There are some particular aspects that should be considered when complete sulfate-reducing bioreactors are operated, as their relatively low gas production rates compared with fully methanogenic reactors, which can lead to a decrease of the mass transfer efficiency. Also the possible precipitation of inorganic metal sulfides is another factor that could cause problems in some reactor configurations, such as the anaerobic filter design. Therefore, the selection of the reactor type has to be in accordance with the aim of the treatment: the removal of organic matter, the removal of sulfate or the removal of both.

Biological sulfate removal canbe carried out in wastewater under-saturated or over-saturated with respect to gypsum, as well as for treatment of acid water directly. However, it is more cost-effective if sulfate is removed with $CaCO_3$ or lime from a over-saturated wastewater (Maree *et al.* 2004)

6.5 RECENT DEVELOPMENTS AND CHALLENGES

6.5.1 Steering of the competition between SRB and MPB

Methods that influence the outcome of the competition between SRB and MPB would be useful to develop fully methanogenic or sulfate-reducing sludge, depending on the desired process application. Moreover, they can prevent potential process failures due to sulfide inhibition. However, adequate methods to steer the competition between ASRB and AMPB that can be applied at full scale are not available.

The best way to steer a reactor toward the predominance of one population over the other involves:

- *Manipulation of the influent composition*: increase of acetate concentration (Omil *et al.* 1996), addition of iron (Isa *et al.* 1986a, b), addition of transition elements (Clancy *et al.* 1992) or availability of the electron donor (Vroblesky *et al.* 1996).
- *Manipulation of inoculums composition*: adding pure cultures of the desirable population or creating unfavorable conditions for the undesirable population during short time intervals (Omil *et al.* 1997a).
- *Manipulation of the environmental conditions*: alteration of pH (Omil *et al.* 1996) or shocks of high temperature (Visser *et al.* 1993; Omil *et al.* 1997b; Vallero *et al.* 2004).
- *Manipulation of the reactor design*: staged sludge bed (USSB) reactor (Lens *et al.* 1998b) or baffled reactor (Lens *et al.* 1999).

However, more research is needed to develop reliable methods that can be used for full-scale applications.

6.5.2 SRB-based bioremediation techniques

SRB are present in anaerobic and even in aerobic wastewater treatment sludges, the development of processes using their capacity to degrade a wide range of organic compounds, opens promising perspectives for environmental biotechnology:

- SRB do not require balanced growth with acetogens, which implies less sensitivity to organic overloads.
- SRB are less sensitive to toxicants.
- Heavy metals are precipitated by sulfide, thus reducing their potential toxic effects.
- SRB can metabolize organic inhibitors such as aromatics, alkanes, chlorinated compounds and long chain fatty acids.

On the other hand, reactors based on organic matter removal by sulfate reduction lack one of the major advantages of methanogenic treatment, the recovery of methane from organic compounds.

6.5.3 Micro-aerophilic treatment

Another way to utilize the capacities of SRB to degrade organic matter, is the use of reactors with combined aerobic/anaerobic conditions to oxidize sulfide. Also, these conditions are needed for a complete mineralization of certain xenobiotics (Field *et al.* 1995), where (facultative) aerobic bacteria utilize the degradation products of the SRB. Besides oxygen, also nitrate can be used in micro-aerophilic systems (Lens *et al.* 1999), nitrate is much more soluble compared to molecular oxygen.

Sulfide oxidation proceeds both chemically and biologically. Biological sulfide oxidation in wastewater treatment systems is typically associated with the activity of colorless sulfur bacteria. Those bacteria utilize the energy derived from the following overall reactions (Kuenen 1975):

$$2HS^- + O_2 \rightarrow 2S^0 + 2OH^- \qquad \Delta G^0 = -169.4 \text{ kJ/mol}$$
$$2HS^- + 4O_2 \rightarrow 2SO_4^{2-} + 2H^+ \qquad \Delta G^0 = -732.6 \text{ kJ/mol}$$

Generation of sulfate probably proceeds via elemental sulfur and sulfite. Alternatively, sulfide can also be biologically oxidized to thiosulfate and tetrathionite (Stratford *et al.* 1995).

In order to obtain feasible sulfide oxidation in micro-aerobically operated wastewater treatment systems, sulfide oxidation should be effective in its competition for oxygen with other oxidative processes like aerobic oxidation of organic COD. In addition, the sulfide oxidation process should be faster than the re-reduction of oxidized sulfur species.

6.6 WASTEWATER TREATMENT

The way in which the industry can aboard the treatment of sulfate-containing wastewater greatly depends, besides the COD/sulfate ratio, on the flow and concentration of the stream. Table 6.6 shows different applications of sulfate reduction bacteria related with sulfate or heavy metal removal, and liquid or solid wastes treatment. According to the characteristic of the wastewaters, the applications can be divided in different categories that are discussed in Table 6.6.

6.6.1 Very high sulfate concentration and low flow streams

When the sulfate concentration in the wastewater is not very high and the flow is low, the wastewater is mixed and treated together with the other wastewater stream generated in the industry. However, if the sulfate concentration is very high, the stream has to be managed separately to avoid the perturbation of the total wastewater treatment.

This kind of waste stream is a very typical case in pharmaceutical and some chemical industries, in which sulfuric acid is used to control pH in bio-chemical processes. In this case one of the better options consists of changing in the operational conditions, replacing sulfuric acid by other inorganic (HCl) or organic acid (acetic or formic acid). The use of organic acids will increase the production cost and the organic load of the wastewater, but will improve highly the treatability of the final effluent. The effluent can be mixed with the other wastewater streams and treated together.

This solution, however, cannot be easily applied in the pharmaceutical industry because medicaments production is under a very strict protocol that made quite

Table 6.6 Biotechnological applications using SRB.

Application	Reference
Biological sulfate removal	
Industrial wastewaters	Särner 1990
AMD	Maree *et al.* 1991
Spent sulfuric acid	Stucki *et al.* 1993
Scrubbing waters SO_2 rich gases	Kaufman *et al.* 1996
Heavy metal removal	
Extensive treatment (wetlands)	Hao *et al.* 1996
High rate reactors	Tichy *et al.* 1998
Process water	
AMD	Barnes *et al.* 1991
Ground water	Scheeren *et al.* 1991
Micro-aerobic treatment	
Treatment domestic sewage	Takahashi and Kyosai 1991
Reduction waste sludge production	Lens *et al.* 1995b
Solid waste treatment	
Gypsum processes	Deswaef *et al.* 1996

difficult to modify the operational conditions. When a change in operational conditions is not possible, the solution adopted is normally the segregation of the high sulfate stream and a separated management. If the flow is no high, the stream is treated externally as a hazardous waste.

6.6.2 High flow streams

There are different options for the treatment of high flow streams with high sulfate concentration. One of the options is removal of sulfate by chemical precipitation, and the others are the use of biological processes.

6.6.2.1 Chemical precipitation

If the sulfate concentration is high, it can be precipitated as gypsum ($CaSO_4$), that has a solubility of 2.4 g/L. With this treatment sulfate concentration can be reduced to values lower than 1200 mg/L, allowing a posterior anaerobic treatment of the wastewater. One of the problems associated with this treatment is the deposition of the gypsum, with the problems associated with solids separation and handling and the possibility of pipe clogging.

6.6.2.2 Aerobic treatment

Aerobic treatment will not generate any problem related with sulfate reduction. Depending on organic matter concentration this alternative can be expensive compared to anaerobic treatment, but could be an appropriated one when sulfate concentration is below discharge limits, typically below 2000 mg/L. This process can be preceded by a chemical precipitation to reduce sulfate concentration to permit limits.

6.6.2.3 *Anaerobic treatment*

This is the more common process employed at industrial level for the treatment of wastewaters with high organic load. The applicability of the anaerobic treatment to high sulfate wastewater depend on the inhibition problems generated by hydrogen sulfide that have been discussed above. To reduce the problems generated by H_2S in the biogas there are some treatment possibilities:

- *Biogas cleaning*. Absorption with a basic solution or oxidation.
- *Aerobic post-treatment*. Organic matter is reduced in anaerobic step together with sulfate reduction. In the aerobic step the dissolved sulfide is reoxidized to sulfate.

However, at industrial level, some of the problems generated in the treatment of wastewaters with high sulfate concentration are related not only to the sulfate, but to the presence of other compounds like calcium in the case of pulp and paper industries or organic nitrogen in the case of fermentation industries, mainly when the COD/SO_4^{2-} is high enough, (>10) to avoid inhibition problems caused by H_2S.

The presence of calcium in wastewaters to be treated anaerobically will lead to biomass mineralization due to the precipitation of $CaCO_3$ inside the reactor. This fact makes necessary to increase reactor control and biomass waste to maintain activity inside the reactor. Calcium carbonate precipitation also occurs, creating scaling problems, in the outlet reactor due to the pH change produced by CO_2 degasification in these points.

Wastewaters with high total Kjeldahl nitrogen (TKN) content. Effluents with high concentrations of organic matter, sulfate and TKN can be found in the fermentation industry. The treatment of these effluents can be carried out in an anaerobic reactor followed by an aerobic one. In the anaerobic reactor biodegradable COD is removed, sulfate is reduced to H_2S and more than 80% of organic nitrogen is transformed to ammonium. In the aerobic reactor ammonia is oxidized to nitrate and the dissolved sulfide is also oxidized again to sulfate. As the hydraulic retention time in the aerobic reactor is usually high enough to obtain nitrification, the aeration can be stopped in some periods (12 h/day), decreasing the dissolved oxygen to levels that permit denitrification of the nitrate formed in the aerated phase. Part of the raw influent can be by-passed directly to the aerobic reactor to bring organic matter to get denitrification.

6.6.2.4 *Wastewaters with high calcium content*

Maree *et al.* (2004) describe a chemical/biological process in which sulfate and sulfide are removed simultaneously during biological treatment. Sulfate is removed partially by chemical pre-treatment and in the biological stage sulfate is reduced to sulfide; the addition of organic matter could be necessary. Sulfide is oxidized by allowing oxygen to enter the system in a controlled way.

This system is employed for the treatment of a coal discharge leachate with a pH of 2.2 and a sulfate concentration of 9200 mg/L. The treatment system consists of:

- Calcium carbonate neutralization. Powder $CaCO_3$ is used to raise the pH to 7.0 and to precipitate some metal (Fe^{3+} and Al^{3+}) as hydroxides. CO_2 generated

Table 6.7 Operational performance of the ADM full-scale anaerobic wastewater treatment plant.

Operational process parameters	Values
Influent sulfate (g/L)	3.43
Influent COD/sulfate ratio	3.61
Hydraulic retention time (h)	1.4
COD removal efficiency (%)	52
Volumetric loading rate (kg COD/m^3-day)	9.0
Biogas production (m^3/h)	1053
CH_4 in biogas (%)	65.5
H_2S in biogas (%)	4.8

during neutralization is used downstream for pH adjustment of the lime treated water.

- Lime treatment/gypsum crystallization/$CaCO_3$ precipitation. Lime is employed to increase pH to for precipitation of metal such manganese and magnesium. Sulfate is partially removed, to less than 1200 mg/L due to gypsum crystallization. Latter on the pH is reduced with the CO_2 generated in the previous step.
- Biological sulfate removal. Sucrose is added as organic carbon and energy source. In this step $CaCO_3$ is produced that can be recycled to the first step.

With this system sulfate is removed to 1200 mg/L in the gypsum crystallization step and to 200 mg/L in the biological step.

6.6.2.5 Citric acid production wastewater treatment

A full scale, fixed bed reactor was employed to treat the wastewater from the ADM citric acid production plant at Ringaskiddy, Cork, Ireland and has been described by O'Flaherty *et al.* (1998b) and Colleran *et al.* (1998). The support material consisted of polypropylene cascade rings. The operational performance of the plant after the start-up is presented in Table 6.7.

One important fact noted in this plant was that the methane observed yield exceeded the theoretically possible yield. This was attributed to the presence of betaine in the digester effluent. Betaine comes from the sugar beet and is incompletely oxidized in the standard COD assay, underestimating the organic matter content of the influent, but can be biologically degraded. In this plant the real organic matter was calculated to be approximately a 35% higher than the value determined analytically. This implies that the COD/sulfate ratio was 5.6:1, rather than 3.6:1, and that the volumetric organic loading rate applied to the reactor was higher than that indicated in Table 6.7. The real COD removal was also higher than that presented in the table.

Analysis of the digester biomass after 5 years of operation indicated that hydrogen and propionate utilizing SRB had out-competed hydrogenotrophic methanogens and propionate syntrophs; however, acetoclastic methanogens played a dominant role in acetate conversion.

Table 6.8 Average wastewater characteristics from the fermentation industry.

Flow (m^3/day)	2100
COD_t (mg/L)	11000
COD_s(mg/L)	8000
BOD_5 (g/L)	8000
TKN (mg/L)	1600
N-NH_4^+ (mg/L)	400
SO_4^{2-} (mg/L)	4000

6.6.2.6 Fermentation industry wastewater treatment

Fermentation industry wastewater usually has a high sulfate concentration, together with high COD and TKN. The treatment of these wastewaters implies the combination of the cycles of carbon, nitrogen and sulfur.

The Burns Philp Food, S.A. yeast production factory located in Cordoba (Spain) generates a wastewater whose characteristics are reflected in Table 6.8. After an evaluation of the alternatives, it was decided to treat the wastewater in the municipal wastewater treatment plant, located 5 km away from the factory.

The wastewater is stored and homogenized in the factory and then pumped to the municipal wastewater treatment plant, where it is pre-treated anaerobically. This treatment is carried out in two upflow anaerobic sludge bed (UASB) systems, with a volume of 1400 m^3 each and equipped with nine BIOPAQ modules. Ferric sulfate is added to the effluent from the anaerobic reactor to reduce the concentration of soluble sulfide, and is treated jointly with the municipal wastewater.

The anaerobic COD removal efficiency is higher than 75% and the efficiency for BOD_5 is higher than 85%. This efficiencies difference is due to the presence in the influent of biodegradable, but non-chemically oxidable organic matter (Section 6.6.2.5). Although the COD/SO_4^{2-} ratio is not very high, inhibition problems in the anaerobic treatment due to H_2S toxicity manifested.

The organic load from the industrial wastewater supposes around 10% of the total organic load that treats the municipal plant. In this plant takes place nitrification and partial denitrification of the ammonia present in the influent.

REFERENCES

Anderson, G.K., Sanderson, J.A., Saw, C.B. and Donnelly, T. (1988) Fate of COD in an anaerobic system treating high sulfate bearing wastewater. In *Biotechnology for Degradation of Toxic Chemicals in Hazardous Wastes* (eds. R.J. Scholze Jr, E.D. Smith, J.T. Bandy, Y.C. Wu and J.V. Basilico). Noyes Data Corporation, Park Ridge, New Jersey, USA, pp. 504–531.

Barnes, L.J., Janssen, F.J., Scheeren, P.J.H., Versteegh, J.H. and Koch, R.O. (1991) Simultaneous microbial removal of sulphate and heavy metals from wastewater. *Trans. Instn. Metall.* (Sect. C: Mineral Process Extr. Metal) **101**, 183–189.

Boshoff, G., Duncan, J. and Rose, P.D. (2004a) Tannery effluent as a carbon source for biological sulphate reduction. *Wat. Res.* **38**, 2651–2658.

Boshoff, G., Duncan, J. and Rose, P.D. (2004b) The use of micro-algal biomass as a carbon source for biological sulphate reducing systems. *Wat. Res.* **38**, 2659–2666.

Buisman, C.J.N. and Lettinga, G. (1990) Sulphide removal from anaerobic waste treatment of a papermill. *Wat. Res.* **24**(3), 313–319.

Buisman, C.J.N., Geraats, B.G., Ijspeert, P. and Lettinga G. (1990a) Optimization of sulphur production in a biotechnological sulphide-removing reactor. *Biotechnol. Bioeng.* **50**, 50–56.

Buisman, C.J.N., Ijspeert, P., Janssen, A. and Lettinga, G. (1990b) Kinetics of chemical and biological sulphide oxidation in aqueous solutions. *Wat. Res.* **24**(5), 667–671.

Burgess, J.E. and Stuetz, R.M. (2002) Activated sludge for the treatment of sulphur-rich wastewaters. *Minerals Eng.* **15**, 839–846.

Carrondo, M.J.T., Silva, J.M.C., Figueira, M.I.I., Ganho, R.M.B. and Oliveira, J.F.S. (1983) Anaerobic filter treatment of molasses fermentation wastewater. *Wat. Sci. Technol.* **15**(8/9), 117–126.

Clancy, P.B., Venkataraman, N. and Lynd, L.R. (1992) Biochemical inhibition of sulfate reduction in batch and continuous anaerobic digesters. *Wat. Sci. Technol.* **25**, 51–60.

Colleran, E., Finnegan, S. and Lens, P. (1995) Anaerobic treatment of sulphate-containing waste streams. *Antonie van Leeuwenhoek* **67**, 29–46.

Colleran, E., Pender, S., Philpott, U., O'Flaherty, V. and Leahy, B. (1998) Full-scale and laboratory scale anaerobic treatment of citric acid production wastewater. *Biodegradation* **9**, 233–245.

Daniels, L., Belay, N. and Rajagopal, B.S. (1986) Assimilatory reduction of sulfate and sulfite by methanogenic bacteria. *Appl. Environ. Microbiol.* **51**, 703–709.

De Smul, A. and Verstraete, W. (1999) Retention of sulfate reducing bacteria in expanded-granular-blanket reactors. *Wat. Environ. Res.* **71**, 427–431.

Deswaef, S., Salmon, T., Hiligsmann, S., Taillieu X., Milande N., Thonart P. and Crine M. (1996) Treatment of gypsum wastewater in a two stage anaerobic reactor. *Wat. Sci. Technol.* **34**, 367–374.

Field, J.A., Stams, A.J.M., Kato, M. and Schraa, G. (1995) Enhanced biodegradation of aromatic pollutants in cocultures of anaerobic and aerobic consortia. *Antonie van Leeuwenhoek* **67**, 47–77.

Genschow, E., Hegemann, W. and Maschke, C. (1996) Biological sulfate removal from tannery wastewater in a two-stage anaerobic treatment. *Wat. Res.* **30**(9), 2072–2078.

Ghigliazza, R., Lodi, A. and Rovatty, M. (2000) Kinetic and process considerations on biological reduction of soluble and scarcely soluble sulfates. *Resour. Conserv. Recy.* **29**, 181–194.

Gupta, A., Flora, J.R.V., Gupta, M., Sayles, G.D. and Suidan M.T. (1994) Methanogenesis and sulfate reduction in chemostats. I. Kinetic studies and experiments. *Wat. Res.* **28**, 781–793.

Habets, L.H.A. and de Vegt, A.L. (1991) Anaerobic treatment of bleached TMP and CTMP effluent in the biopaq UASB system. *Wat. Sci. Technol.* **24**, 331–345.

Hao, O.J., Chen, J.M., Huang, L. and Buglass, R.L. (1996) Sulfate-reducing bacteria. *Crit. Rev. Environ. Sci. Technol.* **26**, 155–187.

Harada, H., Uemura, S. and Momonoi, K. (1994) Interactions between sulphate-reducing bacteria and methane-producing bacteria in UASB reactors fed with low strength wastes containing different levels of sulfate. *Wat. Res.* **28**, 355–367.

Harmsen, H.J.M., Akkermans, A.D.L., Stams, A.J.M. and de Vos, W.M. (1996) Population dynamics of propionate oxidizing bacteria under methanogenic and sulfidogenic conditions in anaerobic granular sludge. *Appl. Environ. Microbiol.* **62**, 2163–2168.

Hilton, B.L. and Oleskiewicz, J.A. (1988) Sulfide-induced inhibition of anaerobic digestion. *J. Environ. Eng.* **114**, 1377–1391.

Hoeks, F.W.J.M.M., Ten Hoopen, H.J.G., Roels, J.A. and Kuenen, J.G. (1984) Anaerobic treatment of acid water (methane production in a sulfate rich environment). *Progr. Ind. Microbiol.* **20**, 113–119.

Huslhoff Pol, L., Lens, P., Stams, A.J.M. and Lettinga, G. (1998) Anaerobic treatment of sulfate-rich wastewaters. *Biodegradation* **9**(3/4) 213–224.

Hulshoff Pol, L., Lens, P.N.L. and Weijma, J. (2001) New development in reactor and process technology for sulphate reduction. *Water Sci. Technol.* **44**(8), 67–76.

Isa, Z., Grusenmeyer, S. and Verstraete, W. (1986a) Sulfate reduction relative to methane production in high-rate anaerobic digestion: technical aspects. *Appl. Environ. Microbiol.* **51**, 572–579.

Isa, Z., Grusenmeyer, S. and Verstraete, W. (1986b) Sulfate reduction relative to methane production in high-rate anaerobic digestion: microbiological aspects. *Appl. Environ. Microbiol.* **51**, 580–587.

Janssen, A.J.H., Ma, S.C., Lens, P. and Lettinga, G. (1997) Performance of a sulfide-oxidizing expanded-bed reactor supplied with dissolved oxygen. *Biotechnol. Bioeng.* **53**(1), 32–40.

Janssen, A.J.H., Meijer, S., Bontsema, J. and Lettinga, G. (1998) Application of the redox potential for controlling a sulfide oxidizing bioreactor. *Biotechnol. Bioeng.* **60**(2), 147–155.

Jensen, A.B. and Webb, C. (1995) Ferrous sulphate oxidation using *Thiobacillus ferrooxidants*: a review. *Process Biochem.* **30**, 225–236.

Kaufman, E.N., Little, M.H. and Selvaraj, P.T. (1996) Recycling of FGD gypsum to calcium carbonate and elemental sulfur using mixed sulfate-reducing bacteria with sewage digest as a carbon source. *J. Chem. Tech. Biotechnol.* **66**, 365–374.

Koster, I.W., Rinzema, A., de Vegt, A.L. and Lettinga, G. (1986) Sulfide inhibition of the methanogenic activity of granular sludge at different pH levels. *Wat. Res.* **12**, 1561–1567.

Kroiss, H. and Plahl-Wabnegg, F. (1983) Sulfide toxicity with anaerobic wastewater treatment. In *Proceedings of the European Symposium on Anaerobic Wastewater Treatment* (AWWT), TNO Corporate Communications Department, The Hague, The Netherlands, pp. 72–78.

Kuenen, J.G. (1975) Colourless sulphur bacteria and their role in the sulphur cycle. *Plant Soil* **43**, 49–76.

Lawrence, A.W. and McCarty, P.L. (1965) The role of sulfide in preventing heavy metals toxicity in anaerobic treatment. *J. Wat. Poll. Cont. Fed.* **37**, 392–409.

Lens, P., Massone, A., Rozzi, A. and Verstraete, W. (1995a) Effect of sulfate concentration and scraping on aerobic fixed film reactors. *Wat. Res.* **29**, 857–870.

Lens, P., De Poorter, M.P., Cronenberg, C.C. and Verstraete, W.H. (1995b) Sulfate reducing and methane producing bacteria in aerobic wastewater treatment. *Wat. Res.* **29**, 871–880.

Lens, P., Visser, A., Janssen, A.J.H., Hulshoff Pol, L.W. and Lettinga, G. (1998a) Biotechnological treatment of sulfate rich wastewater. *Crit. Rev. Environ. Sci. Technol.* **28**, 41–88.

Lens, P., van der Bosch, M., Hulshoff Pol, L.W. and Lettinga, G. (1998b) The use of staged sludge bed reactors for the treatment of sulfate rich wastewaters. *Wat. Res.* **32**, 1178–1192.

Lens, P., Sipma, J., Hulshoff Pol, L.W. and Lettinga, G. (1999) Effect of staging and nitrate addition on sulphidogenic acetate removal. *Wat. Res.* **34**, 31–42.

Lens, P., Vallero, M., Esposito, G. and Zandvoort, M. (2002) Perspectives of sulfate reducing bioreactors in environmental biotechnology. *Rev. Environ. Sci. Biotechn.* **1**, 311–325.

Maree, J.P. and Strydon, W.F. (1985) Biological sulfate removal in an upflow packed bed reactor. *Wat. Res.* **19**(9) 1101–1106.

Maree, J.P., Hulse, G., Dods, D. and Schutte, C.E. (1991) Pilot plant studies on biological sulfate removal from industrial effluent. *Wat. Sci. Technol.* **23**, 1293–1300.

Maree, J.P., Greben, H.A. and de Beer, M. (2004) Treatment of acid and sulphate-rich effluents in an integrated biological/chemical process. *Wat. S.A.* **30**,183–189.

McCartney, D.M. and Oleskiewicz, J.A. (1993) Competition between methanogens and sulfate reducers: effect of COD:sulfate ratio and acclimatization. *Wat. Environ. Res.* **65**, 655–664.

Mendez, R., Lema, J.M. and Soto, M. (1995) Treatment of seafood-processing wastewaters in mesophilic and thermophilic anaerobic filters. *Wat. Environ. Res.* **67**(1), 33–45.

Middleton, A.G. and Lawrence, A.W. (1977) Kinetics of microbial sulfate reduction. *J. Wat. Pollut. Control Fed.* **229**, 1659–1670.

Mohanrao, G.J., Subrahmanyam, P.V., Deshmukh, S.B. and Saroja, S. (1970) Waste treatment at a synthetic drug factory in India. *J. Wat. Pollut. Control Fed.* **42**(8), 1530–1543.

Mulder, A. (1984) The effects of high sulfate concentrations on the methane fermentation of waste water. *Progr. Ind. Microbiol.* **20**, 133–143.

Nanninga, H.J. and Gottschal, J.C. (1986) Anaerobic purification of wastewater from potato-starch producing factory. *Wat. Res.* **20**(1), 97–103.

Oleskiewicz, J.A., Marstaller, T. and McCartney, D.M. (1989) Effects of pH on sulfide toxicity to anaerobic processes. *Environ. Technol. Lett.* **10**, 815–822.

O'Flaherty, V. and Colleran, E (2000) Sulfur problems in anaerobic digestion. In: *Environmental Technologies to Treat Sulfur Pollution. Principles and Engineering*. (eds. P.N.L. Lens and Pol L. Hulshoff). International Water Association, London, pp. 467–489.

O'Flaherty, V., Mahony, T., O'Kennedy, R. and Colleran, E. (1998a) Effect of pH and growth kinetics and sulfide toxicity thresholds of a range of methanogenic, syntrophic and sulfate-reducing bacteria. *Process Biochem.* **33**(5), 555–569.

O'Flaherty, V., Lens, P., Leahy, B. and Colleran, E. (1998b) Long term competition between sulfate-reducing and methane-producing bacteria during the full-scale anaerobic treatment of citric acid production wastewater. *Wat. Res.* **32**(3), 815–825.

O'Flaherty, V., Colohan, S., Mulkerrins, D. and Coleran, E. (1999) Effect of sulfate addition on volatile fatty acids and ethanol degradation in an anaerobic hybrid reactor. II. Microbial interactions and toxic effects. *Biores. Technol.* **68**, 109–120.

Omil, F., Lens, P., Hulshoff Pol, L. and Lettinga, G. (1996) Effect of upward velocity and sulfide concentration on volatile fatty acid degradation in a sulphidogenic granular sludge reactor. *Process Biochem.* **31**, 699–710.

Omil, F., Oude Elferink, S.J.W.H., Lens, P., Hulshoff Pol L. and Lettinga, G. (1997a) Effect of the inoculation with *Desulforhadbus amnigenus* and pH or O_2 shocks on the competition between sulfate reducing and methanogenic bacteria in an acetate fed UASB reactor. *Biores. Technol.* **60**, 113–122.

Omil, F., Bakker, C.D., Hulshhoff Pol, L. and Lettinga, G. (1997b) The effect of pH and low temperature shocks on the competition between sulphate reducing and methane producing bacteria in UASB reactors. *Environ. Technol.* **18**, 255–264.

Omil, F., Lens, P., Visser, A., Hulshoff Pol, L.W. and Lettinga, G. (1998) Long term competition between sulfate reducing and methanogenic bacteria in UASB reactors treating volatile fatty acids. *Biotechnol. Bioeng.* **57**, 676–685.

Oude Elferink, S.J.W.H., Visser, A., Hulshoff Pol, L.W. and Stams, A.J.M. (1994) Sulfate reduction in methanogenic bioreactors. *FEMS Microbiol. Rev.* **15**, 119–136.

Oude Elferink, S.J.W.H., Maas, R.N., Harmsen, H.J.M. and Stams, A.J.M. (1995) *Desulforhabdus amnigenus* gen.nov. sp.nov., a sulfate reducer isolated from anaerobic granular sludge. *Arch. Microbiol.* **164**, 119–124.

Oude Elferink, S.J.W.H., Luppens, S.B.I., Marcelis, C.L.M. and Stams, A.J.M. (1998) Kinetics of acetate utilization by two sulfate reducers isolated from anaerobic granular sludge. *Appl. Environ. Microbiol.* **64**, 2301–2303.

Oude Elferink, S.J.W.H., Akkermans-van Vliet, W.M., Bogte, J.J. and Stams, A.J.M. (1999) *Desulfobacca acetoxidans* gen. nov. sp., a novel acetate-degrading sulfate reducer isolated from sulfidogenic sludge. *Int. J. Syst. Bacteriol.* **49**, 345–350.

Overmeire, A., Lens, P. and Verstraete, W. (1994) Mass transfer limitation of sulfate in methanogenic aggregates. *Biotechnol. Bioeng.* **44**, 387–391.

Park, D., Lee, D.S., Joung, J.Y. and Park, J.M. (2005) Comparison of different bioreactor systems for indirect H_2S removal using iron-oxidizing bacteria. *Process Biochem.* **40**, 1461–1467.

Polprasert, C. and Haas, C.N. (1995) Effect of sulfate on anaerobic processes fed with dual substrates. *Wat. Sci. Tech.* **31**, 101–107.

Rinzema, A. (1988) Anaerobic treatment of wastewater with high concentrations of lipid or sulfate. Ph.D. Thesis, Wageningen Agricultural University, The Netherlands.

Rinzema, A. and Lettinga, G. (1988) Anaerobic treatment of sulfate containing waste water. In *Biotreatment Systems* (ed. D.L. Wise), vol. III. CRC Press Inc., Boca Raton, USA pp. 65–109.

Ritmann, B.E. and McCarty, P.L. (2000) *Environmental Biotechnology Principles and Application*. McGraw-Hill Companies Inc., New York.

Roest, K., Heiling, H.G.H.J., Smidt, H., de Vos, W.M., Stams, A.J.M. and Akkermanms, A.D.L. (2005) Community analysis of a full-scale bioreactor treating paper mill wastewater. *Syst. Appl. Microbiol.* **28**, 175–785.

Särner, E. (1990) Removal of sulphate and sulphite in an anaerobic trickling (ANTRIC) filter. *Water Sci. Technol.* **22**, 395–404.

Scheeren, P.J.H., Koch, R.O., Buisman, C.J.N., Barnes, L.J. and Versteegh, J.H. (1991) New biological treatment plant for heavy metal contaminated groundwater. *Trans. Instn Min. Metall.* (Sect. C: Mineral Process. Extr. Metall.) **101**, 190–199.

Sengül, F. and Müzzinoglu, A. (1991) A laboratory study for biological treatment of industrial wastewaters containing high sulfides. In *New Developments in Industrial Wastewater Treatment* (eds. A. Türkman and O. Uslu). Kluwar Academic Publishers, The Netherlands.

Speece, R.E. (1983) Anaerobic biotechnology for industrial wastewater. *Env. Sci. Technol.* **17**, 416A–427A.

Speece, R.E. (1996) Anaerobic biotechnology for industrial wastewater. *Archae Press*. Nashville, Tennessee.

Steudel, R. (2000) The chemical sulfur cycle. In *Environmental Technologies to Treat Sulfur Pollution. Principles and Engineering*. IWA Publishing.

Stratford, J., Dias, A.E. and Knowles, C.J. (1995) The utilization of thiocyanate as a nitrogen source by a heterotrophic bacterium: the degradative pathway involves formation of ammonia and tetrathionate. *Microbiology* **140**, 2657–2662.

Stucki, G., Hanselmann, K.W. and Hürzeler, A. (1993) Biological sulfuric acid transformation: reactor design and process optimization. *Biotechnol. Bioeng.* **41**, 303–315.

Svardal, K., Götzendorfer, K., Nowak, O. and Kroiss, H. (1993) Treatment of citric acid wastewater for high quality effluent on the anaerobic–aerobic route. *Wat. Sci. Technol.* **28**, 177–186.

Szendry, L.M. (1983) Startup and operation of the Bacardi Corporation anaerobic biofilter. In *Proceedings of the 3rd International Symposium Anaerobic Digestion*. Evans and Falkner Inc. Watertown, Mass, pp. 365–377.

Tabak, H.H., Scharp, R., Burckle, J., Kawahara, F.K. and Govind, R. (2003) Advances in biotreatment of acid mine drainage and biorecovery of metal. 1. Metal precipitation for recovery and recycle. *Biodegradation* **14**, 423–436.

Takahashi, M. and Kyosai, S. (1991) Pilot plant study on microaerobic self-granulated sludge process (multi-stage reversing-flow bioreactor: MRB). *Wat. Sci. Tech.* **23**, 973–980.

Thauer, R.K., Jungermann, K. and Decker, K. (1977) Energy conservation in chemotrophic anaerobic bacteria. *Bacteriol. Rev.* **41**, 100–180.

Thompson, G., Swain, J., Kay, M. and Forster, C.F. (2001). The treatment of pulp and paper mill effluent: a review. *Bioresource Technol.* **77**, 275–286.

Tichy, R., Lens, P., Grotenhuis, J.T.C. and Bos, P. (1998) Solid-state reduced sulfur compounds: environmental aspects and bioremediation. *Crit. Rev. Environ. Sci. Technol.* **28**, 1–40.

Tursman, J.F and Cork, D.J. (1989) Influence of sulfate and sulfite-reducing bacteria on anaerobic digestion technology. In *Biological Waste Treatment*. Alan R. Liss Inc., New York, pp. 273–281.

Utgikar, V.P., Harmon, S.M., Chaudhary, N., Tabak, H.H., Govind, R. and Haines, J.R. (2002) Inhibition of Sulfate-Reducing bacteria by metal sulphide formation in bioremediation of acid mine drainage. *Environ. Toxicol.* **17**, 40–48.

Vallero, M.V.G., Camarero, E., Lettinga, G. and Lens, P.N.L.(2004) Thermophilic (55–65 degrees C) and extreme thermophilic (70–80 degrees C) sulfate reduction in methanol and formate-fed UASB reactors. *Biotechnol. Prog.* **20**, 1382–1392.

Van Houten, R.T., Yun, S.Y. and Lettinga, G. (1997) Thermophilic sulphate and sulphite reduction in lab-scale gas-lift reactors using H_2 and CO_2 as energy and carbon source. *Biotechnol. Bioeng.* **55**, 807–814.

Visser, A. (1995) The anaerobic treatment of sulfate-containing wastewater. PhD Thesis, Wageningen Agricultural University, The Netherlands.

Visser, A., Gao, Y. and Lettinga, G. (1992) The anaerobic treatment of a synthetic sulfate containing wastewater under thermophilic (55°C) conditions. *Wat. Sci. Technol.* **25**, 193–202.

Visser, A., Gao, Y. and Lettinga, G. (1993a) Effects of short-term temperature increases on the mesophilic anaerobic breakdown of sulfate containing synthetic wastewater. *Wat. Res.* **27**, 541–550.

Visser, A., Beeksma, I., van der Zee, F., Stams, A.J.M. and Lettinga, G. (1993b) Anaerobic degradation of volatile fatty acids at different sulfate concentrations. *Appl. Microbiol. Biotechnol.* **40**, 549–556.

Visser, A., Hulshoff Pol, L.W. and Lettinga, G. (1996) Competition of methanogenic and sulfidogenic bacteria. *Wat. Sci. Technol.* **33**, 99–110.

Vroblesky, D.A., Bradley, P.M. and Chapelle, F.H. (1996) Influence of the electron donor and the minimum sulfate concentration required for sulfate reduction in a petroleum hydrocarbon-contaminated aquifer. *Environ. Sci. Technol.* **30**(4), 1377–1381.

Widdel, F. (1988) Microbiology and ecology of sulfate and sulfur-reducing bacteria. In *Biology of Anaerobic Microorganisms*. John Wiley and Sons, New York, p. 469.

Yoda, M., Kitagawa, M. and Miyaji, Y. (1987) Long term competition between sulfate-reducing and methane-producing bacteria for acetate in anaerobic biofilm. *Wat. Res.* **21**(12), 1547–1556.

7

Application of biological treatment systems for food-processing wastewaters

N. Bernet and E. Paul

Implementation of ISO 14001 certification and more stringent environmental legislation have been important drivers for the food industries to improve their process efficiency. To operate a sustainable food industry, energy efficient processes, as well as recycling of water and packing material should be first implemented. Moreover, considering the production of effluents characterized by high volumes and/or high organic contents, actions to reduce the environmental loads of food industries must first be performed (Kirby *et al.* 2003; Pap 2004). In any case however, end of pipe effluent has still to be treated, that is the concern of this chapter.

7.1 CHARACTERISTICS AND SPECIFICITY OF FOOD-PROCESSING WASTEWATERS

7.1.1 General characteristics

Characterization of food-processing wastewaters should take into account: (1) all parameters targeted by the effluent discharge regulations, but also (2) some parameters relevant for (pre-)treatment technology design, such as settleability, or

 ISBN: 1843391147. Published by IWA Publishing, London, UK.

that may directly affect the biological process efficiency, such as the biodegradation rate and nutrient balances (COD/N and COD/P ratios; COD, chemical oxygen demand), micro-nutrient concentrations, alkalinity, among others.

Categorization of agro-industry effluents is helpful to define wastewater qualities allowing to propose general rules for the choice of appropriate treatment technologies. A first classification is given by the type of industry sector: dairy, slaughterhouse, cannery, winery, brewery, etc. Another interesting categorization distinguishes between nutrient (N and/or P) poor and rich wastewaters. The former include mainly wastewaters from vegetal processing (winery, brewery, etc.); typical examples are wastewaters of the beet sugar industry, starch industry, wine producing, and fruit and vegetable processing industries (Prendl and Nikolavcic 2000). The latter comprise mostly wastewaters from animal processing (slaughterhouse, dairy, fish) which are rich in nitrogen (proteins) and sometimes in phosphorus (dairy). Considering this categorization, Tables 7.1 and 7.2 summarize results from literature of conventional chemical characterization for a wide spectrum of food-processing wastewaters.

Care is required in interpreting concentration data which are very dependent of the industrial processes. For example, animal type and size may vary widely and influence greatly the characteristics of a slaughterhouse wastewater. Moreover, concentrations are inevitably influenced by water consumption, production and cleaning procedures, and the efforts for byproduct recovery. The best basis for expressing and comparing pollution output is the use of the specific pollution (i.e. the production based pollution on the unit of transformed/produced material). Unfortunately, such data are rarely available and not recently updated. In Tables 7.1 and 7.2, both specific pollution productions and concentrations in the wastewaters are given when available.

Some major features of the food-processing wastewaters can be underlined from Tables 7.1 and 7.2. Firstly, the quality and quantity of effluents may greatly vary from one industry to another and even within a class of industry. Secondly, as shown by the high values of the BOD_5^*/COD ratios, the majority of food-processing wastewaters are compatible with a biological secondary treatment with or without a pre-treatment step.

7.1.2 Temperature

Temperature of the effluents can be rather high but variable: range for dairy wastewater has been found to be between 3°C and 43°C; in breweries, effluent temperature is included in a range of 18°C and 40°C. The wastewater temperature is variable from one plant to another but within the same plant, effluents can be submitted to seasonal and diurnal temperature fluctuations. Temperature can be a significant issue in the choice and economics of wastewater treatment operations. For example, it can influence greatly the choice and performances of anaerobic systems, nitrification process and fat emulsification.

---

*The soluble biochemical oxygen demand of the effluent is generally below 5 mg/L which is represented as BOD_5.

Table 7.1 Characteristics of food-processing wastewater from animal origin.

Industry	Slaughterhouse	Dairy	Fish canning	Fish processing
Specific flow	5–10 m^3/t living weight	1–3 m^3/t milk	4 m^3/t fish	5–10 m^3/t fish
Specific COD	4–18 kg/t live carcass[a]	8 kg/m^3 milk intake[a]	42–116 kg/t fish intake[a]	13–39 kg/t unloaded fish
Specific BOD_5	10–20 kg/t slaughtered animal	0.8–2.5 kg/t milk		
TCOD (g/L)	1.5–11.2	0.19–7.6	2.2–10.6	4–6
SCOD (g/L)	0.11–10		0.5–1.4	
BOD_5 (g/L)	0.49–8.0	0.38–5.7	4.5–7.3	
TSS (g/L)	0.22–6	0.1–14.2	1.5–2.2	0.28
TKN (mg/L)	90–700	6% BOD_5, 14–140	240–890	540
Total P (mg/L)	12–120	0.2–181		
Alkalinity as $CaCO_3$ (mg/L)	350–1014	225–1550		1200
BOD_5/DCO		0.25–0.78		
pH	4.9–7.2	3–13	5.8–7.6	4.5
Reference	Johns (1995); Massé *et al.* (1998); World Bank Group (1998); [a]UNEP (2000a)	Danalewitch *et al.* (1998); World Bank Group (1998); [a]UNEP (2000b); Andreottola *et al.* (2002)	GTZ (1997); [a]UNEP (2000c); Palenzuela-Rollon *et al.* (2002)	Aspé *et al.* (1997)

TCOD and SCOD: total and soluble chemical oxygen demand; BOD_5: biochemical oxygen demand; TSS: total suspended solids; TKN: total Kjeldahl nitrogen.

Table 7.2 Characteristics of food-processing wastewater from vegetal origin.

Industry	Winery	Brewery	Fruit	Vegetable	Potato	Sugar	Vegetable oil
Specific flow		2–8 hL/hL beer	2.5–6.5 m^3/t	3.5–8.5 m^3/t		0.2–1.8 m^3/t	2–25 m^3/t product
Specific load (g COD/100 kg of processed raisin-day)	6–30						
Specific BOD_5 (kg/unit of production)			1.3–44	1.2–30			
Specific TSS (kg/unit of production)			0.3–7.5	0.2–19			
COD (g/L)	5.7–18	2–6			5.2–13.9	2.3–10	30–60
BOD_5 (g/L)		1.2–3.6			2.0–6.0	1.7–7	20–35
TSS (g/L)	0.8–3	0.2–1			<7.1	<5	10
TKN (mg/L)		25–80			120–210		500–800
Total P (mg/L)		10–50			28–34		
pH		4.5–12			6–7	4–7	
Reference	Rozzi *et al.* (1998)	Driessen and Vereijken (2003)	World Bank Group (1998)	World Bank Group (1998)	El-Gohari *et al.* (1999)	World Bank Group (1998); Kansal *et al.* (1998)	World Bank Group (1998)

COD: chemical oxygen demand; BOD_5: biochemical oxygen demand; TSS: total suspended solids; TKN: total Kjeldahl nitrogen.

Table 7.3 Daily wastewater flow in a fruit juice factory in regard to seasonal and production differences (m^3/day). (Austermann-Haun *et al.* 1997a).

Season	Minimum	Maximum	Average
Autumn campaign	264–800	1280–1624	641[a]–1061
Winter and spring (only bottling)	200–420	360–620	299–285
Summer (treatment of berries, stone fruit and bottling)	300–480	500–825	279–587

[a]First campaign without bottle washing, bottling and sanitary wastewater.

Table 7.4 COD in the wastewater at different seasonal and production times (mg/L). (Austermann-Haun *et al.* 1997a).

Season	Minimum	Maximum	Average
Autumn campaign	740–1296[a]	1721–3395	1520–2411
Winter and spring (only bottling)	428–1147	1039–2258	820–1602
Summer (treatment of berries, stone fruit and bottling)	375–691	1670–2310	896–1418

[a]First campaign without bottle washing, bottling and sanitary wastewater.

7.1.3 High flow rate and concentration variability

Food-process industry has often to manage seasonal production, with important consequences on the variability of the wastewater (flows, concentrations, characteristics). Tables 7.3 and 7.4 illustrate this variability in flows and COD concentration respectively, in the case of a fruit juice factory in Germany (Austermann-Haun *et al.* 1997a).

Due to the very high variability, *in-situ* experimental information is required to define properly, quantitatively and qualitatively, the wastewater to be treated.

7.1.4 Lipid content

Wastewater from slaughterhouses, dairy and vegetable oil processing contain lipids that can cause problems, especially in anaerobic digestion, but also in aerobic processes.

The anaerobic degradation of fat-rich wastewaters is slower than that of fat-poor wastewaters due to the low rate of the fat hydrolysis step. However, this prevents accumulation of volatile fatty acids (VFA) and favors the overall process (Vidal *et al.* 2000).

Lipids are hydrolyzed in glycerol and long-chain fatty acids (LCFA) that are reported to exert an acute toxic effect on the microorganisms involved in the β-oxidation and methanogenic pathways (Pereira *et al.* 2004). Unsaturated LCFA seemed to have a greater inhibitory effect than saturated LCFA, especially on methane production from acetate. Thus, they should be saturated to prevent inhibition in anaerobic processes.

Moreover, formation of scum layers and wash-out of sludge, due to the low specific density of the lipids can occur (Zeeman and Sanders 2001).

Lipid degradation and inhibition have been assessed in one- and two-phase anaerobic digestion processes. The two-phase systems have been shown to yield higher efficiencies than a corresponding single-phase system for substrates with greater lipid content (Demirel and Yenigün 2002). However, other authors show that, unlike the hydrolysis of proteins and carbohydrate, lipid hydrolysis is hardly occurring in the absence of methanogenesis (Zeeman and Sanders 2001). Moreover, Lettinga (1995) states that there does not exist any need for phase separation when treating non-acidified or slightly acidified wastewaters. Phase separation even is detrimental, in case the acidogenic organisms are not removed from the effluent of the acidogenic reactor, because they deteriorate the settlability of granular sludge and also negatively affect the formation and growth of granular sludge.

In the case of aerobic treatment, the presence of grease may lead to bulking and to a high oxygen demand. Depending on temperature, a change in lipid physical state can provoke some clogging, especially when using fixed biomass reactors.

7.1.5 Protein content

Food-processing industries such as abattoir, dairy, whey, cheese, fish and certain vegetable processing produce wastewater containing significant amounts of proteins (Ramsay and Pullammanappallil 2001). Proteins are hydrolyzed to amino acids and further degraded to VFA, carbon dioxide, hydrogen, ammonium and sulfide, either through anaerobic oxidation linked to hydrogen production or via fermentation according to Stickland reaction (McInerney 1988).

Depending on the pH value of the system and the initial protein concentration, ammonia and hydrogen sulfide can cause inhibition of the anaerobic digestion process (Vavilin *et al.* 1995). For non-adapted methanogenic sludges, ammonia inhibition has been observed to start at concentrations of 0.7 g N/L with a complete stop of methanogenesis at approximately 1.7–2 g N/L at a pH value maintained between 7.6 and 7.95 (Koster and Lettinga 1984). However, the toxicity threshold level can be strongly increased after adaptation of the microbial populations (Koster and Lettinga 1988).

7.1.6 Sulfate content

Some processes use sulfuric acid or sulfate-rich feed stocks (e.g. sea-food processing industry, baker yeast industry) and generate wastewaters containing sulfate (Hulshoff Pol *et al.* 1998). Under anaerobic conditions, dissimilatory sulfate reduction bacteria (SRB) use sulfate as a terminal electron acceptor for the degradation of organic compounds and hydrogen. Sulfate is reduced to sulfide which is toxic for anaerobic digestion (Hulshoff Pol *et al.* 1998). Therefore, SRB compete with methanogens for electron donor and they produce sulfide which is potentially toxic for the anaerobic ecosystem. Anaerobic digestion of sulfate-rich wastewaters is treated more extensively in the Chapter 6 of this book, including inhibition aspects.

7.1.7 Suspended solids

Organic suspended solids (SS) are often present in the wastewaters produced in the food industry (potato processing, fish meal processing, etc.). This is a serious drawback for the use of high-rate anaerobic reactors such as upflow anaerobic sludge blanket (UASB) or anaerobic filters, in which case the implementation of a previous hydrolysis–acidification step would be required (Guerrero *et al.* 1999). The performance of UASB reactors for the treatment of wastewaters with a high SS concentration can be affected when entrapped solids accumulate in the sludge bed. Consequently, the amount of excess sludge will increase, leading to a decrease of the sludge age and a deterioration of the process.

In the case of energetic valorization of methane, it may be useful to digest the volatile suspended solids (VSS) which can contribute significantly to biogas production.

7.2 PRELIMINARY TREATMENTS

Pre-treatment and primary treatments of food-industry effluents mainly consist in screening, flow equalization, neutralization, air flotation and settling. Trickling filters (TF) and anaerobic processes used prior to a conventional aerobic treatment process are considered latter in this chapter. Preliminary treatments aim to remove fats, proteins and solids (coarse particles, fibers, fat, bone, hair, meat, etc.) because these elements induce extremely high BOD_5 and SS loads as it is the case for blood-rich wastewaters (Johns 1995) or can cause adverse effects to secondary treatments. Austermann-Haun *et al.* (1999) underlined the importance of efficient pre-treatments when fat and SS-sensitive secondary treatment processes, such as expanded granular sludge bed (EGSB), are used.

To reduce the biochemical oxygen demand (BOD) load fed to secondary treatments (solids, lipids) or to recover proteins from processing wastewaters, dissolved air flotation (DAF) can be interestingly used (Rusten *et al.* 1990; 1993). As an example, Schneider *et al.* (1995) reported on the recovery of proteins suspended in the effluent of a soybean protein plant for the production of a product suitable for animal feed enrichment. DAF refers to a physical/chemical treatment with water–solid separation by the injection of fine gas bubbles (air). However, a lot of systems are subjected to considerable operational problems. Sensitivity to flow variations and solid settling are often observed, leading to bulking and to sludge with poor dewaterability. With chemical addition (polymers and coagulants), DAF units can achieve COD reductions ranging from 32% to 90% and are capable of removing large amounts of nutrients (Rusten *et al.* 1990; Kiepper 2001; Mittal 2005). It also increases protein clumping and precipitation as well as fat flotation. The use of chitosan, which is food grade, has been proposed to replace the synthetic anionic and cationic polymers traditionally used for the DAF operation (Selmer-Olsen *et al.* 1996). With such a process, the proteins retain their value as food product and can be recycled within the food industry.

Treatability tests such as plain settling, flotation and coagulation–flocculation–settling should be used to define the requirements for a primary settling of the effluent. Orhon *et al.* (1999) gave an example of this type of physical fractionation for three industrial wastewaters.

7.3 ANAEROBIC DIGESTION

7.3.1 General considerations

Microbiology of anaerobic digestion and a description and design of anaerobic processes being applied to the treatment of industrial wastewater have been developed in Chapters 3 and 4 of this book. Anaerobic processes generate energy in the form of biogas and produce sludge in an amount that is significantly lower than that resulting from aerobic processes. It is why anaerobic digestion is generally an adapted process for the treatment of food-processing wastewaters which contain a high concentration of biodegradable organic matter. As a consequence, about 3/4 of the digesters built in the world treat food-processing wastewaters (Frankin 2001; Kleerebezem and Macarie 2003). Table 7.5 gives for each industrial sector, the number of anaerobic processes built in the world for treating food-processing wastewaters. From these 1222 processes, about 2/3 are granular processes of which 78% are UASB and 22% are EGSB/IC (IC, internal circulation). Concerning the industrial sector of brewery and malt, it represents 23% of the built processes. Vegetal-processing industries are largely

Table 7.5 Number of commercial scale anaerobic reactors built in the world by January 2003 for treating food-process wastewaters. (From Kleerebezem and Macarie 2003).

Type of wastewater	Type of reactor						
	Low rate[a]	Activated carbon	Fixed bed[b]	Moving bed[c]	UASB	EGSB/IC	Total
Brewery and malt	2	–	6	4	185	88	285
Distillery and ethanol	25	31	40	–	76	9	181
Other beverages	–	3	11	2	88	15	119
Sugar production	–	49	7	1	32	3	92
Potato processing	14	4	2	–	46	10	76
Dairy, ice-cream and cheese	12	10	10	2	27	6	67
Starch production	2	9	10	2	34	7	64
Yeast production	7	8	6	–	25	8	54
Candy and confectionery	4	–	3	–	15	2	24
Citric acid production	2	3	1	1	3	5	15
Coffee processing	–	–	7	–	4	1	12
Wine processing	–	–	6	1	3	1	11
Fish and sea-food processing	1	4	–	–	2	1	8
Miscellaneous	10	22	40	5	112	25	214

[a]Including continually stirred tank reactor (CSTR), lagoons and bulk volume fermenter (BVF) reactors from the firm ADI.
[b]Including upflow, downflow and hybrid reactors.
[c]Including upflow fluidized bed and mobilized film reactors.

the more represented sector to use anaerobic digestion for the treatment of their wastewater.

7.3.2 Specificities of food-processing wastewaters for anaerobic digestion

Design of anaerobic processes must take into account specificities that can affect sludge properties, especially in the case of granular processes where the efficiency of the process is directly linked to the settling capacities of the sludge. When high settling capacities are obtained, solid retention times of over 200 days can be achieved at hydraulic residence time (HRT) of only 6 h (Hulshoff Pol *et al.* 2004).

Besides effluent composition in lipids, proteins, VSS (see Section 7.1.2), flow and concentration variations and temperatures are the main parameters to be considered in the choice of a process: granular *vs.* non-granular sludge or biofilm, high-rate *vs.* low-rate process, one-step *vs.* two-step process, mesophilic *vs.* thermophilic.

Albagnac and Verrier (1983) give choice criteria based on the COD and VSS concentrations of the effluent to be treated. High-rate processes in which it is possible to operate at a high solid retention time and a low hydraulic retention time are recommended for low-strength wastewaters whereas conventional technologies such as anaerobic contact can be used for the treatment of high-strength wastewaters (more than 50 g COD/L).

In the case of high-rate processes, Weiland and Rozzi (1991) resume the typical operation regimes with respect to the total suspended solids (TSS) and COD concentrations (Figure 7.1.).

The presence of oil and grease (O&G) and/or proteins must be considered as an important parameter in the choice of a process, as it is mentioned in Part 7.1.

In anaerobic systems such as UASB reactors, the O&G content of the wastewater has generally to be decreased below 50 mg/L. Over this value, flotation of the granular sludge will be observed resulting in a progressive but irremediable washout of the active biomass and a total failure of the process (Maat and Gorur 1990). As a consequence, an efficient pre-treatment must be implemented to remove

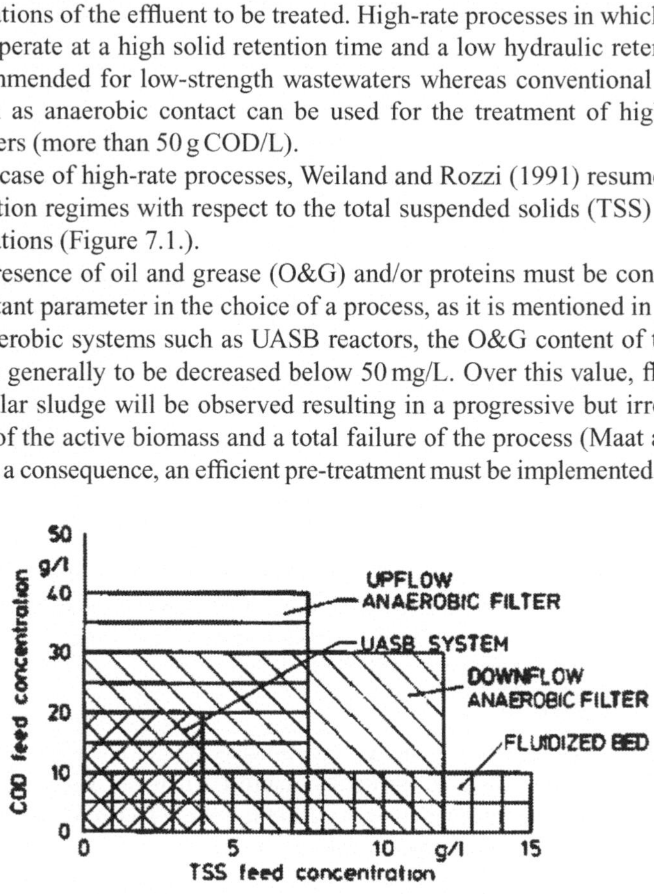

Figure 7.1 Typical operation regimes of high-rate systems with respect to the TSS content and COD load of the wastewater (Weiland and Rozzi 1991).

O&G if a high-rate system has to be applied. On the contrary, it has been shown that low-rate anaerobic reactors such as the ADI–BVF system can deal with high amounts of O&G without the need of pre-treatment. This is probably due to the long hydraulic retention time of these systems, which is compatible with the hydrolysis of the lipids and the subsequent digestion of the LCFA. These reactors operate also at much lower liquid and gas superficial velocities than the high rate ones. Since both parameters contribute to the flotation of the sludge attached to the lipids, a decrease of their value should also reduce the potential of sludge flotation and the associated loss of active biomass by washout (Grant *et al.* 2002).

It must be noted that an option can be to decrease the TSS and/or O&G content before the anaerobic treatment by an adequate pre-treatment.

The choice of a process can also be dependent on the available surface on the industrial site where the process should be built: from a high-rate and low-footprint process (FP, IC, EGSB) to an anaerobic pond.

Last but not the least, the complexity of the process may be an important criteria, especially in developing countries where simple processes should be preferred. It is one of the reason of the success of the UASB process in Latin-America and Southeast Asia (Macarie 1992).

7.3.2.1 Flow and concentration variation

Food-processing wastewaters can be submitted to important flow and concentration variations. It is suggested that high-performance reactors such as UASB cannot scope with high-load-rate variations (Leitão *et al.*, in press). Schmidt and Ahring (1997) investigated the treatment of wastewaters from a multiproduct food-processing company generating four different types of wastewaters, using a UASB concept. They concluded that problems could occur if no precautions are taken, especially when changing between wastewaters with high and low organic loading rate (OLR) and when changing between wastewaters with high and low content of lipids and proteins. On the other hand, Austermann-Haun *et al.* (1997b) demonstrated that a UASB reactor is suitable for campaign industries like fruit-juice producers, running only 2 months a year with higher concentrated wastewater, but also for the time outside the campaign when the wastewater concentrations are low (average COD 1000 mg/L). It can be necessary to set up an equalization tank at the head of the treatment plant or after the pre-treatment in order to smooth the variations. If the equalization tank provides a long enough retention time, it can also play the role of acidogenic reactor (e.g. see Comeau *et al.* 1996).

7.3.2.2 Temperature

Most of anaerobic digesters treating food-process wastewaters are operated in mesophilic conditions. It is of particular interest when the wastewater is discharged at a high temperature: in this case, all the energy recovered from methane can be used for other purpose than heating the reactor.

Table 4.7 (Chapter 4 of this book) gives suggested organic volumetric loads in granular sludge UASB reactors as a function of temperature treating mainly dissolved COD.

Temperature-phased systems incorporate thermophilic and mesophilic digestion processes into one system. Due to the increased energy requirement, this system can be recommended only when high solid and pathogen removal is required (Dugba and Zhang 1999). Indeed, significant inactivation of pathogens occurs during the anaerobic digestion (Dohanyos *et al.* 2004) and thermophilic conditions have been shown to be more effective for inactivating pathogenic bacteria than mesophilic ones (Watanabe *et al.* 1997).

In the case of a protein-rich wastewater, thermophilic conditions may favor methanogenesis inhibition by free-ammonia (Guerrero *et al.* 1999).

7.3.3 Case studies

7.3.3.1 UASB reactor in a fruit juice factory (Austermann-Haun et al. 1997b)

The anaerobic treatment plant was built in 1989 at the fruit juice factory Lindavia Fruchtsaft at Lindau, Germany, to treat the wastewater before it is discharged into the municipal wastewater treatment plant of Lindau.

Design data Influent flow 1920–2400 m^3/day; COD 2.5–5.0 g/L; maximum daily load 10 ton COD/day; COD/BOD_5 ratio 1.5; pH 4–12; temperature 30°C.

Design criteria Hydraulic retention time in the acidification tank 5.5–7 h; volumetric loading rate of the UASB reactor 10–17 kg COD/m^3-day; COD removal efficiency 80–90%.

Treatment plant (Figure 7.2) It consists of a rotary sieve (1.5 mm) and a sand grip chamber (30 m^3) – both located in the factory, two buffer tanks (65 m^3) with the possibility of pH regulation, the dosage of nutrient salts (N and P) and an anti-foaming agent, a washing liquor tank (151 m^3), a wastewater filtration unit (gap

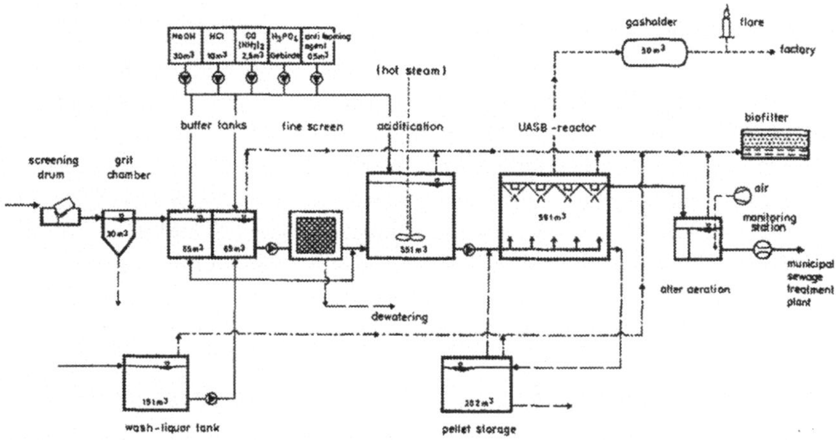

Figure 7.2 Flow diagram of the anaerobic wastewater treatment plant in the fruit juice factory (Austermann-Haun *et al.* 1997b).

width 0.1 mm) running with the dosage of 10 g/m^3 coagulation aid, belt-type filter-press for dewatering of the solids separated in the filtration unit, an acidification tank (551 m^3), a UASB reactor (591 m^3, A = 131.4 m^2), an aerated tank (8.5 m^3), a pellet storage (202 m^3), a gasholder (50 m^3) with a flair, a burner to produce steam from the biogas, a biofilter for treatment of the exhaust air coming of the acidification tank, the top of the UASB reactor, the operation building, the aerated tank and the buffer tanks.

The wastewater treatment plant was initiated in September 1990, the investment costs were 4.5 million DM (€ 2.3 million).

Characteristics of the wastewater Due to the seasonal production, wastewater flows and concentrations are very variable during the year (Tables 7.3 and 7.4).

Running experiences

Acidification Depending on the wastewater composition, 10–16% of the COD is based on organic acid (mainly acetic acid). Due to the lower amount of wastewater, the design HRT was never reached. It varied between 12 and 15 h during the campaign, up to 20 h outside the campaign. At an HRT of 13 h, 20–27% of the COD were based on organic acids.

UASB reactor The reactor was inoculated with 130 m^3 pellet sludge from a UASB reactor used in the paper industry. This large amount of inoculum (21 kg SS/m^3) shortens the start-up period.

During the first campaign, the average COD elimination was between 82% and 88% at a volumetric OLR of about 3.5 kg COD/m^3-day. Outside the campaign, it ranged from 87% to 90% at a very low loading of 0.4–1.3 kg COD/m^3-day. The UASB reactor proved to be very reliable throughout the year. Problems occurred only at a sudden start of a campaign, when the wastewater flow rose to an amount which was four times as high as the normal flow rate within a week (maximum load 8.5 kg COD/m^3-day). In this case, the COD removal efficiency was about 70% during the first three weeks. After that, it rose to over 90%.

Running expenses During the first year of operation, the specific running expenses were 2.52 DM/m^3 (€1.29/m^3), including all costs for materials, personnel, etc., but excluding capital costs. According to the authors, these running expenses could be reduced by routine operation in the following years down to 0.99 DM/m^3 (€0.51/m^3). Energy recovery by the use of biogas was estimated to be 0.177 DM/m^3 (€0.09 m^3).

This example shows that a UASB reactor can be applied in campaign industries running only 2 months a year with concentrated wastewaters as well as for the time outside the campaign when COD concentrations are low.

7.3.3.2 Two-step anaerobic filter in a Winery (Moletta 2005)

This example is at a winery producing 35,000 hL/year and located at Goult Lumières, France. The annual wastewater production is 1200 m^3. During vintage

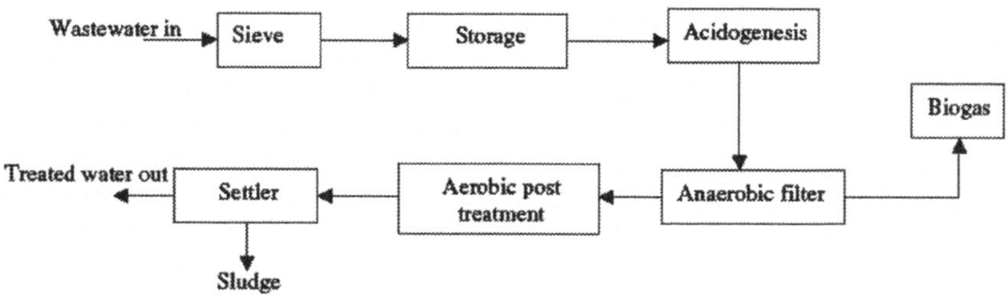

Figure 7.3 Flow-chart of the two-step anaerobic treatment (Moletta 2005).

the COD wastewater content is between 8 and 16 g/L with a 1 g/L of suspended solid. The flow sheet of the process is shown in Figure 7.3.

The wastewater is first sieved before storage. The tank used can store the total volume produced per day. The wastewater goes to 5 m^3 acidogenic reactor with temperature and pH control (37°C and 6.5, respectively). The acidogenic wastewater reaches a 4 m^3 anaerobic filter and then goes to a 15 m^3 aerobic post treatment. The biogas is burnt by a flare. This strategy, with the full storage of the wastewater, allows a small volume of reactor and treatment during all the year.

In the acidogenic phase, the soluble COD removal yield is 24% and the HRT is 20–38 h. The OLR of the anaerobic filter is between 4.6 and 11 kg COD/m^3-day with an average yield of 70%. The OLR of the aerobic post-treatment is 1.3 kg COD/m^3-day. The COD removal yield of the overall process was between 88% and 98%. The investment cost was €120,000 and the running cost is €4500/year (1998).

7.3.3.3 Anaerobic treatment of potato-processing wastewater (Zoutberg and Eker 1999)

Zoutberg and Eker (1999) reported three full-scale applications of anaerobic granular processes (two UASB and one EGSB) for the treatment of potato-processing wastewaters. Wastewater in the potato-processing industry contains substantial amounts of SS. The UASB process is characterized by longer hydraulic retention times than that of the EGSB process. Therefore, use of the UASB process results in a higher removal of SS and higher overall COD removal efficiencies. The EGSB process has been designed for removal of soluble COD mainly. The use of the EGSB in the potato-processing industry is focused for those applications where the anaerobic effluent will be discharged to sewer or to a final aerobic post-treatment.

In the three plants presented in the paper, coarse solids are removed in screens and most of the SS in a pre-clarifier. Then, the wastewater is sent to a buffer tank, followed by a conditioning tank where pH and temperature are controlled.

Total COD concentrations after screening and pre-settlement varied from 4500 to 7500 mg/L, soluble COD between 1425 and 6000 mg/L. Average flows after buffering ranged from 37 to 90 m^3/h. Total COD removal of 79–80% was obtained in the UASB processes. The UASB were coupled to an aerobic process to achieve the targeted COD discharge limit.

7.4 AEROBIC TREATMENT

As described in the previous paragraph, great advantages may be obtained by running an anaerobic process for the treatment of a food-processing wastewater. However, when the wastewater contains a lower COD concentration than that required for running an anaerobic system with a reasonable economical efficiency, or when stringent regulations on COD and nutrient effluent discharge have to be faced, an aerobic treatment must be chosen. Among the existing aerobic treatment technologies, the activated sludge process is widely used, above all for large treatment plants. Other possible biological treatment systems include land treatment or pond systems when space is available, trickling filters (TF), sequencing batch reactor (SBR) and rotating biological contactors (RBC). Some results on the performances of aerobic processes are synthesized on review papers for the treatment of wastewaters from meat industry (Johns 1995), more recently from abattoirs (Mittal 2005).

Selection of a biological treatment system should be based on (1) the required effluent quality, (2) the characteristics of the wastewater to be treated, (3) the size of the plant, (4) simplicity and easiness of the operation, (5) operational stability faced to variations of wastewater characteristics, flow rates and loads, (6) the presence of a previous (anaerobic) pre-treatment when the wastewater COD concentration is high and (7) investment and operating costs.

7.4.1 Activated sludge systems

Activated sludge (AS) system consists of two main process units: the aeration basin where biodegradation occurs and the clarifier allowing the activated sludge to be separated from the treated wastewater, concentrated and recycled to the aeration basin. The conventional model for AS design has been already presented in the Chapter 4 of this book. However, specificities of AS design procedure treating wastewaters from food industries must be underlined.

7.4.1.1 AS tank design

Process modeling is useful and widely used for activated sludge design. Conventional methods generally consider various simplifying hypotheses, such as: (1) the reactor is perfectly mixed and operates under steady state, (2) biomass growth in the sedimentation tank is negligible and so no reaction takes place in this settler, (3) biomass is represented by VSS and all the influent biodegradable COD is soluble. Under these conditions, the mass balance on the biodegradable COD leads to Equation (7.1).

$$r_S \cdot V = q_S \cdot X \cdot V = Q_e(S_e - S) \tag{7.1}$$

where V is the volume of the reactor, Q_e is the influent flow rate, S_e the influent biodegradable COD and S the reactor (and so the effluent) biodegradable COD.

The substrate consumption rate (r_S) can be decomposed into a substrate specific consumption rate (q_S) multiplied by the active biomass concentration (X).

For a given temperature, T, the q_S value strongly depends on the concentration S in the reactor and a Monod kinetic relation is usually used to describe this dependence.

$$q_{S,T} = q_{S\,max,T} \cdot \frac{S}{K_S + S} \tag{7.2}$$

where $q_{S,T}$ is the specific substrate consumption rate at the temperature T and $q_{S\,max,T}$ is the maximum specific substrate consumption rate at the temperature T.

Stringent regulations impose a low S concentration in the effluent (S concentration smaller than K_S value) and consequently, $q_{S,T}$ will be low compared to $q_{S\,max,T}$. In Equation (7.2), q_S can be substituted by μ/Y (Equation (7.3)) with the expression of μ given in Equation (7.4).

$$q_S = \frac{1}{X} \cdot \frac{dS}{dt} = \left(\frac{1}{X} \cdot \frac{dX}{dt}\right) \cdot \frac{dS}{dX} = \mu \cdot \frac{1}{Y} \tag{7.3}$$

where Y is the biomass yield and μ the specific growth rate.

$$\mu = \frac{1}{\theta_X} + k_d = \frac{1 + k_d\theta_X}{\theta_X} \tag{7.4}$$

where $\theta_X = (V \cdot X)/(Q_w X_w + Q_e X_e)$, which is the mean biomass residence time (or sludge age) and k_d the endogenous decay coefficient; Q_w and Q_e are the waste sludge and treated effluent flow rates and X_w and X_e are the concentrations of VSS in the waste sludge and in the treated effluent respectively.

Hence, it yields to Equation (7.5) where the total mass of active biomass required to obtain a required S value is given as a function of θ_X:

$$V \cdot X = \frac{Y\theta_X Q_e (S_e - S)}{(1 + \theta_X k_d)} = Y_{obs} Q_e (S_e - S)\theta_X \tag{7.5}$$

where Y_{obs} is the observed yield

Hence,

$$X = Y_{obs}(S_e - S)\frac{\theta_X}{\theta_H} \tag{7.6}$$

where θ_H is the hydraulic retention time.

Various conclusions can be drawn from Equations (7.5 and 7.6).

Increasing X is of crucial importance for minimizing the aerated tank volume. It can be achieved by increasing θ_X (thus the recycled sludge ratio) for a given θ_H and S_e.

However, settling is adversely affected by sludge concentration. Therefore, biomass concentration in the aerated tank must be maintained below a critical value that allows the sludge settling under the higher hydraulic flow. The designed biomass concentration in the activated sludge tank is thus a compromise between two opposite objectives: a high degradation rate, consequence of a high AS biomass concentration (leading to a small AS reactor volume) and a high settling velocity due to a low AS biomass concentration (leading to a small settler surface area). This compromise is well documented in Ekama *et al.* (1997). In the case of food-industry wastewater treatment, the sludge settling and thickening capacities are generally poor and may greatly fluctuate during operation. The design value of the AS biomass concentration should not be too high and some security factor should be applied for the design of the settler surface area. Table 4.2 in Chapter 4 gives some design parameter values for protein- or vegetable-rich wastewaters. A simple estimation of this surface area (A) is obtained using the following equation:

$$A = \frac{Q_e}{q_{A\,\max}} \tag{7.7}$$

where $q_{A\,\max}$ is the maximal surface flow rate.

Values of 0.25 and 0.35 m/h in the case of dairy and slaughterhouse respectively are preferred to the 0.6 m/h commonly used for urban wastewater. High variations of Q_e have generally to be considered. Consequently a 70 or 80 or 95 percentile value of the maximum flow rate should be adopted for an appropriate design.

X will increase with S_e and with Q_e for a given θ_X. In the case of food-industry wastewater treatment, the influent COD is often rather high compared to the urban wastewater COD. Consequently, the design θ_H value should be higher than that used for the design of a conventional AS plant treating urban wastewater. In some cases, S_e is so high that biomass recycling is no more necessary (or feasible). A continuous process without biomass recycling (chemostat-like process) may be chosen. Post-treatment to remove SS is thus required. A better alternative solution would be the implementation of an anaerobic pre-treatment.

7.4.1.2 Specificities of food-processing wastewater treatments

More insights on COD

Hydrolyzable COD fractions The use of process modeling for design is pertinent and accepted only if effluent and biomass characteristics are correctly defined for each case considered. For treatment of food-processing wastewaters, the hypothesis that all the biodegradable COD is soluble and readily biodegradable may not be valid. The high content in slowly hydrolyzable COD should be considered in the design because its biodegradation is often the major rate limiting process. The biodegradable COD is therefore generally fractionated into a readily biodegradable COD (S_S) and a hydrolyzable COD (X_S). Moreover, there is also in many cases experimental evidence that two hydrolysis rates should be considered to represent correctly the COD removal (Orhon *et al.* 1999). These authors proposed a dual hydrolysis model to improve the design and operation procedures. It is described in the process kinetic and stoechiometry matrix shown in Table 7.6.

Values of the kinetic parameters associated to the biological degradation of these COD fractions are given by Orhon *et al.* (1999) for various treatments of food-processing wastewaters, including from slaughterhouse, poultry and meat industries.

Table 7.6 Process kinetic and stoichiometry for COD removal. (Orhon *et al.* 1999).

	Component (I)							
Process (j)	1 S_S	2 S_H	3 X_S	4 X_H	5 X_P	6 S_P	7 S_O	Process rate, β_j (M/L^3T^{-1})
1 Growth	$-\frac{1}{Y_H}$			1			$\left(-\frac{1-Y_H}{Y_H}\right)$	$\mu_{max}\frac{S}{K_S+S}X_H$
2 Rapid hydrolysis	1	−1						$k_{hS}\frac{\frac{S_H}{X_H}}{K_{XS}+\frac{S_H}{X_H}}X_H$
3 Slow hydrolysis	1		−1					$k_{hX}\frac{\frac{X_S}{X_H}}{K_{XX}+\frac{X_S}{X_H}}X_H$
4 Decay				−1	f_{EX}	f_{ES}	$-(1-f_{EX}-f_{ES})$	$b_H X_H$
Parameter($M\backslash L^3$)	COD	COD	COD	COD	COD	COD	−COD (O_2)	

S_S: readily biodegradable COD; S_H: rapidly hydrolyzable COD; X_S: slowly hydrolyzable COD; X_H: heterotrophic biomass; S_P: inert soluble microbial products; X_P: inert particulate microbial products; S_O: dissolved oxygen; Y_H: heterotrophic yield coefficient; f_{EX}: particulate inert fraction of metabolic product; f_{ES}: soluble inert fraction of metabolic product; b_H: endogenous decay rate.

COD inert fractions Influent settleable inert COD (X_I) is not included in the matrix. However, due to its accumulation at a given θ_X, X_I, as X_P (the inert particulate COD fraction produced by the microorganisms), influences greatly the active biomass fraction in the sludge and consequently the aerobic volume calculation. Both inert particulate COD fractions participate also directly to the sludge production. It is therefore recommended to correctly evaluate X_I, and X_P COD fractions.

As described in the matrix given in Table 7.6, care must be given to the inert soluble COD coming either from the industrial wastewater or produced during the aerobic treatment (soluble microbial products (S_P) originating from growth and lysis processes). This inert soluble COD will be found of course in the effluent of the treatment plant leading to difficulties in complying some stringent effluent discharge regulations. Methods to quantify S_I (soluble inert COD) and S_P were proposed by Germirli *et al.* (1993) and by Orhon *et al.* (1992). A study on S_P generation was conducted by Goorany and Oztürk (2000) on an effluent from the fermentation industry. For COD concentrations of 1540–6645 mg/L, S_P were expected to vary in the range of 12.5–8.6% of the initial COD respectively, while the S_I/initial COD ratio was found to be around 0.12. Values of S_P produced during the treatment of dairy, cheese whey, citric acid wastewaters can be found in Germirli *et al.* (1993).

Solid–liquid separation Solid leakage from the secondary clarifier possibly results in non-compliance with treatment objectives for TSS and BOD_5. It is certainly the major problem when operating the water chain of an AS system treating a food-industry wastewater. Origins of the settling problems are the presence of a high concentration of easily biodegradable compounds, of grease and sulfur, of a low dissolved oxygen concentration in the aeration tank and shock loads. The following problems are frequently encountered: (1) viscous bulking, where large amounts of exocellular slime are produced by microorganisms; (2) filamentous bulking, where filamentous organisms in the sludge flocs extend into the bulk liquid and interfere with settling and compaction; (3) foaming/scum formation, caused by non-degradable surfactants or by specific microorganisms (actinomycetes) in the sludge (Jenkins 1992; Jenkins *et al.* 1993). The extent of the problem stood out from a survey performed on 23 poultry processing facilities. This survey showed that the majority of the AS treatment plants questioned faced bulking problems (Kiepper 2001). Other origins of settling problems are dispersed growth, where the sludge does not flocculate, pin point floc, where small, compact, relatively weak flocs are formed and blanket rising, where denitrification releases nitrogen gas which floats the sludge blanket.

Solutions to remediate to these problems include the sludge partial chloration or ozonation, sludge ballasting (Clauss *et al.* 1998), the design of a selector, nutrient supplementation and choice of the reactor hydraulic (Pujol and Canler 1994). Reactor volume may be arranged in different ways that can greatly influence the effectiveness of the treatment, primarily by inducing significantly different sludge structures. In the case of dairy-wastewater treatment, at constant total reactor volume, an in-series reactor arrangement proved to be more efficient, in respect to the effluent COD and sludge characteristics, than a parallel arrangement (Jobbágy *et al.* 1994).

Matsché *et al.* (2002) indicate an original solution to avoid bulking problems when the effluent from the dairy industry can be brought to the urban-wastewater treatment plant. They propose to adsorb the easily biodegradable fraction of the dairy wastewater onto the excess sludge of the urban wastewater on a separate contact tank (solid retention time of 2 days; storage yield coefficient Y_{STO} in the range of 0.8–0.9 g COD, X_{STO}/g COD, SS). Then the sludge is dewatered (6% dry weight) and transferred to an anaerobic digester where it yielded a subsequent increased gas production.

7.4.2 Other treatment processes

7.4.2.1 Sequencing batch reactor

SBR is a time-oriented system operating over repeated cycles of five phases – fill, react, settle, decant, and idle. Mace and Mata-Alvarez (2002) gave an excellent overview on the use of the SBR technology for urban- and industrial-wastewater treatment. Performances are mainly controlled by the choice of the solid retention time, the hydraulic retention time and by operational parameters (sludge settleability, dissolved oxygen, influent characteristics). As SBR technology treats wastewater in small batches, it fits well with most food-processing wastewater collection systems. In particular, it offers a flexible and efficient mode of wastewater treatment for plants of small size. Dairy wastewaters from small cheese-making plants in the Jura Mountains (France) were successfully treated by a SBR reactor (Torrijos *et al.* 2001). Pollution from these cheese-making units is generally at the level of 200–300 pe (population equivalent). In that case, the SBR technology led to purification levels of 97.7% for total COD and 99.8% for BOD_5. The process was also financially advantageous, as the cost of treatment represented less than 1% of the price paid by the cooperative for its milk. Similar conclusions were drawn by Torrijos and Moletta (1997) in a winery producing 7300 hL annually located at Domaine du Mouton (Bordeaux, France) and by Houbron *et al.* (1998). Another example of the treatment of dairy wastewaters can be found in the paper of Samkutty *et al.* (1996). Table 7.7 gives some results on the removal performances of SBR processes treating food-processing wastewater.

7.4.2.2 Lagoons and wetlands

Aerobic lagoons are large shallow earthen basins that use algae in combination with other microorganisms for wastewater treatment. Slow water flow causes SS to settle. Oxygen is supplied naturally by mass transfer through the liquid–air interface and through photosynthesis. Aeration devices may also be used to intensify the biological degradation. The advantages of constructed wetlands include low operational cost, their "green appeal" and simplicity, and low energy requirements (Johns 1995). The feasibility of constructed wetlands however varies with wastewater characteristics and climate. Constructed wetlands are also being installed as polishing technology prior to final discharge of wastewater.

Due to a high complexity of these treatment systems their optimal design and operation is usually determined by the experience and intuition of the designer/operator. Cronk (1996) gave a comprehensive description on constructed wetlands designed to treat wastewater from dairy and swine operations. In this paper, the author mentions design parameters proposed by US EPA: HRT: 3–10 day; 4.5–135 kg BOD_5/acre-day or by NRCS: 73 kg BOD_5/ha-day, HRT >12 day, 0.15–0.20 m depth. The k–C* Model from CH2M HILL and Payne Engineering is also recalled. It is useful to determine the surface area required for a given load and is based on an area loading equation:

$$S = -\frac{Q}{k} \cdot \ln \frac{C_o - C^*}{C_i - C^*} \tag{7.8}$$

where, Q: annual flow (m^3/year); k: rate constant (m/year); C_i: inflow concentration (mg/L); C_0: outflow concentration (mg/L); C^*: background concentration (mg/L). Background concentration can be defined as concentration representing a natural state of the environment prior to any anthropogenic impact.

Table 7.8 presents some removal performances for constructed wetlands. High BOD_5 reductions are generally observed (up to 98%), but effluent SS concentrations are often high because of poor sludge settling.

7.4.2.3 Case study: Boiry Sainte Rictrude factory for sugar production (Fonade et al. 2000)

The plant produces sugar from sugar beet and the effluents which must be treated are essentially the waters used to wash the beets. During the harvest, about 2 months a year, the waters are recycled to the plant, resulting in an accumulation of the organic load.

The objectives required for the treatment lagoon were to avoid anaerobic processes causing bad smells and to obtain a 50% abatement of the organic load in order to reduce COD to admissible levels in the recycling. The effluent flow rate was 500 m^3/h and the COD load 70 ton/day. The lagoon was designed from biotreatability tests and its volume was found to be 72,000 m^3; 152 hydroejectors (1225 kW) were used to provide the 36 ton O_2/day required. They are located along the bank. Their relative positions (Figure 7.4) were studied at laboratory scale (1:55 linear scale model).

The results obtained show that the oxidoreduction parameter, rH, could be maintained at a value higher than 10 which is necessary to avoid bad smells. The system was characterized by a good oxygen transfer, close to 1.6 kg O_2/kWh.

7.4.2.4 Trickling filter

It is mainly used as a pre-treatment for industrial effluents. Bough *et al.* (1987) proposed the use of single-stage and a two-stage trickling filter for the direct treatment of dairy wastewater and then for the post-treatment of an anaerobic low-rate system.

Table 7.7 Removal performances of SBR processes treating food-processing wastewater.

Reference	Wastewater type	COD (mg/L)	Operating parameters	η_{COD} (%)	η_{TSS}(%)	η_{TKN} (%)
Li and Zhang (2002)	Dairy	10,000	HRT 1–3 day HRT 4 day	80–85	63	76 >95
Torrijos *et al.* (2001)	Cheese industry	2500	HRT 6 day	TCOD: 97.7, BOD_5: 99.8	High	–
Torrijos and Moletta (1997)	Winery			TCOD: 93, BOD_5: 97.5		50
Hamoda and Al Awadi (1995)	Dairy	3200	–	TCOD: 95	96	–
Torrijos *et al.* (2004)	Goat cheese dairy	12,000–15,000	0.4–0.9 kg COD/m^3-day	TCOD > 99	–	95
Thayalakumaran *et al.* (2003)	Meat processing	800–2000	SRT: 15 day, HRT: 2.5 day 0.56–0.72 kg COD/m^3-day	TCOD > 99	–	99

η_{COD}: COD removal rate; η_{TSS}: TSS removal rate; η_{TKN}: TKN removal rate; HRT: hydraulic retention time.

Table 7.8 Some design guidelines and removal performances for constructed wetland.

Reference	Wastewater type	COD (mg/L)	Operating parameters	η_{BOD5} (%)	η_{TSS} (%)	η_{TKN} (%)
Schaafsma *et al.* (2000)	Milkhouse, 170 cows	4000	HRT = 42 day; 0.055 ha; 0.15 m depth	97	96	98
Newman *et al.* (2000)	Dairy farm 100 dairy cows	3000	HRT = 42 day; Q_e = 2.65 m^3/day; 7.3 g BOD_5/m^2-day	85	94	53
French Water Agency (1993)	Ice-cream	4650	Q_e = 100 m^3/day; HRT = 40 day + polishing HRT = 20 day	97.8	88	83

η_{BOD5}: BOD_5 removal rate; η_{TSS}: TSS removal rate; η_{TKN}: TKN removal rate; HRT: hydraulic retention time.

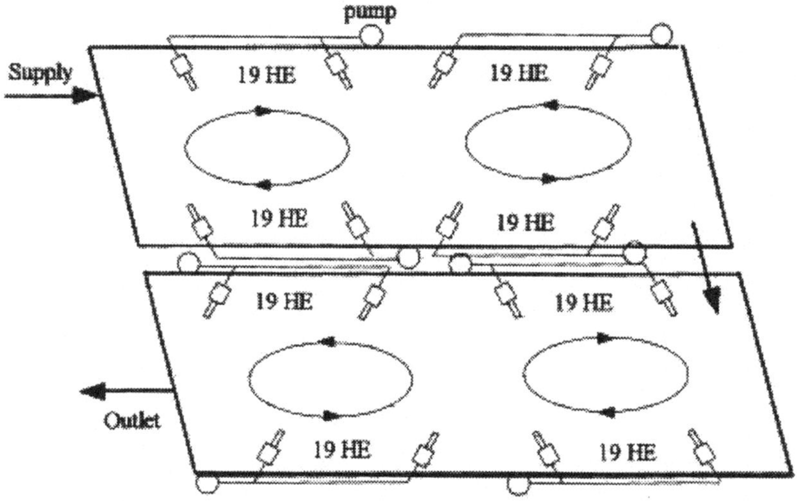

Figure 7.4 Scheme of the lagoon and aeration system (HE: hydroejector) (Fonade *et al.* 2000).

More recently, Austermann-Haun *et al.* (1999) designed pre-treatment systems for the highly loaded wastewater streams of a brewery, a fish factory and two vegetable-processing plants (with seasonal wastewater flow and composition variations). They reported the utilization of a high-rate plastic media biological filter as a buffer system placed after a fixed-film methane reactor to obtain a stable COD concentration in the activated sludge system.

7.4.2.5 Rotating biological contactor

The RBC operation is based on discs on a rotating shaft that allows the discs to be partly submerged in the effluent and support the biofilm responsible for the degradation of the pollution. The RBC is flexible to fluctuation of the wastewater organic load, rather cheap to run, and does not require too much land. Parameters that influence its removal performances are hydraulic retention time, disc rotational speed and disc submergence. A recent review on RBC is proposed by Patwardhan (2003).

7.4.2.6 Moving beds

Particle-based biofilm reactors provide the potential to develop compact and high-rate processes (Nicolella *et al.* 2000a). In these reactors, a large biomass content can be maintained (up to 30 g/L), and the large specific surface area (up to 3000 m^2/m^3) seems to ensure that the conversions are not strongly limited by the biofilm liquid mass-transfer rate. However, high shear stress from water circulation and carrier collision is necessary to control biofilm development and stabilize the reactor system (van Loosdrecht *et al.* 1995). A new generation of high-load, efficient biofilm reactors are operating throughout the world with several full-scale applications. A review on applications of particulate biofilm reactors can be found in Nicolella *et al.* (2000b).

Beside this moving-bed technology, air lift reactors were also developed. The first airlift reactor (type CIRCOX) in the brewing industry was applied at the Grolsch brewery in Enschede in the Netherlands in 1996 (Driessen *et al.* 1997).

7.5 TWO-STAGE ANAEROBIC–AEROBIC TREATMENT

Direct treatment by aerobic processes of strong wastewaters is not recommended because of inherent difficulties such as high aeration costs, oxygen transfer problems, high waste sludge produced and settling problems. Anaerobic pre-treatment is therefore an efficient and cost-effective solution in that case. Moreover, there is the possibility to valorize the organic biodegradable matter as energy and the space requirement is lower than that of an aerobic process alone.

Optimum design of the two stages must be determined. Indeed, according to Del Pozo *et al.* (2003), overcoming 80% of the degradation of a slaughterhouse wastewater by anaerobic pre-treatment resulted in a significant decrease in the removal rate. This was explained by the low-effluent biodegradable COD concentration that involved anaerobic kinetic limitation.

If nitrification and denitrification must be performed in the aerobic stage, it is necessary to run a bypass (fresh or acidified wastewater) around the methane reactor to provide sufficient biodegradable COD for denitrification (Austermann-Haun *et al.* 1999). However, this solution may lead to bulking problems on the AS plant. Another limitation for the use of an anaerobic pre-treatment is when struvite or other salts may precipitate. This problem and their solutions are described by Austermann-Haun *et al.* (1999).

An evaluation of the advantages of the two-stage anaerobic–aerobic technology over a purely aerobic system was given by Driessen and Vereijken (2003). The example is based on brewery with an annual beer production of 1,000,000 hL, operating 5 days a week, having a water to beer ratio of 7 hL/hL and 15% loss of water. The wastewater has a biodegradable COD of 3000 mg/L and 250 mg/L of inert solids. The wastewater to beer ratio is calculated to be approximately 0.51 m^3/hL beer and specific COD production amounts approximately 1.53 kg COD/hL beer. Results of the comparison are presented in Table 7.9.

7.5.1 Case study: Anaerobic–aerobic treatment of cheese wastewater in Mexico (Monroy *et al.* 1995)

The plant of "El Sauz", located at Cortazar, Guanajuato, produces 1500 ton of milk derivatives per month and 500 m^3 of wastewater per day. The mean concentrations of the wastewater are as follows: COD 4430 mg/L, BOD_5 3000 mg/L, TSS 1100 mg/L, O&G 754 mg/L, PO_4-P 14 mg/L, NH_4-N 18 mg/L, pH 7.3.

The plant had a treatment system consisting in a degreasing tank and three ponds connected in series (total volume: 10,500 m^3). But the performances of the system were not satisfactory. Therefore, it was decided to redesign the plant in four steps:

(1) *Reduction of wastewater pollution*: Including a segregation of plant effluents (recycling of cleaning chemicals such as H_3PO_4 and NaOH), removal of salts

from effluent by concentration and drying of the brine produced during the regeneration of the ion exchange resins, and substitution of cleaning chemicals to reduce P concentration in the wastewater and the formation of foam.

(2) *Pre-treatment*: Increase of the degreasing capacity (mechanical de-emulsification, flotation using coarse bubble diffusers, gravity separation).

(3) *Secondary biological treatment*:
   - Anaerobic pond of 4000 m^3 fed at a loading rate of 0.55 kg COD/m^3-day. The influent distribution was optimized compared to the previous configuration and based on a homogenous distribution of the influent at the bottom of the pond as in a UASB reactor.
   - Aerobic pond fed with the effluent from the anaerobic pond. Three surface-mixer-diffusers were added in the pond with a total capacity of 71 kW. The pond was designed to reduce the water COD from 800 to 112 mg/L using an aeration capacity of 45 kW. This capacity was calculated to maintain a mean dissolved oxygen concentration of 2 mg/L and a mixed liquor suspended solids (MLSS) concentration of 1.2 g/L. The HRT is 7 days, without solid recirculation. With respect to N, P and O&G, the actual design would allow to obtain 1, 7 and 10 mg/L, respectively, in the treated wastewater.

(4) *Tertiary biological treatment*: The third initial pond was converted in a water hyacinth pond (WHP) fed with the effluent from the aerobic pond. The TSS

Table 7.9 Comparison between completely aerobic and combined anaerobic–aerobic treatment systems from the points of view of energy and sludge production. (From Driessen and Verijken 2003).

	Completely aerobic	Combined anaerobic–aerobic	Savings
Energy production (MJ/hL)	0.0	+13.8	+13.8
Energy consumption (MJ/hL)	−4.6	−1.5	+2.8
Energy balance (MJ/hL)	**−4.6**	**+12.3**	**+16.9**
Biosolids (aerobic) (kg TS/hL)	0.25	0.05	0.20 (80%)
Inert solids (kg TS/hL)	0.15	0.15	0 (0%)
Total sludge (kg TS/hL)	**0.40**	**0.20**	**0.20 (50%)**

Table 7.10 Performance of the different stages for the treatment of cheese wastewater. (Monroy *et al.* 1995).

	pH	O&G (mg/L)	TCOD (mg/L)	SCOD (mg/L)	TSS (mg/L)	TKN (mg/L)	P (mg/L)
Influent	7.3	950	4426		1110	737	13
O&G tank	6.5	300	3436	3000	1087	29	15
Anaerobic pond	6.4		2246	1111	469	71	20
Aerated pond	7.3		1100	344	963	5	21
WHP	7.3	70	700	211	65	16	17

from the aerobic pond are pumped back to the anaerobic pond for stabilization. Water hyacinth must be harvested every week to maintain a density of 8 kg (as fresh weight)/m^2 with a high growth rate (0.24 kg/m^2-day). The performance of the different stages is given in Table 7.10.

7.6 BIOLOGICAL TREATMENT FOR NUTRIENT REMOVAL

As shown in Table 7.1, effluents from the animal processing industries are rich in nutrients. For example, the meat-processing effluents contained 75–200 mg TKN/L and 20–40 mg P/L for COD concentrations in the range of 800–2000 mg/L after primary treatment. To treat these nutrient flows, conventional nitrogen and phosphorus treatment systems can be applied. These systems are already widely described in the literature (e.g. see Henze *et al.* 2002) and will not be further detailed in this chapter.

7.6.1 Specificities of nitrogen removal in the context of treatment of food-industry wastewaters

Conventional nitrification–denitrification systems can be applied to remove nitrogen from the food-industry effluents. SBR technology can also be easily adapted for nitrogen removal (Mace and Mata-Alvarez 2002).

One original application of SBR for that purpose was described by Keller *et al.* (1997) who performed with success a simultaneous nitrification and denitrification (SND) from meat-processing wastewater. Page *et al.* (1997) used a full-scale SBR for nitrification of the effluent of a BVF low-rate anaerobic reactor. In the case of wetlands and ponds, nitrogen removal is enhanced during the growing season when high temperatures stimulate plant and microbial population growth (Gambrell and Patrick 1978). Processes adapted for the treatment of nitrogen-rich wastewaters are presented in Chapter 5 of this book.

When an anaerobic pretreatment is operated, the anaerobic effluent has a low COD/N ratio, which can lead to an incomplete denitrification. It is possible either to operate a conventional nitrification–denitrification process with the addition of an external carbon source or to combine the anaerobic digestion process with an aerobic nitrifying reactor and to recycle the nitrified effluent in the digester (Bernet *et al.* 2001). Mosquera-Corral *et al.* (2001) proposed this strategy for the treatment of pretreated effluents from a fish-canning industry. This configuration is still in development and has not been implemented at full-scale.

When the food industry is closed to an urban-wastewater treatment plant, it is worth testing a combined treatment. Huhtamäki and Huhtamäki (2003) presented a full-scale test using food-industry wastewater to increase the nitrogen treatment efficiency. A suspended carrier biofilm process, which utilizes industrial wastewater as carbon source was chosen to rebuild the conventional wastewater treatment plant. Over 70% nitrogen removal was achieved with redox-controlled activated sludge process and BOD_5/N ratio of 20/100.

7.6.2 Examples of phosphorus removal systems

Comeau *et al.* (1996) and Keller *et al.* (1997) achieved enhanced biological phosphorus removal (EBPR) in pilot-scale SBR systems applied to the treatment of a fermented dairy-processing wastewater and of a slaughterhouse wastewater, respectively. More recently, Merzouki *et al.* (2005) showed the importance of a pre-fermentation on the efficiency of a denitrifying EBPR process.

Chemical precipitation from dairy processes wastewater was applied with success by Bickers *et al.* (2003).

Struvite precipitation has been reported by Austermann-Haun *et al.* (1999) when treating wastewaters of wheat and potato starch factories that contain high concentrations of magnesium, nitrogen and phosphate. This process can be controlled by reducing pH, reducing the ion concentrations to a low value or by inhibiting the struvite formation. One other method for controlling fouling is to form struvite where it may be controlled, giving the option for recovering it (Doyle and Parsons 2002).

REFERENCES

Albagnac, G. and Verrier, D. (1983) Méthanisation des effluents d'industries agro-alimentaires. *Biomasse Actualités.*

Andreottola, G., Foladori, P., Ragazzi, M. and Villa, R. (2002) Dairy wastewater treatment in a moving bed biofilm reactor. *Wat. Sci. Technol.* **45**(12), 321–328.

Aspé, E., Marti, M.C. and Roeckel, M. (1997) Anaerobic treatment of fishery wastewater using a marine sediment inoculum. *Wat. Res.* **31**(9), 2147–2160.

Austermann-Haun, U., Seyfried, C.F. and Rosenwinkel, K.H. (1997a) Full scale experiences with anaerobic pre-treatment of wastewater in the food and beverage industry in Germany. *Wat. Sci. Technol.* **36**(2–3), 321–328.

Austermann-Haun, U., Seyfried, C.F. and Rosenwinkel, K.H. (1997b) UASB-reactor in the fruit juice industry. *Wat. Sci. Technol.* **36**(6–7), 407–414.

Austermann-Haun, U., Meyer, H., Seyfried, C.F. and Rosenwinkel, K.H. (1999) Full scale experiences with anaerobic/aerobic treatment plants in the food and beverage industry. *Wat. Sci. Technol.* **40**(1), 305–312.

Bernet, N., Akunna, J.C., Delgenès, J.P. and Moletta, R. (2001) Denitrification in methanogenic reactors: state of art. *Proceedings of the 9th World Congress, Anaerobic Digestion 2001*, Antwerpen, Belgium, September 2–6. Part 2, pp. 227–229.

Bickers, P.O., Bhamidimarri, R., Shepherd, J. and Russell, J. (2003) Biological phosphorus removal from a phosphorus-rich dairy processing wastewater. *Wat. Sci. Technol.* **48**(8), 43–51.

Bough, W.A., Burk, E.D., Bybee, E.C., Eckert, R.W., Clark-Thomas, D. (1987) The evolution of a treatment system for high strength dairy processing waste. *Proceedings of the Food Processing Waste Conference*, Atlanta, GA, USA, September 1–2, 25 pp.

Clauss, F., Helaine, D., Balavoine, C. and Bidault, A. (1998) Improving activated sludge floc structure and aggregation for enhanced settling and thickening performances. *Wat. Sci. Technol.* **38**(8–9), 35–44.

Comeau, Y., Lamarre, D., Roberge, F., Perrier, M., Desjardins, G., Hade, C. and Mayer, R. (1996) Biological nutrient removal from a phosphorus-rich pre-fermented industrial wastewater. *Wat. Sci. Technol.* **34**(1–2), 169–177.

Cronk, J.K. (1996) Constructed wetlands to treat wastewater from dairy and swine operations: a review. *Agric. Ecosys. Environ.* **58**(2–3), 97–114.

Danalewich, J.R., Papagiannis, T.G., Belyea, R.L., Tumbleson, R.E. and Raskin, L. (1998) Characterization of dairy waste streams, current treatment practices, and potential for nutrient removal. *Wat. Res.* **32**(12), 3555–3568.

Del Pozo, R., Tas, D.O., Dulkadiroglu, H., Orhon, D. and Diez, V. (2003) Biodegradability of slaughterhouse wastewater with high blood content under anaerobic and aerobic conditions. *J. Chem. Technol. Biotechnol.* **78**, 384–391.

Demirel, B. and Yenigün, O. (2002) Two-phase anaerobic digestion processes: a review. *J. Chem. Technol. Biotechnol.* **77**(7), 743–755.

Dohanyos, M., Zabranska, J., Kutil, J. and Jenicek, P. (2004) Improvement of anaerobic digestion of sludge. *Wat. Sci. Technol.* **49**(10), 89–96.

Doyle, J.D. and Parsons, S.A. (2002) Struvite formation, control and recovery. *Wat. Res.* **36**(16), 3925–3940.

Driessen, W. and Vereijkent, T. (2003) Recent developments in biological treatment of brewery effluent. *Proceedings of The Institute and Guild of Brewing Convention*, Livingstone, Zambia, March 2–7. www.environmental-expert.com/technology/paques/PAPER%20268.pdf

Driessen, W., Habets, L. and Vereijken, T. (1997) Novel anaerobic–aerobic process to meet strict effluent plant design requirements. *Fermentation* **10**(4), 243–250.

Dugba, P.N. and Zhang, R.H. (1999) Treatment of dairy wastewater with two-stage anaerobic sequencing batch reactor systems – thermophilic versus mesophilic operations. *Biores. Technol.* **68**(3), 225–233.

Ekama, G.A., Barnard, J.L., Gunthert, F.X., Krebs, P., McCorquodale, J.A., Parker, D.S. and Wahlberg, E.J. (1997) *Secondary Settling Tanks. Theory, Modelling, Design and Operation.* IWA Publishing, 232 pp.

El-Gohary, F.A., Nasr, F.A. and Aly, H.I. (1999) Cost-effective pre-treatment of food-processing industrial wastewater. *Wat. Sci. Technol.* **40**(7), 17–24.

Fonade, C., Rols, J.L., Goma, G., Doubrovine, N., Bermejo, M. and Grasa, J.P. (2000) Improvement of industrial wastewater treatment by aerated lagoon: case studies. *Wat. Sci. Technol.* **42**(5–6), 193–200.

Frankin, R.J. (2001) Full-scale experiences with anaerobic treatment of industrial wastewater. *Wat. Sci. Technol.* **44**(8), 1–6.

French Water Agency – Adour-Garonne (1993) Le fonctionnement des lagunes aérées dans l'industrie agro-alimentaire. Study Report. 189 pp.

Gambrell, R.P. and Patrick, W.H.J. (1978) Chemical and microbiological properties of anaerobic soils and sediments. In: *Plant Life in Anaerobic Environments* (eds. D.D. Hook and R.M.M. Crawford). Ann Arbor Science, MI, USA, pp. 375–423.

Germirli, F., Orhon, D., Artan, N., Ubay, E. and Görgun, E. (1993) Effect of two-stage treatment on the biological treatability of strong industrial wastewaters. *Wat. Sci. Technol.* **28**(2), 145–154.

Goorany, O. and Ozturk, I. (2000) Soluble microbial product formation during biological treatment of fermentation industry effluent. *Wat. Sci. Technol.* **42**(1–2), 111–116.

Grant, S., Landine, R., Wilson, D., Molina, J., Norton, S., Qiu, Z. and Cocci, A. (2002) Case studies of low-rate anaerobic treatment of dairy processing wastewaters. *Proceedings of the Water Environment Federation's 8th Annual Industrial Wastes technical and Regulatory Conference*, Atlantic City, NJ, USA, August 11–14, 20 pp.

GTZ (1997) Environmental management guideline for the fish canning industry. http://www.gtzth.org/library/files/gtzth_2000_2266_5_2.zip

Guerrero, L., Omil, F., Mendez, R. and Lema, J.M. (1999) Anaerobic hydrolysis and acidogenesis of wastewaters from food industries with high content of organic solids and protein. *Wat. Res.* **33**(15), 3281–3290.

Hamoda, M.F. and Al-Awadi, S.M. (1995) Wastewater management in a dairy farm. *Wat. Sci. Technol.* **32**(11), 1–11.

Henze, M., Harremoes, P., La Cour Jansen, J.I. and Arvin, E. (2002) *Wastewater Treatment. Biological and Chemical Processes*. 3rd edn. Springer Verlag, 420 pp.

Houbron, E., Torrijos, M. and Moletta, R. (1998) Application du procédé SBR aux effluents vinicoles. Résultats de trois années de suivi. *Proceedings of the 2nd International Specialized Conference on Winery Wastewaters*, Bordeaux, France, May 5–7, Cemagref Editions, pp. 189–196.

Huhtamäki, M. and Huhtamäki, L. (2003) Full-scale tests using food industry wastewater for nitrogen removal and a solution to increase the effluent treatment efficiency. *Proceeedings of the 9th IWA International Specialised Conference on Large Wastewater Treatment Plants*, Prague, Czech Republic, September 1–4.

Hulshoff Pol, L.W.H., Lens, P.N.L., Stams, A.J.M. and Lettinga, G. (1998) Anaerobic treatment of sulphate-rich wastewaters. *Biodegradation* **9**(3–4), 213–224.

Hulshoff Pol, L.W., de Castro Lopes, S.I., Lettinga, G. and Lens, P. (2004) Anaerobic sludge granulation. *Wat. Res.* **38**(6), 1376–1389.

Jenkins, D. (1992) Towards a comprehensive model of activated sludge bulking and foaming. *Wat. Sci. Technol.* **25**(6), 215–230.

Jenkins, D., Richard, M.G. and Daigger, C.T. (1993) *Manual on the Causes and Control of Activated Sludge Bulking and Foaming*, 2nd edn. Lewis Publishers, Chelsea, MI, USA.

Jobbágy, A., Grady Jr, C.P.L., Morsanyi, G., Nyeste, L. and Simon, J. (1994) Comparative studies of differently arranged activated sludge systems. *Wat. Sci. Technol.* **30**(11), 263–269.

Johns, M.R. (1995) Developments in wastewater treatment in the meat processing industry: a review. *Biores. Technol.* **54**(3), 203–216.

Kansal, A., Rajeshwari, K.V., Balakrishnan, M., Lata, K. and Kishore, V.V.N. (1998) Anaerobic digestion technologies for energy recovery from industrial wastewater – a study in Indian context. *TERI Inform. Monitor Environ. Sci.* **3**(2), 67–75.

Keller, J., Subramaniam, K., Gosswein, J. and Greenfield, P.F. (1997) Nutrient removal from industrial wastewater using single tank sequencing batch reactors. *Wat. Sci. Technol.* **35**(6), 137–144.

Kiepper, B. (2001) A survey of wastewater treatment practices in the broiler industry. *Proceedings of the WEF Annual Conference*, Atlanta, GA, USA.

Kirby, R.M., Bartram, J. and Carr, R. (2003) Water in food production and processing: quantity and quality concerns. *Food Cont.* **14**, 283–299.

Kleerebezem, R. and Macarie, H. (2003) Treating industrial wastewater: anaerobic digestion comes of age. *Chem. Eng.* **110**(4), 56–64.

Koster, I.W. and Lettinga, G. (1984) The influence of ammonium-nitrogen on the specific activity of pellitized methanogenic sludge. *Agricult. Waste.* **9**, 205–216.

Koster, I.W. and Lettinga, G. (1988) Anaerobic digestion at extreme ammonia concentrations. *Biol. Waste.* **25**, 51–59.

Leitão, R.C., van Haandel, A.C., Zeeman, G. and Lettinga, G. (2006) The effects of operational and environmental variations on anaerobic wastewater treatment systems: a review. *Biores. Technol.* **97**(9), 1105–1118.

Lettinga, G. (1995) Anaerobic digestion and wastewater treatment systems. *Ant. van Leeuwen.* **67**, 3–28.

Li, X. and Zhang, R. (2002) Aerobic treatment of dairy wastewater with sequencing batch reactor systems. *Bioproc. Biosyst. Eng.* **25**, 103–109.

Maat, D.Z. and Gorur, S.S. (1990) Start-up and performance of a full scale UASB anaerobic wastewater treatment facility. *Proceedings of the 44th Purdue University Industrial Waste Conference*, Chelsea, MI, USA, Lewis Publishers, Inc., pp. 209–214.

Macarie, H. (1992) *Anaerobic Treatment of the Wastewater from a Petrochemical Plant Producing Terephthalic Acid (1,4-Benzenedicarboxylic Acid).* PhD Thesis, Université de Provence, France.

Macarie, H. (2002) Les procédés de traitement biologique aérobie et anaérobie en sucrerie de canne. *Actes de la 3ème Rencontre Internationale en Langue Française sur la Canne à Sucre,* 5–8 juin 2001, Yaoundé, Cameroun, AFCAS Nogent sur Marne, France, ISBN 2-9506270-1-3, pp. 222–236.

Mace, S. and Mata-Alvarez, J. (2002) Utilization of SBR technology for wastewater treatment: an overview. *Ind. Eng. Chem. Res.* **41**, 5539–5553.

Massé, D.I., Masse, L. and Bourgeois, N. (1998) Anaerobic processing of slaughterhouse wastewater in a SBR. *Proceedings of the 8th International Conference on the FAO ESCORENA Network on Recycling of Agricultural, Municipal and Industrial Residues in Agriculture*, Rennes, France, May 26–29, pp. 375–387.

Matsché, N., Winkler, S., Prendl, L., Dornhofer, K. and Wandl, G. (2002) Treatment of easily biodegradable wastewater avoiding bulking sludge. *Wat. Sci. Technol.* **46**(1–2), 503–506.

McInerney, M.J. (1988) Anaerobic hydrolysis and fermentation of fats and proteins. In: *Biology of Anaerobic Microorganisms*, ed. A.J.B. Zehnder. John Wiley and sons, New York, pp. 373–415.

Merzouki, M., Bernet, N., Delgenès, J.P. and Benlemlih, M. (2005) Effect of prefermentation on denitrifying phosphorus removal in slaughterhouse wastewater. *Biores. Technol.* **96**, 1317–1322.

Mittal, G.S. (2006) Treatment of wastewater from abattoirs before land application – a review. *Biores. Technol.* **97**(9), 1119–1135.

Moletta, R. (2005) Winery and distillery wastewater treatment by anaerobic digestion. *Wat. Sci. Technol.* **51**(1), 137–144.

Monroy, O., Vazquez, F., Derramadero, J.C. and Guyot, J. P. (1995) Anaerobic–aerobic treatment of cheese wastewater with national technology in Mexico: the case of "El Sauz". *Wat. Sci. Technol.* **32**(12), 149–156.

Mosquera-Corral, A., Sanchez, M., Campos, J.L., Mendez, R. and Lema, J.M. (2001) Simultaneous methanogenesis and denitrification of pretreated effluents from a fish canning industry. *Wat. Res.* **35**(2), 411–418.

Newman, J.M., Clausen, J.C. and Neafsey, J.A. (2000) Seasonal performance of a wetland constructed to process dairy milkhouse wastewater in Connecticut. *Ecol. Eng.* **14**, 181–198.

Nicolella, C., van Loosdrecht, M.C.M. and Heijnen, S.J. (2000a) Particle-based biofilm reactor technology. *Trend. Biotechnol.* **18**, 312–320.

Nicolella, C., van Loosdrecht, M.C.M. and Heijnen, J.J. (2000b) Wastewater treatment with particulate biofilm reactors. *J. Biotechnol.* **80**(1), 1–33.

Orhon, D., Artan, N., Büyükmurat, S. and Görgun, E. (1992) The effect of residual COD on the biological treatability of textile wastewaters. *Wat. Sci. Technol.* **26**(3–4), 815–825.

Orhon, D., Tash, R. and Sözen, S. (1999) Experimental basis of activated sludge treatment for industrial wastewaters – the state of the art. *Wat. Sci. Technol.* **40**(1), 1–11.

Page, I.C., Grant, S.R., Landine, R.C., Landine, R.C. and Brown, G.J. (1997). Abattoir wastewater treatment plant nitrifies at low temperatures: a case study. Paper 975048, *ASAE Annual International Meeting*, Mineapolis, MN, USA, August, 12 pp.

Palenzuela-Rollon, A., Zeeman, G., Lubberding, H.J., Lettinga, G. and Alaerts, G.J. (2002) Treatment of fish processing wastewater in a one- or two-step upflow anaerobic sludge blanket (UASB) reactor. *Wat. Sci. Technol.* **45**(10), 207–212.

Pap, N. (2004) Industrial ecology and recycling course. http://cc.oulu.fi/~polamwww/FoodIE.pdf

Patwardhan, A.W. (2003) Rotating biological contactors: a review. *Ind. Eng. Chem. Res.* **42**(10), 2035–2051.

Pereira, M.A., Sousa, D.Z., Mota, M. and Alves, M.M. (2004) Mineralization of LCFA associated with anaerobic sludge: kinetics, enhancement of methanogenic activity, and effect of VFA. *Biotechnol. Bioeng.* **88**(4), 502–511.

Prendl, L. and Nikolavcic, B. (2000) Aerobic treatment of industrial waste water – experiences with the dosage of nitrogen and phosphorus. *Wat. Sci. Technol.* **41**(9), 241–249.

Pujol, R. and Canler, J.P. (1994) Contact zone: French practice with low F/M bulking control. *Wat. Sci. Technol.* **29**(7), 221–228.

Ramsay, I.R. and Pullammanappallil, P.C. (2001) Protein degradation during anaerobic wastewater treatment: derivation of stoichiometry. *Biodegradation* **12**(4), 247–257.

Rozzi, A., Malpei, F. and Padoani, L. (1998) Estimate of polluting loads in effluents of Italian North East wineries. In *Proceedings of the 2nd International Specialized Conference on Winery Wastewaters*, Bordeaux, France, May 5–7, Cemagref Editions, pp. 33–40.

Rusten *et al.* (1990) Coagulation as pretreatment of food industry wastewater. *Wat. Sci. Technol.* **22**(9), 1–8.

Rusten *et al.* (1993) Chemical pretreatment of dairy wastewater. *Wat. Sci. Technol.* **28**(2), 67–76.

Samkutty, P.J., Gough, R.H. and McGrew, P. (1996) Biological treatment of dairy plant wastewater. *J. Environ. Sci. Health Part A-Toxic/Hazard. Subst. Environ. Eng.* **31**(9), 2143–2153.

Schaafsma, J.A., Baldwin, A.H. and Streb, C.A. (2000) An evaluation of a constructed wetland to treat wastewater from a dairy farm in Maryland, USA. *Ecol. Eng.* **14**, 199–206.

Schmidt, J.E. and Ahring, B.K. (1997) Treatment of waste water from a multi product food-processing company, in upflow anaerobic sludge blanket (UASB) reactors: the effects of seasonal variation. *Pure Appl. Chem.* **69**(11), 2447–2452.

Schneider, I.A.H., Manera Neto, V., Soares, A., Rech, R.L. and Rubio, J. (1995) Primary treatment of a soybean protein bearing effluent by dissolved air flotation and by sedimentation. *Wat. Res.* **29**(1), 69–75.

Selmer-Olsen, E., Ratnaweera, H.C. and Pehrson, R. (1996) A novel pretreatment process for dairy wastewater with chitosan produced from shrimp-shell waste. *Wat. Sci. Technol.* **34**(11), 33–40.

Thayalakumaran, N., Bhamidimarri, R. and Bickers, P.O. (2003) Biological nutrient removal from meat processing wastewater using a sequencing batch reactor. *Wat. Sci. Technol.* **47**(10), 101–108.

Torrijos, M. and Moletta, R. (1997) Winery wastewater depollution by sequencing batch reactor. *Wat. Sci. Technol.* **35**(1), 249–257.

Torrijos, M., Vuitton, V. and Moletta, R. (2001) The SBR process: an efficient and economic solution for the treatment of wastewater at small cheesemaking dairies in the Jura mountains. *Wat. Sci. Technol.* **43**(3), 373–380.

Torrijos, M., Sousbie, P., Moletta, R. and Delgenès, J.P. (2004) High COD wastewater treatment in an aerobic SBR: treatment of effluent from a small farm goat's cheese dairy. *Wat. Sci. Technol.* **50**(10), 259–267.

UNEP (2000a) Cleaner production assessment in meat processing. http://www.agrifood-forum.net/publications/guide/meatguide.zip

UNEP (2000b) Cleaner production assessment in dairy processing. http://www.agrifood-forum.net/publications/guide/dairyguide.zip

UNEP (2000c) Cleaner production assessment in fish processing. http://www.agrifood-forum.net/publications/guide/fishguide.zip

van Loosdrecht, M.C.M., Eikelboom, D., Gjaltema, A., Mulder, A., Tijhuis, L. and Heijnen, J.J. (1995) Biofilm structures. *Wat. Sci. Technol.* **32**(8), 35–43.

Vavilin, V.A., Vasiliev, V.B., Rytov, S.V. and Ponomarev, A.V. (1995) Modelling ammonia and hydrogen sulfide inhibition in anaerobic digestion. *Wat. Res.* **29**(3), 827–835.

Vidal, G., Carvalho, A., Mendez, R. and Lema, J.M. (2000) Influence of the content in fats and proteins on the anaerobic biodegradability of dairy wastewaters. *Biores. Technol.* **74**(3), 231–239.

Watanabe, H., Kitamura, T., Ochi, S. and Ozaki, M. (1997) Inactivation of pathogenic bacteria under mesophilic and thermophilic conditions. *Wat. Sci. Technol.* **36**(6–7), 25–32.

Weiland, P. and Rozzi, A. (1991) The start-up, operation and monitoring of high-rate anaerobic treatment systems: discusser's report. *Wat. Sci. Technol.* **24**(8), 257–277.

World Bank Group (1998) *Pollution Prevention and Abatement Handbook: Toward Cleaner Production*. Washington, DC.

Zeeman, G. and Sanders, W. (2001) Potential of anaerobic digestion of complex waste(water). *Wat. Sci. Technol.* **44**(8), 115–122.

Zoutberg, G.R. and Eker, Z. (1999) Anaerobic treatment of potato processing wastewater. *Wat. Sci. Technol.* **40**(1), 297–304.

8

Application of biological treatment systems for chemical and petrochemical wastewaters

E. Razo-Flores, H. Macarie and F. Morier

8.1 INTRODUCTION

The chemical and petrochemical industry (CPI) produces worldwide thousands of compounds each year and is still in expansion. The global market of the CPI is huge and only during the first 9 months of 2004 the sales in the USA summed up US $110.4 billion for 26 companies of the chemical sector (Storck *et al.* 2005). The CPI developed strongly in the USA in the middle of the 20th century and expanded afterwards to Europe, Japan and some developing countries. Despite all the economical benefits that CPI has produced, it must be emphasized that it also created serious environmental problems. CPI generates a high volume of wastewaters containing organic and/or inorganic compounds including metals. Until the 1970s, effluent discharge requirements for the CPI industry usually considered only biochemical oxygen demand (BOD_5) and suspended solids (SS). As a consequence, wastewater management in this industrial sector was rather rudimentary

© 2006 IWA Publishing. *Advanced Biological Treatment Processes for Industrial Wastewaters* edited by F.J. Cervantes, S.G. Pavlostathis and A.C. van Haandel. ISBN: 1843391147. Published by IWA Publishing, London, UK.

and often limited to mechanical and chemical primary treatment (Trobisch 1992). When applied, biological treatment was frequently performed in unlined lagoons. Activated sludge, occasionally replaced by trickling filters, was also applied when land availability was restricted (Eckenfelder and Englande 1996). However, due to the discovery of groundwater contamination caused by the unlined lagoons and a more stringent environmental legislation controlling the discharge of specific compounds potentially harmful to aquatic forms of life or human health, CPI has been under increased scrutiny during the last three decades. Over this period, several new technologies have been developed and applied for treating CPI wastewaters (CPIW). Most of them oxidize organic compounds using either biological (aerobic or anaerobic) or physicochemical (chemical oxidation, incineration, wet oxidation (WO)) methods or a combination of both (Trobisch 1992).

It is clear that the "end-of-pipe treatment" approach that CPI used for several decades is no longer affordable. Currently, thanks to new regulatory considerations, a holistic integrated environmental protection approach is a prerequisite for CPI in order to comply with present and future regulations and still remain competitive achieving the goal of sustainable development. Particularly, efforts must be concentrated towards the development of "cleaner production technologies" allowing a reduction of pollution discharge at the source (Trobisch 1992; Strotmann and Weisbrodt 1994; Ding *et al.* 1996). It must be emphasized that very often waste minimization can be achieved through simple adjustments of the existing processes and that these adjustments translate frequently into economic benefits for the companies that implement them (Alzamora-Rumazo *et al.* 2000). Water recycling, at least for cooling purposes, must be also applied as much as possible in order to reduce the amount of water intake (Huber 1967; Baron *et al.* 2000; Wong 2000). A general review of the "state of the art" for treatment of CPIW is presented in this chapter. Specifics on characteristics of CPIW, toxicity, biodegradability and case studies are also presented. Special emphasis will be drawn to the more novel anaerobic systems.

8.2 CHARACTERIZATION OF CPI WASTEWATER

8.2.1 Source and characteristics

The raw materials used by CPI are almost all petroleum-based products that are transformed using a wide variety of processes producing a wide range of waste streams. Consequently, due to the complexity of CPI, it is very difficult to make general statements for the wastewater generated by this industrial sector (Carmichael and Strzepek 1987). This is reinforced by the fact that even when the same final product is manufactured, the raw materials and the production technology used may vary from one installation to another resulting in different product mixtures and so in wastewaters with diverse characteristics. A general feature of CPIW, however, is the strong fluctuations in pH, temperature and strength, that may occur due to maintenance activities or upsets in the production process (Kleerebezem 1999). This trait leads to the need of large equalization capacities to smooth the variations when a biological treatment is employed. The three basic uses for water in CPI are cooling water, steam generation and process water that account approximately for 80%, 5% and 15%, respectively, of intake water (Carmichael and Strzepek 1987).

CPI processes have been classified into four groups on the basis of water use within the process (Carmichael and Strzepek 1987):

(a) Non-aqueous processes.
(b) Processes with process water contact as steam diluent or adsorbent.
(c) Aqueous liquid-phase reaction systems.
(d) Batch and semi-continuous processes.

Except for spills, storm-water run-off and sanitary systems, the above-listed four process groups are the main source of contaminants in the CPI.

Before designing a general treatment for CPIW, it is necessary to know where the contaminated streams are generated, the levels of contamination and whether they can be segregated with their own pre-treatment if needed (Berné and Cordonnier 1995). The parameters that define potential contamination in CPIW include parameters common in municipal wastewater on the one hand (SS, BOD_5, chemical oxygen demand (COD), nitrogen compounds, etc.), and parameters specific for the CPIW on the other (hydrocarbons, phenols and alkylphenols, sulfur compounds, salts, alkalinity, etc.). The characteristics of some CPIW are presented in Table 8.1.

Table 8.1 Characteristics of wastewater from the production of some organic chemicals or from refinery operations.

Product or waste stream	SS (mg/L)	COD (mg/L)	BOD_5 (mg/L)
Acetaldehyde	150–300	40,000–60,000	15,000–25,000
Acrylates	50–100	2000–3200	1000–2000
Acrylonitrile	80–150	600–1200	200–500
Aspartame	200–2000	10,000–30,000	7000–20,000
Dimethylterephthalate (DMT)	NA	17,000–142,000	NA
Esters	20–100	10,000–20,000	5000–12,000
Ethanolamine	~0	199–424	110–236
Ethylene and propylene	20–40	800–1200	400–600
Isocyanate	40–75	900–1600	300–600
Ketones	50–100	20,000–40,000	10,000–20,000
Methacrylic acid	6000–12,000	7000–12,000	NA
Methyl and ethyl parathion	50–100	4000–6000	2000–3500
Nylon intermediates	~0	9000–16,000	5000–9000
Oil from coal[a]	NA	8000–18,000	NA
Organic acids	100–200	5000–15,000	300–600
Organic phosphate compounds	200–400	1500–3000	500–1000
Polyethylene terephthalate	~0	12,000–18,600	7800–13,400
Phthalate plasticizers[b]	92–164	5000–13,900	3400–9500
Phthalic and maleic anhydrides	20–50	150–300	NA
Phenol (from toluene)	NA	18,000–30,500	NA
Spent caustic liquors	NA	240,000–400,000	NA
Sour waters	NA	1200–1600	NA
Styrene–divinylbenzene polymer	NA	8000–10,000	NA

SS: suspended solids; NA: not available; [a]Fisher-Tropsch process; [b]Mainly dioctyl phthalate, batch process. Adapted from Jørgensen (1979); Subrahmanyam *et al.* (1982); Olmos *et al.* (2004) and authors' files.

These wastewaters are often characterized by high COD concentrations, which may range from few hundred to several thousands mg/L, high total dissolved solids and contamination by heavy metals. The SS concentration is usually much lower and even nil in some instances. The quality of the water required by the CPI in their manufacturing processes being generally of the deionized grade, the resulting CPIW is normally devoid (or at least greatly unbalanced) of macro- (N and P) and micronutrients necessary for biological treatment (Kleerebezem 1999). This is obviously not the case for the streams generated during the production of nitrogen or phosphorus containing compounds for which, either nitrogen or phosphorus may be in excess and need to be removed from the wastewater (Subrahmanyam *et al.* 1982; Strotmann and Weisbrodt 1994; Cheng *et al.* 2004).

One of the major issues in the treatment of CPIW is the toxicity and biodegradability of the contaminants present in the waste streams. Consequently, it is very important to know as much as possible the chemical composition of the wastewater. As can be observed from the COD/BOD_5 ratio of the wastewaters shown in Table 8.1, which is a measure of their biodegradability, some of them will be amenable to a direct biological treatment (aerobic or anaerobic). In other cases, where the strength of the wastewater is too high or potentially toxic, the waste stream will have to be treated either by a combined physicochemical–biological process or by a physicochemical process (Gulyas 1997). The proper CPIW treatment will depend on the final goal to be achieved: recycle, treatment for discharging to a sewer or to a water body, etc. A diagram indicating the general available technologies for CPIW treatment is shown in Figure 8.1.

8.2.2 Toxicity

An important point, which has to be considered to estimate if a given CPIW can be or not a good candidate for biological treatment corresponds to the concentration of its individual organic and inorganic components. Indeed, by comparison with published data (e.g. Blum and Speece 1991) it will be possible to evaluate if these concentrations have reached critical levels which make the wastewater toxic towards the different microbial groups necessary to its purification.

It is noteworthy that contrarily to the still common believe among environmental engineers, anaerobic microorganisms, particularly methanogens, are not more susceptible to toxic compounds than the aerobic ones, except in the case of chlorinated aliphatic hydrocarbons and alcohols (Blum and Speece 1991). This means, that a toxic wastewater will be probably as hard to treat aerobically as anaerobically and will require special pre-treatments (Gulyas 1997) before the biological step, or acclimation of the biomass and/or specific reactor designs. Thanks to this, wastewaters containing for instance the extremely toxic formaldehyde or hydrogen cyanide at concentrations as high as 10 g/L and 125 mg/L, respectively, can be treated very efficiently anaerobically (Zoutberg and de Been 1997; Gijzen *et al.* 2000).

It must be recognized however that due to the slower growth rate of the anaerobic biomass, a toxic shock may have a greater impact on an anaerobic treatment system than an aerobic one because the anaerobes would take longer to recover and reach their original population level. This has to be tempered by the fact that

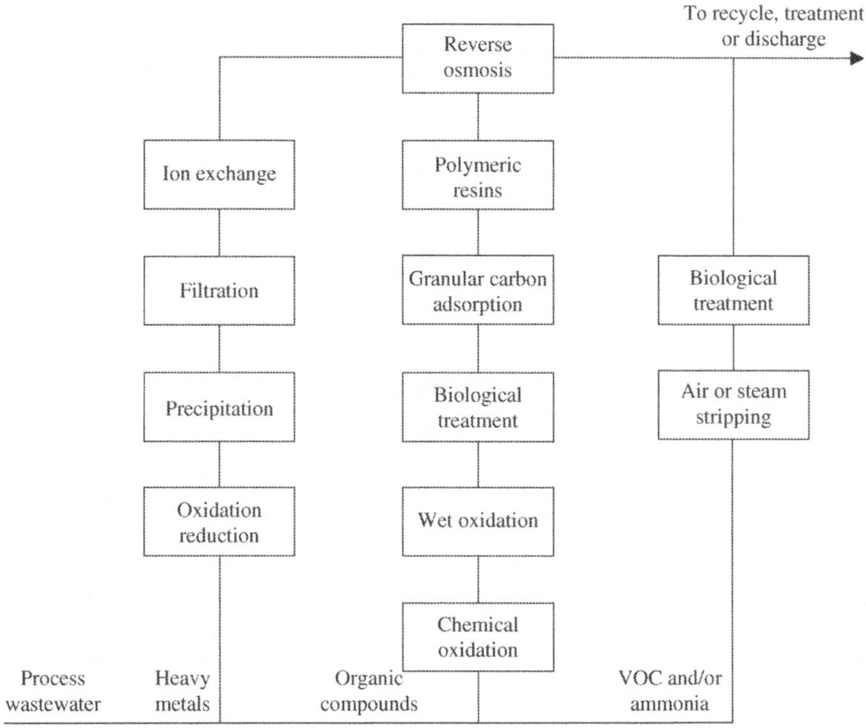

Figure 8.1 General available technologies for chemical and petrochemical wastewater treatment (Modified from Eckenfelder and Englande 1996).

the biomass concentration in modern high-rate anaerobic reactors is in general 10–20 times higher than in conventional activated sludge and that the impact of a toxicant is obviously related to the ratio of the amount of toxicant to the amount of biomass (Moreno-Andrade and Buitrón 2003).

No precise guidelines can be given in order to determine the potential toxicity of a wastewater from the chemical structure of its organic components. Notwithstanding, it can be said that esters, ketones and carboxylic acids are generally relatively non-toxic contrarily to aldehydes, nitro and chlorinated compounds which are chemically much more reactive. It can be said also, that within a class of compounds, the toxicity should increase with the presence of unsaturated carbon structure which increments the reactivity, a benzene ring and the hydrophobicity of the molecule since this last characteristic increases the ability of the compound to solubilize in the lipid bacterial membranes which may damage the membrane functions like ion transport and may cause cellular lysis. On the opposite, the toxicity appears to decrease with an increase in the carbon length in the case of aliphatic compounds and with the number of hydroxyl groups in the case of the aromatic ones (Chou *et al.* 1978b; Sierra-Alvarez and Lettinga 1991; Donlon *et al.* 1995). Several authors have reported the concentrations that reduced in 50% the acetoclastic methanogenic activity (IC_{50}) of anaerobic granular sludge when exposed to several kinds of phenolic compounds (Sierra-Alvarez and Lettinga 1991; Donlon *et al.* 1995; Olguín-Lora *et al.* 2003). In the specific case of phenol, the IC_{50} values using non-acclimated and phenol-acclimated granular sludge are in the range of 470–7802 mg/L (Razo-Flores *et al.* 2004).

With respect to the inorganic compounds, safe concentrations below which anaerobic treatment is applicable without fearing toxicity are: 5 g Na^+/L, 3.5 g K^+/L, 2.8 g Ca^{2+}/L, 1.8 g NH_4^+/L (pH 7), 50 mg H_2S-S/L and a sulfate concentration giving a COD/SO_4^{2-} ratio (g/g) higher than 10 (Kugelman and Chin 1971; Rinzema and Lettinga 1988). These values represent however very rough guidelines and wastewaters with higher concentrations should be still treatable anaerobically after biomass acclimation.

8.2.3 Biodegradability and recalcitrance

The biodegradability of an organic compound is known to depend upon a variety of physical and biochemical factors, including chemical structure, sorption characteristics, volatility, ionic character, solubility, concentration (not too low, not too high), environmental conditions (pH, temperature, etc.) and availability of terminal electron acceptors (Fewson 1988; Providenti *et al.* 1993; Field 2002). The biodegradation of organic compounds can proceed under aerobic or anaerobic conditions depending on the electron acceptor available: oxygen for aerobic respiration, or nitrate, sulfate, oxidized metal ions (e.g. Fe^{3+} and Mn^{4+}), protons and bicarbonates for anaerobic respiration. Proton and bicarbonate reduction are the primary electron acceptors utilized under methanogenic conditions (Schink 1997). The mechanisms of biodegradation are completely different under aerobic and anaerobic conditions. For example, under aerobic conditions, aromatic compounds are transformed by monooxygenases and dioxygenases into a few central intermediates incorporating oxygen into the aromatic ring prior to ring fission while under anaerobic conditions, the aromatic ring is reductively attacked (Fuchs *et al.* 1994). The aerobic and anaerobic biodegradation of organic compounds present in CPIW has been reviewed extensively (Ludzack and Ettinger 1960; Matsui *et al.* 1975; Pitter 1976; Chou *et al.* 1978a; Gerike and Fischer 1979; Horowitz *et al.* 1982; Ghisalba 1983; Lund and Rodriguez 1984; Evans and Fuchs 1988; Zitomer and Speece 1993; Field *et al.* 1995).

As a first approximation, the COD/BOD_5 ratio used to determine the biodegradability of wastes under aerobic conditions will be also a good indicator of their anaerobic biodegradability and all effluents with a ratio below 3 should be in principle good candidates for anaerobic treatment. Of course, depending of the nature of the organic compounds present in the wastewater, some differences will exist between the potential to apply aerobic or methanogenic treatment. For instance, the anaerobic option will not be feasible in the case of wastewaters containing molecules without oxygen in their carbonaceous skeleton, as are the aliphatic (alkanes, alkenes, alkynes) and aromatic (e.g. benzene, *o*-, *m*- and *p*-xylene) hydrocarbons (Field *et al.* 1995). Indeed, even if some of these compounds can be biodegraded to methane (e.g. benzene, hexadecane), the rate of the reaction will be extremely low compared to their attack in the presence of molecular oxygen and so their removal is doubtful under practical conditions of anaerobic reactor operation.

Restrictions to the anaerobic degradation of other classes of compounds exist also despite the presence of oxygen in their structure. That is the case of several molecules containing tertiary substituted carbon and/or ether bonds (e.g. *tert*-butanol, metyl *tert*-butyl ether (MTBE), *tert*-amyl methyl ether, ethyl *tert*-butyl ether, butyl ether, diethyl ether, etc.). Nevertheless, it must be recognized that several of these compounds, even if known to be aerobically biodegradable (Fiorenza and Rifai 2003), will equally be difficult to degrade under such conditions. This is mainly due to the steric hindrance caused by the tertiary substituted carbon, which will not facilitate their attack by oxygenases. In order to overcome this difficulty, the aerobic microorganisms need to invest a high amount of energy to metabolize them. This, results in a low efficiency of biomass production (see for instance the low biomass yields measured for different aerobic consortia growing on MTBE; Fortin *et al.* 2001). As a consequence, the retention of the microorganisms degrading these compounds in conventional aerobic treatment units will not be easy, and such units will hardly acquire the capacity to remove them.

Beside its limitations, the range of molecules amenable to anaerobic treatment remains very large and includes a great variety of aliphatic and aromatic carboxylic acids, aldehydes, alcohols, esters, ketones, etc. (Macarie 2000). Moreover, it must be emphasized that in the same way aerobic treatment will be the only answer for some wastewaters, anaerobic treatment will be the sole answer for others. That is the case of wastewaters containing certain polyols (Harvey and Rubiano 1983), azo dyes (Razo-Flores *et al.* 1997), nitroaromatics (Razo-Flores *et al.* 1999), and polychlorinated aromatic and aliphatic compounds (van Eeckert and Schraa 2001) which are usually persistent in aerobic environments but can be completely converted to CH_4 and CO_2 in anaerobic ones or at least transformed into products which are then mineralizable in the presence of oxygen (Zitomer and Speece 1993; Field *et al.* 1995).

8.3 CASE STUDIES

8.3.1 Styrene monomer and propylene oxide production

Styrene monomer (SM) and propylene oxide (PO) are basic petrochemical compounds whose worldwide production accounts for several million of tones per year. The chemical structures of SM and PO are shown in Figure 8.2.

SM is an aromatic hydrocarbon which, under normal conditions, is a clear, colorless, flammable liquid. The derivatives of SM are used in the manufacture of plastic

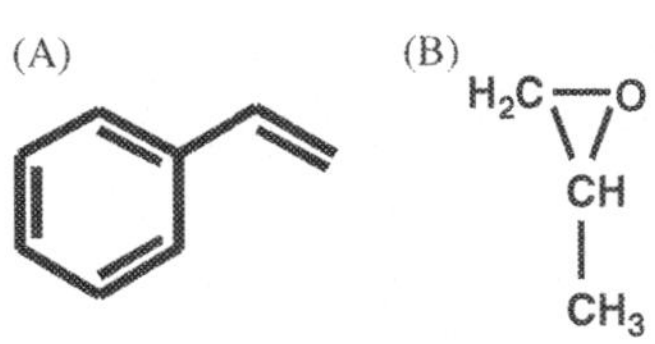

Figure 8.2 Chemical structures of (A) SM and (B) PO.

and rubber products, including polystyrene. The conventional method for producing SM is the alkylation of benzene with ethylene followed by dehydrogenation.

PO is a colorless liquid with an ether-like odor and is used as a chemical building block for the production of urethane, coatings, adhesives, sealants, elastomer, etc. PO is usually obtained by the "Chlorohydrin" or the "PO/MTBE" processes. In the first process, the propylene to PO reaction goes together with chlorine to chlorinated hydrocarbons reaction. In the second process, isobutene is converted into *tert*-butyl alcohol (TBA) parallel to the propylene to PO reaction. TBA is the main raw material for the fuel additive MTBE.

A notable exception for the production of both compounds is the so-called "styrene monomer propylene oxide (SMPO)" process developed by Shell. In this process organic peroxides are first produced by liquid-phase air oxidation of ethylbenzene, followed by epoxidation in the presence of a transition metal catalyst and the formation of PO and SM. The SMPO process has some advantages over the two other processes (e.g. avoids the use of expensive, toxic and corrosive reagents that also produce highly toxic chlorinated organic byproducts). However, all of them produce high volumes of wastewater that have to be treated before discharge.

Shell Chemicals is the second larger producer of SM and one of its main petrochemical plants is located in Moerdijk, The Netherlands (Shell Nederland Chemie, SNC). SNC has two SMPO plants producing 1000 and 460 kton/year of SM and PO, respectively (http://www.shellchemicals.com). Due to the characteristics of the wastewater generated from these plants a combined physicochemical–biological treatment process was applied (Frankin *et al.* 1994) as shown in Figure 8.3.

The main objectives of the combined treatment were: (a) the reduction of toxicity as well as the partial oxidation of organic compounds by WO in order to make the waste stream more suitable for biological treatment; and (b) anaerobic biological oxidation of the organic compounds. In the following sections the main features of the WO and anaerobic processes as well as the main bottlenecks faced when treating SMPO wastewater from SNC and the main strategies applied to overcome these difficulties are discussed.

8.3.1.1 WO process

WO is the oxidation of soluble or suspended components in an aqueous environment using oxygen as the oxidizing agent. When air is used as the source of oxygen the process is referred as wet air oxidation. The oxidation reactions occur at

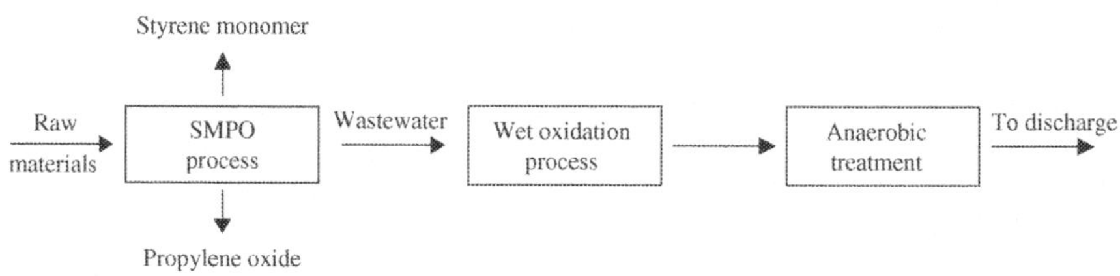

Figure 8.3 SMPO process and wastewater treatment scheme at SNC (Modified from Frankin *et al.* 1994).

temperatures of 200–325°C, pressures of 20–210 bar (2–21 × 10^3 kPa), residence time of 10–90 min with conversion efficiencies between 80% and 99% (Levec 1997). Since the oxidation reactions are exothermic, sufficient energy may be released in the reactor to allow the WO system to operate without any additional heat input. This autothermal operation is crucial in the scale-up as it determines the overall economics of waste destruction (Levec 1997). The main products of the reactions are: CO_2, H_2O, N_2, salts and organic acids. Wastewaters that are most effectively treated by WO have strength in the range of 20–200 g COD/L and flow rates above 1 m^3/h (Levec 1997). This process can be applied to satisfy a variety of objectives such as: elimination of toxicity or reactivity, destruction of specific compounds, pre-treatment to produce readily biodegradable residual organics, process liquor treatment for recycle/recovery and gross reduction in COD.

The WO system selected by SNC for the pre-treatment of its SMPO wastewater was supplied by USFilter and is known under the trademark Zimpro®. In 1997, nearly 200 units based on this technology were operating worldwide on different industrial wastewaters or for the dewatering or volatile solid destruction of sludges from municipal wastewater treatment plants (Levec 1997). At Moerdjik, the Zimpro effluent (720 m^3/day) was characterized by a pH, temperature and pressure of 9, 130°C and 50 bar (5 × 10^3 kPa), respectively. Around 80% of its total COD was in the form of benzoate and acetate, along with a high concentration of salt and bicarbonate. Other characteristics of this effluent are given in Table 8.2.

8.3.1.2 Anaerobic treatment

Before applying a biological process for the treatment of the Zimpro effluent, a 2-year feasibility study was executed to assess the design and performance criteria for the full-scale anaerobic reactor (Frankin *et al.* 1994). Special attention was given to the high levels of salt, benzoate as well as the presence of numerous other chemicals at lower concentrations, such as PO, SM, benzaldehyde, phenol, toluene, propanol, ethylbenzene, cumene, methylphenylcarbinol, methylphenylketone, etc. (Enger *et al.* 1989) that could hinder the anaerobic treatment. Results of the laboratory research indicated that a 1:1 dilution of the Zimpro effluent with an external source of water was necessary to guarantee optimal treatment efficiency. The design of the anaerobic process provided by Biothane Systems International had the following characteristics: an upflow anaerobic sludge bed (UASB) reactor with a volume of

Table 8.2 Composition of the SMPO wastewater after treatment with the Zimpro process.

Constituent	Concentration (g/L)
Sodium	10–15
Benzoate	6–7
Acetate	4–6
Bicarbonate	20–30
COD	20

Modified from Frankin *et al.* (1994).

1430 m^3, volumetric loading rate of 10 kg COD/m^3-day, hydraulic retention time (HRT) of 1 day, biogas production rate of 7200 m^3/day and a COD removal efficiency of 80%. A 1:1 feed to recycling ratio was also considered in order to further dilute the wastewater. Figure 8.4 shows a diagram of the Biothane process design.

Before the biological treatment, the Zimpro effluent is cooled and the pressure decreased to atmospheric. Afterwards, CO_2 is released and vented through a knock out vessel. The wastewater is finally conditioned (pH adjustment to 7–7.5 with HCl, micro- and macronutrients addition, dilution, etc.) in a rapid mix tank. In March 1988, the UASB reactor was seeded with 200 m^3 of granular sludge obtained from an UASB reactor treating sugar-refinery wastewater and operated at 35°C. The operation of the process started with a 1:3 dilution of the wastewater and after 1 month of operation the dilution was reduced to a 1:1 ratio. After 2 years of operation the COD removal efficiency was between 80% and 95% even at volumetric loading rates of 20 kg COD/m^3-day. The biodegradation of benzoate and acetate as well as all the other organic compounds present in the wastewater was also highly efficient.

Up to date information indicates that Shell Chemicals is actively working in a new SMPO process that allows wastewater to be recycled back into the production process. Consequently, the intake of fresh water is reduced (108,000 ton/year for two SMPO plants located in Singapore), as is the production of wastewater and the amount of energy used (http://www.shellchemicals.com).

According to this case study, the combined physicochemical–biological process applied by SNC for the treatment of SMPO wastewater was very successful and 17 years after start-up it is still operating effectively (Personal communication of Arnold Mulder from Amecon).

8.3.2 Natural gas and multi-fine-chemical production site

The chemical industrial complex of LACQ, located in the Southwest part of France, originates from the production of natural gas wells which began in the 1950s.

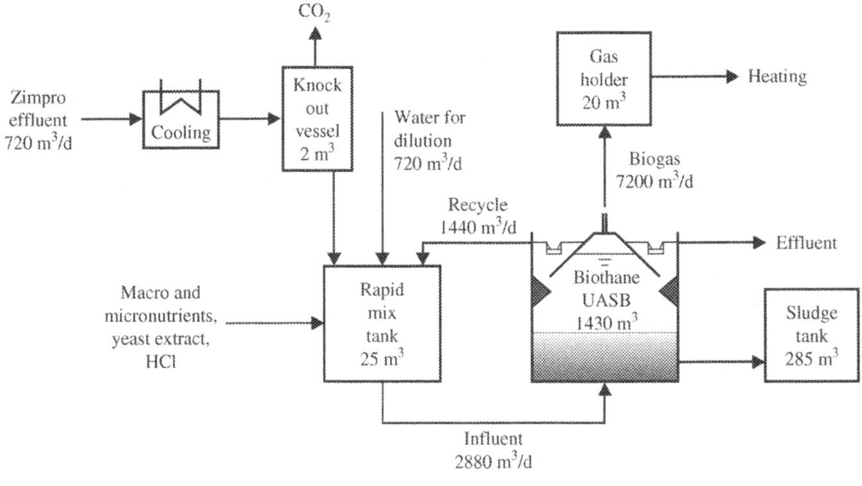

Figure 8.4 Diagram of the Biothane process for the biological treatment of Shell Moerdijk Zimpro effluent (Modified from Frankin *et al.* 1994).

Due to the high content of sulfur in the gas, a complete chemical industry of thiocompounds was established on site and lately, due to the decline of gas production, an active policy has attracted several medium size fine chemical industries. Among the active companies on site are TOTAL, a giant petrochemical company, ARKEMA, a chemical company and SOBEGI, a company which includes various fine chemical facilities belonging to other companies. SOBEGI also provides utilities (including wastewater treatment) to all the other companies and is in charge of the chemical platform development.

Until 2003, the site effluent was treated through a physicochemical plant (neutralization + flotation) and a biological lagoon without a complete sludge removal system. The treatment plant as such, implemented some 20 years ago, was unable to comply with current discharge limits due to the evolution of the industrial activities over the same period. Moreover the floated sludges were incinerated in a site incinerator of which flue-gas treatment had to be improved to conform to new European regulations.

The purpose of the complete project was therefore: revamp the physicochemical plant, build an aerobic biological reactor in replacement of the existing lagoon, and install a new incinerator to eliminate not only the floated sludges but also the biological sludges from the site plant and the nearby city sewage treatment plant. Proserpol, a French wastewater engineering firm, was in charge of the biological plant on a turn-key basis.

8.3.2.1 The environmental constraints

The LACQ chemical complex is located near to a sensitive to pollution Salmon river, Gave de Pau, 100 km from the Atlantic Ocean. Consequently, the discharges are under strict scrutiny by the French environmental regulatory agencies. For refineries, a specific regulation is established according to the processes involved as shown in Table 8.3.

Table 8.3 Specific regulations for refineries in France, maximum allowed values.

Parameters	Unit[a]	Category of refineries			
		1 Standard	2 1 + Cracking	3 1 or 2 + Vapo-cracking + oil production	4 1 or 2 or 3 + Desulfurization
Water consumption	m^3/ton	1	2	4	8
SS	g/ton	2	5	10	15
COD	g/ton	10	15	30	60
BOD_5	g/ton	5	5	10	15
TN	g/ton	5	5	10	15
Grease and oil	g/ton	1	25	5	2
Phenol	g/ton	0.01	0.05	0.05	1

[a]Ton means ton of product; SS: suspended solids; TN: total nitrogen.

The chemical industrial complex of LACQ has been classified under category 4. In addition to this, the LACQ discharge limits are also regulated for the following chemical compounds (gram of toxic per ton of specific chemicals produced): hexachlorocyclo-hexane, carbon tetrachloride, benzene, butadiene, trichloromethane, 1,2-dichloroethane, trichloroethene, perchlorethylene and trichlorobenzene. All these substances can be present in the effluent either as traces or at higher concentrations in case of an accidental, uncontrolled discharge.

8.3.2.2 Effluent characteristics

The main characteristics of the different waste streams generated at LACQ are shown in Table 8.4. As it can be seen the natural gas, thiochemical and polymer production activities contribute more or less in the same proportion to more than 94% of the total flow, but only 58% of the COD load because the corresponding streams are relatively diluted. Almost all the rest of the load comes from the highly concentrated streams generated at very small flow rate by the different fine chemical facilities. Part of these facilities produces also an effluent containing more than 100 g COD/L. Due to its

Table 8.4 Characteristics of the wastewaters produced by the different industrial plants located at LACQ complex and treated on site. The data correspond to design values and to those, which are expected after the physicochemical primary treatments.

Parameters	Production activities						
	Natural gas	Thio-chemical	Polymer	Specialty chemical facilities	Lab.	Total observed	Design over capacity
Pollutants to be treated	HC, amines, CH_3OH	Thio-organics	Lactam SO_2	Detergent alcohol esters		Salinity <5 g/L	
Mean flow rate (m^3/h)	130	100	120	<5	15	370	+80
Maximum flow rate (m^3/h)	210	150	180	5	30	575	
Peak flow rate (m^3/h)	500 to 1500	500	200	5	30	1200	
Mean COD (mg/L)	450	450	<540	<25,000	10	510	
Maximum COD (mg/L)	1000	1000	540	36,000	100		
COD/BOD_5	3–3.5	3–3.5	1.4	3	2		
Mean COD load (ton/day)	1.4	1.1	<1.55	3	0.005	7	+1.4
Peak COD load (ton/day)	3.1	2.4	1.55	4.3	0.05	11.4	
Mean SS (mg/L)	10	10	20	<500	20		
Peak SS (mg/L)	50	50	50	500	50		

HC: hydrocarbons; Lab.: central laboratory; other abbreviations as in Table 8.3.

poor biodegradability, this stream is not managed on site. Indeed, for the moment, it is still more economical to send it for treatment outside compared to the cost which would be required to install and operate a WO process to increase its biodegradability and make it amenable to biological treatment. The last stream treated on site comes from the chemical platform analysis of the central laboratory and represents less than 4% and 0.072% of the total flow rate and organic load, respectively.

8.3.2.3 Flow treatment scheme

Water used in the gas and chemical plants is pumped from the river, screened at 200 μm and clarified under 40 μm by hydro-cyclones before utilization. The wastewater produced at the LACQ complex is treated according to the scheme shown in Figure 8.5.

Pretreatment The SO_2 rich effluent from the polymer plant is oxidized in a special pressurized reactor with pure oxygen under pH control in order to convert SO_2 into sulfate. SO_2, which is a bactericidal compound, would otherwise negatively

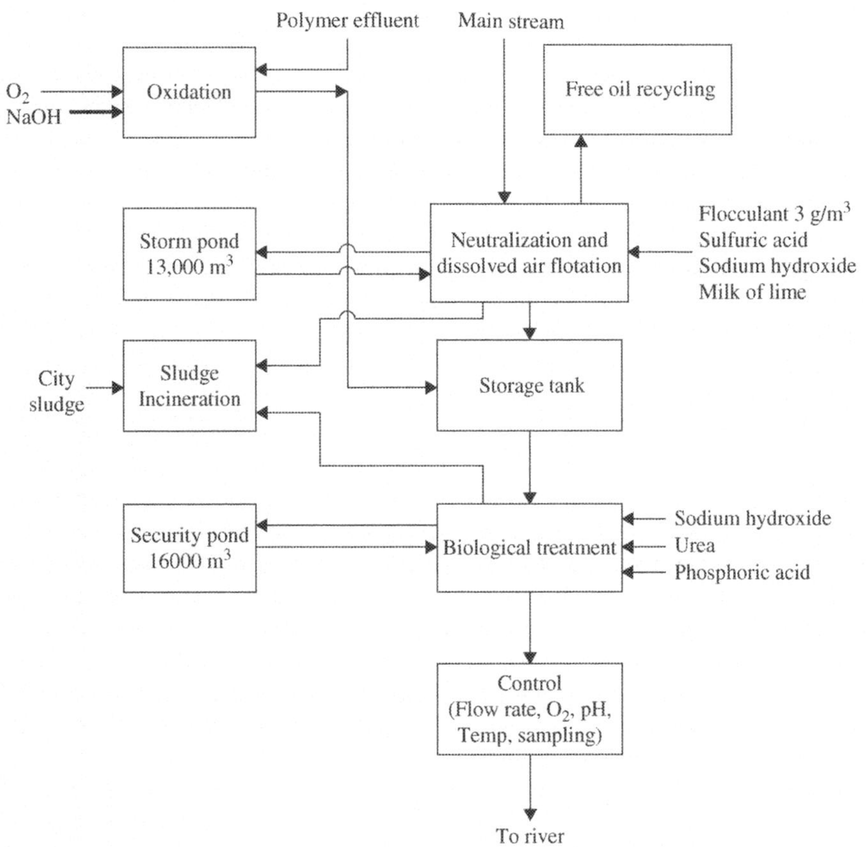

Figure 8.5 LACQ wastewater treatment plant process flow scheme.

affect the operation of the downstream biological process. After oxidation, this stream is fed directly to the biological unit without further pre-treatment.

All the other waste streams are blended and sent to a common pre-treatment unit where the slowly biodegradable and insoluble hydrocarbon solvents are eliminated. These compounds must be removed because they have several detrimental effects in the aeration tanks. They favor the formation of foam, decrease the aeration efficiency and cause sludge flotation due to their adsorption to the biomass. In this treatment unit, the free oil is recovered by a drum skimmer and recycled. The combined effluent is then neutralized in two reactors in series ($100\,m^3$ and $200\,m^3$) by lime or sulfuric acid addition. Sodium hydroxide is used as a back-up in case of lime distribution failure. A cationic polymer is added in line before the dissolved air flotation cell, 7 m diameter, designed for an original flow rate of $900\,m^3/h$. Due to gas production decline which has resulted in a water consumption decrease, the flow rate which is presently treated has been reduced to $250\,m^3/h$. Primary sludge production is $20\,m^3/h$ at 30–60 g/L solids concentration of which 50% (w/w) correspond to the hydrocarbon solvents.

As the river is very pollution sensitive, any risk of pollutant discharge from the wastewater treatment plant must be eliminated. This is achieved by three levels of storage tanks:

(1) Storm pond ($13{,}000\,m^3$) at pre-treatment level, used in case of overflow.
(2) Storage tank ($1800\,m^3$) before the biological plant with pH adjustment which allows a constant feed in the biological process.
(3) Security pond ($16{,}000\,m^3$) normally for treated effluent storage and recycling in case of pollutants concentration over the discharge limits.

Biological treatment An anaerobic pre-treatment of the very concentrated fine chemical facility stream was not considered because, due to the highly changing production cycles, this stream has fluctuating characteristics (both in terms of load and chemical composition) which would have made the performance of an anaerobic treatment unpredictable.

Consequently, it was decided to treat all selected waste streams aerobically. Among the different alternatives, which could have been implemented, sequential batch reactors were not selected because until now, they are not common practice in France. The highly loaded moving-bed biological reactor was also not considered because a final treatment was required and this system is usually used as a pre-treatment. Moreover, space availability was not a limitation at LACQ. Membrane reactors could also have been a good option since they produce an effluent of excellent quality. Due to their cost, they are however usually restricted to cases where water recycling is an objective of the treatment. At LACQ, water reuse was not considered since the treated water must be discharged in order to bleed the salt formed as a byproduct of the chemical activities.

For all these reasons, a conventional activated sludge process was finally selected. Notwithstanding, it was decided to design it with two aeration tanks in series because such scheme is able to produce an effluent of constant quality even during overloading. Oxygen is supplied by three blowers and a fine bubble organic

Table 8.5 Design parameters of the biological process at LACQ[a].

Parameters	Value
BOD_5/N/P ratio	100/5/1
MLSS (g/L)	5
MLVSS (g/L)	3
F/M ratio (kg BOD_5/kg MLVSS-day)	0.06
BOD_5 oxygen demand (kg O_2/kg BOD_5)	0.65
MLVSS oxygen demand (kg O_2/kg MLVSS-day)	0.07
TKN oxygen demand (kg O_2/kg TKN)	4.3
Oxygen transfer ratio	0.5
Settling velocity (m/h)	0.33
Sludge production (kg MLVSS/kg BOD_5)	0.3
Sludge underflow concentration (g MLSS/L)	8

[a]MLSS: mixed liquor suspended solids; MLVSS: mixed liquor volatile suspended solids; F/M: food to microorganism ratio; TKN: total Kjeldahl nitrogen.

Figure 8.6 Partial aerial view of the chemical industrial complex of LACQ together with the wastewater treatment plant build by Proserpol.

membrane distribution system, and regulated by an oxygen probe. The main design parameters used for the operation of the aeration tanks are shown in Table 8.5 and an aerial view of LACQ industrial platform and wastewater treatment system is shown in Figure 8.6.

As it is shown in Table 8.5, the aeration tanks are operated at a very low loading rate (low food to microorganism (F/M) ratio). This was selected to maintain a long

Table 8.6 Performance of the biological system at LACQ during year 2004 (mean values; all in mg/L).

		Jan	Feb	Mar	Apr	May	Jun	Jul	Aug	Sep	Oct	Nov	Dec
In	COD	800	400	500	650	570	470	415	332	660	800	531	657
Out	COD	64	76	65	101	92	177	81	94	76	184	115	82
	BOD_5	10	5	5	15	17	28	57	27	33	101	13	84
	SS	18	22	16	25	31	76[a]	52[a]	36[a]	41[a]	34	42	59[a]

[a]Poor settling characteristics.

sludge retention time (SRT) in order to allow the degradation of slowly biodegradable compounds and to obtain an effluent of better quality. With these operating conditions it is also possible to reduce the sludge production and to obtain a more stabilized sludge. In this particular case, an extra and direct pH control in the biological reactors is required as an acidification reaction takes place during the oxidation of organic sulfur compounds. Nitrogen and phosphorus have to be added using urea and phosphoric acid. For safety reasons, two clarifiers work in parallel, biological sludge is recycled and clarified overflow is controlled before river discharge.

The aeration tanks were seeded with sludge of the aerobic lagoon and started up in the late 2003. The performance of the system during the first year of operation is presented in Table 8.6. During the first 5 months of operation the aeration tanks performed well producing an effluent which was always below the discharge limits. Starting in June 2004, a higher SS concentration was observed in the plant effluent indicating a deterioration of the sludge settling characteristics. This was related to mechanical problems of the mixing and aeration devices, which were soon resolved. In the meantime, flocculant was dosed to the clarifiers in order to comply with the SS discharge limits. The plant has performed well since this problem was solved.

Sludge treatment Floated and biological sludges are fed in a scraped thickener before centrifugation. Cationic flocculant (15 g/kg TS) is used to reach a 18% (w/w) dry centrifuged sludge which is pumped to the incinerator. Industrial sludge is mixed with the publicly owned wastewater treatment plant sludge before incineration in a fluidized bed kiln. Exhaust gas is purified with lime and activated carbon before filtration.

The treatment arrangement applied at LACQ is very classical within the chemical and petrochemical industrial platforms and remains often the most economical option for wastewaters containing a low concentration of contaminants. The activated sludge is the heart in this scheme. Depending on the characteristics of the different streams, it can be complemented by a chemical oxidation pre-treatment when biologically inert COD is present or anaerobic pre-treatment for those streams, which contain a high concentration of biodegradable compounds.

8.3.3 Purified terephthalic acid production

Purified terephthalic acid (PTA) (1,4-benzenedicarboxylic acid) is used together with ethylene glycol for the synthesis of polyethylene terephthalate (PET), the basic saturated polyester which has many applications in our modern life in the form of fibers for the textile industry, films for the photographic audio and video industry and molding resins for the manufacture of food packaging items, including the well known bottles for soft drinks (Park and Sheehan 1996). Due to the massive use of these end products, PTA is produced in huge amounts worldwide. In 1996, its production was estimated to be around 12.5 million metric tons/year (Fligg 1996) and by the time of publication of this chapter should reach 30–35 million tons/year (personal communication of Fernando Varela-Fuentes from "Tereftalatos Mexicanos"). About 66% of this production is located in Asia. The remainder is mostly distributed between North America (22%) and Europe (11%).

8.3.3.1 PTA manufacturing technology and resulting wastewater

Terephthalic acid (TA) became commercially available, with a sufficient purity for polymerization, only in 1965 thanks to a process developed by Amoco Chemical Co. Until today, this process remains the technology of reference. It consists of two steps: oxidation and purification (Park and Sheehan 1996). In the oxidation step, TA is obtained by a liquid-phase air oxidation of *p*-xylene at 175–225°C and $1.5–3.0 \times 10^3$ kPa (15–30 atm), using acetic acid as solvent, Co and Mn as catalysts and bromine as a renewable source of free radicals. The crude TA (CTA) recovered at this stage is already over 99% pure, but this is not sufficient for PET production. CTA is further purified by dissolution in water at 250°C under a pressure of 4×10^3 kPa (40 atm) followed by a hydrogenation in the presence of a palladium catalyst in order to convert 4-carboxybenzaldehyde (4-CBA), the most problematic impurity into *p*-toluic acid (4-methylbenzoic acid) (reduction of the aldehyde group to a methyl) and decrease its concentration from 2000–5000 to below 25 ppm (Khachane *et al.* 2003). During this step, various colored impurities are also hydrogenated to colorless products.

Approximately 2.5–4.5 m^3 of wastewater and 20–40 kg COD are generated per ton of PTA produced (Bushway and Gilman 1986; Shelley 1991; Vandufell 1993; Author's files). Most of this volume and COD load (80–90%) comes from the purification unit. The main characteristics of the global wastewater resulting from the combination of the different streams (oxidation + purification + leak's collecting canals) are presented in Table 8.7. The large fluctuations shown for most of the parameters result from the different qualities of the *p*-xylene feed stocks and oxidation and purification technologies used from one PTA installation to another.

The structure and solubilities of the four main wastewater organic pollutants, which represent 80–90% of the wastewater COD, are given in Table 8.8. As can be seen from these data, TA is very poorly soluble in water. Its solubility is pH dependent and increases with this parameter as exemplified by the high solubility of its disodium salt. Nevertheless, at the temperature and pH of the wastewater, TA remains almost completely protonated and undissolved (Kleerebezem 1999) contrarily to the other compounds, the concentrations of which are below their maximum solubility at the same conditions (Tables 8.7 and 8.8). As a consequence, TA is mostly present in a particulate form in the wastewater and constitutes almost exclusively its SS, as this has been confirmed experimentally (Varela-Fuentes *et al.* 1998).

Table 8.7 Characteristics of PTA plant wastewaters as reported in the literature[a].

Parameter	Values	COD contribution (%)[b]
Temperature (°C)	25–100	
pH	3–5	
Total solids (g/L)	5–8.6	
SS (g/L)	0.3–6	
COD total (g/L)	4.4–20	
COD/BOD_5	1.2–1.7	
TA (g/L)	0.6–6.9	7.4–76.5
p-toluic acid (g/L)	0.05–1	0.8–25.8
Benzoic acid (g/L)	0.05–2.33	1.4–59
Acetic acid (g/L)	0.25–12	2–71
Co (mg/L)	13–57	
Mn (mg/L)	41–76	
Br (mg/L)	36	

[a]Ely and Olsen 1989; Liangming *et al.* 1991; Macarie 1992; Pereboom *et al.* 1994; Cheng *et al.* 1997; Young 1997; Page *et al.* 1998; Varela-Fuentes *et al.* 1998; Primelles *et al.* 1999; Young *et al.* 2000.
[b]Minor organic contaminants, which can be present beside the indicated ones correspond mostly to trimellitic acid (1,2,4-benzenetricarboxylic acid) and 4-CBA as well as to the *o*- and *m*-isomers of terephthalic and *p*-toluic acids. More complex compounds may be found also but at much lower concentrations (for a list, see Roffia *et al.* 1984).

Table 8.8 Structure, COD exerted and solubilities of the main organic contaminants found in PTA plants' wastewaters.

Common name		Terephthalic acid	*p*-toluic acid	Benzoic acid	Acetic acid
IUPAC name		1,4-Benzene-dicarboxylic acid	4-Methyl-benzoic acid	Benzene-carboxylic acid	Ethanoic acid
Chemical structure		$C_8H_6O_4$	$C_8H_8O_2$	$C_7H_6O_2$	$C_2H_4O_2$
Developed structure					CH_3—COOH
COD exerted[a]		1444.65	2115.46	1967.21	1066
Solubility (g/1000 g water)					
Protonated forms	25°C	0.017	0.345	3.45	∞
	35°C	*0.039*	*0.55*	*5.07*	∞
	100°C	0.33	11.6	58.75	∞
Sodium salts	25°C	140	Vs[b]	*628.8*	*505.8*

[a]In mg COD/g compound. Information taken from Seidell (1941); Dean (1992); 1990/1991-Merck Chemical Catalogue, Park and Sheehan (1996) and Han *et al.* (1999). Solubilities in italic have been estimated according to Han *et al.* 1999 for TA, and from the values reported by Seidell (1941) for benzoic and *p*-toluic acids, and from Dean (1992) for sodium acetate and benzoate.
[b]Vs: very soluble.

It must be pointed out that the wastewater is practically devoid of nitrogen and phosphorus and does not contain sulfate. This is the result of the use of very high-quality deionized water for all the production operations which give raise to the waste streams (Wong 2000).

8.3.3.2 Waste minimization and pre-treatments

As can be seen from the previous paragraph, several of the characteristics of PTA wastewater are (absence of N and P) or may be (pH, temperature) unsuitable for a biological treatment and need to be corrected beforehand. One point, which needs particular attention, corresponds to the SS, which are at a concentration that could affect most biological systems. This problem may be solved in different ways.

First, the recovery of TA in the production process can be improved both at the oxidation and purification levels. This may be achieved, for instance, by the replacement of the conventional separation devices (drum or centrifugal disc filters) by a microfiltration one (Kim *et al.* 2002). If they have not been retained in the factory, the TA SS can be physically removed from the wastewater by sedimentation taking advantage of the high specific gravity of TA (1.522 at 25°C). Such settling may be performed in a specially designed primary settler (Page *et al.* 1998; Primelles *et al.* 1999), but it is most often commonly done, in a more or less controlled form, in the equalization and/or cooling basins, which are necessary to smooth the COD peak loads and reduce the wastewater temperature upstream of the biological unit (Bushway and Gilman 1986; Noyola *et al.* 2000). The SS removal can be further improved through the use of flocculants (e.g. Ferric chloride) and/or acidification (Xu *et al.* 1988). The TA recovered after settling is of very poor quality (purity $< 88\%$) and cannot be used for PET production or recycled to the production process, but several other end uses exist (Bushway and Gilman 1986; Wang *et al.* 2002).

The last option to solve the TA SS problem consists to dissolve them with a base (e.g. NaOH). This is often practiced, but it should not be the first choice because it increases considerably the cost of treatment. In any case, even if the SS are removed by sedimentation, the consumption of NaOH to neutralize the acids in solution, in order to reach a pH compatible with a biological treatment, may remain very large. Beside the SS, the organic and inorganic loads to the treatment plant can also be greatly reduced by improvements in the production process (Kelly 1999; Khachane *et al.* 2003).

8.3.3.3 Aerobic biodegradability and treatment

The COD/BOD_5 ratio of PTA wastewater (Table 8.7) indicates that it is easily biodegradable aerobically. Then, it is logical that the first systems implemented to treat PTA wastewater were aerobic (Shelley 1991). Amoco Co developed an original medium loaded activated sludge adapted to this particular case (Lau 1978; Bushway and Gilman 1986; Noyola *et al.* 2000). This system consisted of three identical size aeration tanks operated in series but with 66% of the total aeration capacity installed in the first tank, 20% in the second and 14% in the third. Aeration was apparently provided through mechanical surface aerators (Bushway and Gilman 1986) although

Table 8.9 Operating conditions and performance of the different aerobic systems applied at full-scale for treating PTA wastewater.[a]

Parameters	Three-stage activated sludge	Extended aeration	Deep shaft
Operating conditions			
F/M (kg COD/kg MLSS-day)	0.3–0.5	<0.1	NA
OVLR (kg COD/m^3-day)	3	0.5	11–27
HRT (days)	2–5	19	3–6 h
Aeration tank MLSS (g/L)	5–6	4–9	NA
Aeration tanks DO (mg/L)	2–6	NA	NA
Oxygen demand (kg O_2/kg COD fed)	1.13	NA	NA
Performance			
COD removal (%)	>97	99	91–98
Sludge yield (kg SS/kg COD removed)	0.032–0.152	0.03–0.07	NA
COD out (mg/L)	100–200	40	NA
BOD_5 out (mg/L)	<30	5	50–300
SS out (mg/L)	<50	20	NA

[a]Lau (1978); Primelles *et al.* (1999); Author's files.
F/M: food to microorganism ratio; OVLR: organic volumetric loading rate; HRT: hydraulic retention time; MLSS: mixed liquor suspended solids; DO: dissolved oxygen; NA: not available. For the deep shaft, the OVLR and removal are BOD_5 based.

they may have been replaced later by submerged diffusers (Noyola *et al.* 2000). No precise data seems to be available in the literature on the performances of such system at full scale. However, a laboratory study published by Amoco (Lau 1978) and conducted with synthetic PTA wastewater shows that it was able to eliminate at least 97% of the COD when operated as indicated in Table 8.9. This study has shown that this system could continue to produce an effluent of constant and excellent quality in the presence of 1–3 days peak organic loading rates up to twice the normal loading contrarily to single- or two-stage systems operated under the same conditions. Moreover, the sludge of the three-stage unit always had better settling properties than the sludge of the less staged units.

Since the development of the Amoco treatment design, other aerobic concepts have been applied to PTA wastewater. Depending probably on the surface available on site for the installation of equipment and the disposal of excess sludge, these systems have ranged from the low loaded extended aeration (Primelles *et al.* 1999) to the very-high-loaded deep shaft process. The operating conditions and performance of these systems are compared to those of the Amoco process in Table 8.9.

It must be emphasized that all the previously described aerobic methods have the ability to treat the acid PTA wastewaters without the need for neutralization and that the systems operated with surface aerators tolerate higher TA SS concentrations and wastewater temperature than those using submerged diffusers.

As can be seen from Table 8.9, the different aerobic systems are very efficient for the treatment of PTA wastewaters. Nevertheless, despite their undeniable qualities they present the following disadvantages:

- Long HRTs, at least for the low- and medium-loaded units, which result in large volumes of aeration tanks and corresponding investment.

- High oxygen (energy) requirement. In the case of the Amoco three-stage activated sludge design, the reported oxygen demand corresponds to an electricity consumption of 0.47–0.94 kWh/kg COD fed (considering that 1.2–2.4 kg O_2 can be transferred per kWh according to the usual efficiency of the aeration systems).
- Important consumption of nitrogen and phosphorus which must be added at a COD/N/P ratio of 100/2/0.125 according to Lau (1978).
- Possibility of sludge bulking or at least development of a poorly settling sludge (Brugnaro and Polo 1985).
- Production of large amounts of waste biological sludge despite the fact that the observed sludge yield for most of the applied systems is much lower than the normally expected value (0.5 kg VSS/kg COD removed). This sludge contains a large quantity of metals, particularly Co and Mn and must be stabilized and disposed off under appropriate conditions (Bushway and Gilman 1986; Wang *et al.* 2002).

8.3.3.4 Anaerobic biodegradability and treatment

Amoco experience Due to the increase of energy cost and the restrictions to the disposal of surplus activated sludge in many countries, in the 1980s Amoco undertook a research effort to determine if anaerobic treatment, which is a well known answer to the above problems, could be applied to PTA wastewater. This research effort resulted in the development of a methanogenic process which was first implemented at full scale in 1989 at Capco, a subsidiary of Amoco in Taiwan, and the world's largest PTA manufacturing facility (907,200 ton/year) at that time (Shelley 1991). Amoco claimed that the application of anaerobic treatment at this site allowed cutting the treatment cost by nearly US$4 million/year compared to an equivalent capacity aerobic system.

From the very scarce information which has been disclosed by Amoco about its anaerobic process, it appears that it consists in a downflow anaerobic filter filled with a vertically oriented packing material and operated under the following conditions: pH 7, 37°C, organic loading rate 3–4 kg COD/m^3-day, HRT 3–4.5 days, recycle to feed ratio 1–6, stripping of CO_2 in the recycle line. Under these conditions, the system has seemingly the capacity to remove 80–85% of the total organic carbon (TOC) present in the wastewater (Ely and Olsen 1989; Shelley 1991). A very long start-up period (6 months to 1 year) is however necessary before reaching such performance (Vanduffel 1993). The recycle and stripping of CO_2 are claimed to cut the caustic usage necessary to control the pH and dissolve the TA SS at the entrance of the system by 40–60%.

More or less within the limits of the previous description, the Amoco downflow filter process has been reported to be applied in at least eight different PTA production facilities, but only four of the corresponding locations have been disclosed (Macarie 2000; Kleerebezem and Macarie 2003).

Early full-scale mesophilic anaerobic plants not based on the Amoco process During the 1990s, in a search to increase their business profitability and remain competitive in the market, several PTA producers decided to follow the example of Amoco and to implement anaerobic treatment (Pereboom *et al.* 1994; Page *et al.*

1998; Noyola *et al.* 2000). Amoco being reluctant to commercialize its anaerobic process independently of its PTA production technology, these producers turned towards engineering companies specialized in anaerobic systems to design their treatment plants. It appeared however very rapidly that, due to the almost inexistent public knowledge of the anaerobic biodegradability of PTA wastewater, the performance of most of these early installations was limited to 55–65% COD removal (Pereboom *et al.* 1994; Page *et al.* 1999) except in those cases where acetic acid represented most of the wastewater COD (Young *et al.* 2000).

The data published on the UASB reactor implemented by Grontmij, a Dutch engineering company, for Tuntex in Taiwan (Pereboom *et al.* 1994) suggested that these low efficiencies were without any doubt the result of the absence of TA and *p*-toluic acid degradation, even after prolonged operation (1.5 years in the case of Tuntex), which should have resulted in biomass acclimation. The Tuntex experience revealed also that TA and *p*-toluic acid removal could in fact be achieved, but only at low loading rates. Indeed, TA degradation was finally obtained 2 years after start-up when the organic loading rate was decreased from 10 to 5 kg COD/m^3-day (1.1 to 0.5 kg COD/kg VSS-day), a value closer to the one used by Amoco for its downflow filters (Kleerebezem 1999).

Improved understanding of PTA wastewater anaerobic biodegradability through laboratory scale experiments The need to understand the factors limiting the degradation of TA and *p*-toluic acid in anaerobic reactors and so their treatment performance stimulated a great amount of research which demonstrated (or confirmed) that under mesophilic conditions:

- Most of the organic compounds (phthalate isomers, *p*-toluate, trimellitate, 4-CBA) present in PTA wastewater can be converted to CH_4 and CO_2 (Kleerebezem 1999; Noyola *et al.* 2000).
- Common anaerobic reactor seed sources, such as digested sewage sludge and granular sludge, acquire quite rapidly the ability to degrade all phthalate isomers, but not *p*-toluate (Kleerebezem 1999).
- Terephthalate, phthalate (PA) and benzoate (BA) can be removed at high rates (0.67 g COD-TA/g VSS-day, 0.85 g COD-PA/g VSS-day, 1 g COD-BA/g VSS-day, respectively) when fed to an anaerobic reactor individually as sole carbon and energy sources (Li *et al.* 1995; Tur and Huang 1997; Kleerebezem 1999), contrarily to *p*-toluate, the removal of which is limited to 0.012 g COD-*p*-toluate/g VSS-day (Macarie 1992). As a consequence, *p*-toluate removal efficiencies of 60–95% (COD based) can only be obtained at very low volumetric loading rates ($\leq$1.3 kg COD/m^3-day; Wu *et al.* 2001) while TA, PA and BA removal over 95% (COD based) can be obtained at volumetric loading rates as high as 15–30 kg COD/m^3-day.
- Terephthalate, phthalate, benzoate, *p*-toluate and 4-CBA are not toxic for the bacterial partners involved in their methanogenic breakdown at the concentrations normally found in PTA wastewater (Fajardo *et al.* 1997; Kleerebezem *et al.* 1997).
- Acetate and/or benzoate inhibit the methanogenic breakdown of TA and *p*-toluate, while TA does not affect that of BA and PA that of TA (Pereboom *et al.* 1994; Fajardo *et al.* 1997; Kleerebezem 1999).

- The doubling time of the bacteria responsible for the primary attack of TA (7–28 days) and *p*-toluate (58 days) are extremely long compared to those of the BA (~4 days) and acetate users (~6 days) (Kleerebezem 1999).
- The growth rate and the specific removal activities of the bacteria using TA decrease sharply outside the optimum temperature of 37°C (e.g. 43% of activity lost between 37°C and 30°C) and the pH range of 6.1–7.1 (Kleerebezem 1999).
- Granulation of disperse sludge is very hard to achieve on PTA real wastewater as well as on TA and *p*-toluate when fed as single components (Macarie 1992; Cheng *et al.* 1997; Kleerebezem 1999).

Single-stage over two-stage reactors As can be seen from the previous results, TA and *p*-toluate methanogenic degradation can only occur at a low concentration of acetate and benzoate. In a single-stage reactor of the packed bed or UASB type, this condition seems to be fulfilled solely at low loading rates of the order of 3–5 kg COD/m^3-day. Even at such loading, due to its extremely slow degradation kinetic, *p*-toluic acid will remain almost refractory to the treatment and will have to be eliminated in an aerobic post-treatment unit. This means also that the COD removal which can be achieved anaerobically will mostly depend on the concentration of this compound in the wastewater.

High-rate treatment (⩾10 kg COD/m^3-day) of PTA wastewater with these types of reactors without sacrificing COD removal efficiencies, which means including TA degradation, will be possible at the condition to operate two reactors in series. In this scheme, the first reactor is used to remove acetate and benzoate in order to allow the development of a TA-degrading population in the second stage. The viability of such concept has been demonstrated at laboratory scale (Kleerebezem 1999). Despite of using two reactors, thanks to the higher loads which can be applied, the global HRT of the staged system (and so the volume of the installation) is always smaller than that required by a single-stage unit producing an effluent of similar final quality (identical global COD removal). This is illustrated in Figure 8.7.

It can be seen from this figure that for a loading rate of 14 kg COD/m^3-day in each stage the required HRT is already reduced by 46–57%. Higher HRT reduction (>70%) could be achieved because the first reactor has been found at laboratory scale to be operable up to 40 kg COD/m^3-day, which corresponds to more than twice the highest loading rate at which the second stage can be operated (Kleerebezem 1999). Nevertheless, the first stage should not be loaded too much to always provide an effluent with low acetate and benzoate concentration, even during COD peaks, in order to protect the second stage. Otherwise, the TA-degrading capacity of this stage could be greatly affected by such events reducing the global COD removal of the two-stage system. As a consequence, under field conditions, the organic loading of the first stage should be limited to 20–25 kg COD/m^3-day. Surprisingly, in the same laboratory scale experiments (Kleerebezem 1999), the first stage reactor finally acquired the capacity to remove TA, but 300 days of operation were necessary and this occurred only after the loading rate was reduced from 40 to 9 kg COD/m^3-day. TA degradation in this stage remained however easily influenced by acetate and benzoate shock loads which means that a two-stage reactor configuration should remain a better option.

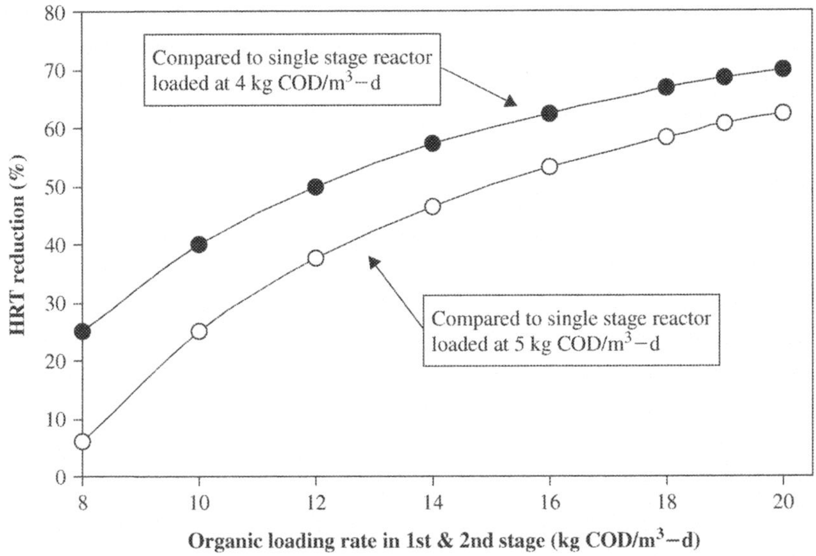

Figure 8.7 Reduction of HRT, which can be obtained with a high-loaded two-stage reactor compared to a low-loaded single-stage unit producing an effluent of same quality. The calculations are for: COD concentration 10 g/L; 50% of COD due to acetate and benzoate, 100% acetate and BA removal in first stage, first and second stage operated at the same loading rate.

Due to the very slow growth rate of the TA degraders, efficient biomass retention is crucial for the second-stage reactor since it will be the only way to obtain the required long SRT and short start-up. This indicates that this stage should be preferentially inoculated with granular biomass and that, if such seeding material is not available, a reactor with some kind of packing or another device improving biomass retention (e.g. anaerobic filters, hybrid reactors) should be used instead of an UASB. It will be also of utmost importance to operate the second-stage reactor very close to the optimal conditions for the growth of the TA users (37°C, pH 7).

By the time of publication of this chapter, the two- or multi-stage reactor in series concept would have been applied at full scale in at least five PTA plants. Four of these units (hybrid reactors) have been installed by the Canadian engineering company ADI (Page *et al.* 1998; Macarie 2000) and the last one (upflow pond) by the Mexican engineering company IBTech (Noyola *et al.* 2000). It must be noted however, that all these plants remain operated at low loading rates ($<$6.5 kg COD/m^3-day). Moreover, the ADI units were originally not designed according to the biodegradability of the different organic compounds present in the PTA wastewaters, but to operate under the patented two-stage cyclic mode described by Howerton and Young (1987). Detailed information on the performance of the staged anaerobic reactors is only available for one ADI unit, a picture of which is presented in Figure 8.8.

Unfortunately, the published information does not include the start-up of the system (April 1995) and covers a period (December 1997 to April 1999) for which the reactors could be considered to be already mature. Thus, it is impossible to confirm

Figure 8.8 Wastewater treatment plant of the Sam Nam Korean PTA producer which is made of an Anamet® anaerobic contact reactor (2000 m^3) designed by the Swedish engineering company Purac in 1990 (background), two in series circular anaerobic hybrid reactors of 1100 m^3 each designed by ADI in 1994 (center) and an aerobic activated sludge post-treatment unit (foreground) (picture courtesy of ADI Systems Inc.).

from this example that the two-stage configuration allowed a faster start-up compared to a single-stage one. Nevertheless, the data provided for this period (Young *et al.* 2000) show that the two-stage system removed more efficiently TA, PA and *iso*-phthalic acid (IA) than a companion single-stage anaerobic contact reactor operated at the same site and at the same global loading rate (4.5 kg COD/m^3-day). A significant removal of all phthalate isomers was achieved in the first stage of the hybrid unit (>50% for PA and IA; <50% for TA) at double the loading rate of the anaerobic contact reactor and the second stage was very efficient leading to their almost complete removal.

Present state of application of anaerobic treatment in the PTA industry By January 2003, mesophilic anaerobic treatment followed by aerobic polishing was applied in at least 24 PTA plants (Macarie 2000; Kleerebezem and Macarie 2003). This suggests that such treatment was already, or was about to become the conventional form of treatment in this industrial sector. The tendency in the last years has been apparently towards the progressive abandonment of low-loaded systems (e.g. anaerobic contact) and their replacement by higher-loaded ones. This is particularly evident in the case of Amoco which does not implement anymore in its new facilities (or when increased treatment capacities are required) its downflow filter, but prefers to contract on a turn-key basis the services of engineering companies specialized in the design of anaerobic systems.

Since the year 2000, the very-high-loaded expanded granular sludge bed (EGSB) and internal circulation (IC) reactors commercialized by the Dutch companies

Biothane and Paques, respectively, have been introduced in the PTA anaerobic market (Kleerebezem and Macarie 2003) and not less than seven PTA plants should be currently using these technologies (four use EGSB, three IC). The IC units are claimed to be single-stage reactors. According to the information provided about one of these units (Driessen *et al.* 2004), they seem to be operated at loading rates around or over 10 kg COD/m^3-day (3.64 kg TOC/m^3-day) and to attain TOC removal efficiencies as high as 90%. The wastewater COD of the described case being due for 40–50% to acetate and benzoate and 30% to TA (Personal communication of Jaap Vogelaar from Paques), this suggests that an effective TA removal can be achieved by the IC reactors at such loading rates contrarily to what was observed previously with single-stage UASB and packed bed systems. This high TA removal rate capacity is said to be the result of the improved solid retention and increased mixing of the IC reactors compared to the other anaerobic reactor designs (Driessen *et al.* 2004). The improved retention of solids would allow such system to retain much more efficiently the slow-growing TA users, while the improved mixing due to the IC of water within the reactor and the higher gas and liquid superficial velocities would improve the contact between the wastewater and the biomass and allow to maintain in the liquid phase very low concentrations of acetate, benzoate and hydrogen which would otherwise inhibit TA degradation. A good upfront buffering in order to avoid big organic load variations at the entrance of the reactor is also stressed by Paques as one of the key factors to achieve TA removal at high loading rates with single-stage IC reactors. Contrarily to the IC units, no information has been published about the Biothane EGSB systems designed for PTA plants. It seems however that both single- and two-stage units have been installed and that single-stage EGSB reactors are able to remove TA at loading rates similar to those reported for the IC system (Personal communication of Marc de Pijper from Biothane Systems International) and probably for the same reasons.

8.3.3.5 Perspectives

As can be seen from this case study, anaerobic treatment can now be considered as an established technology in the PTA industry. Further improvements could be done by looking at the possibility to operate the reactors under thermophilic conditions since PTA wastewater is most often produced at a temperature close to the optimum (55°C) of the thermophilic microorganisms. It has been shown at laboratory scale with synthetic PTA wastewater that such thermophilic treatment is possible at relatively high rates (16 kg COD/m^3-day) with single-stage hybrid reactors (Chen *et al.* 2004). This needs however to be confirmed at pilot scale with real wastewater before a full-scale application is undertaken.

Another interesting option would be to characterize thoroughly the composition of the streams generated by the oxidation and purification steps in order to assess if it would be more convenient to treat them separately. Indeed, acetate is supposed to be mostly present in the oxidation stream. Then, this could allow to avoid (at least partly) the problem of TA degradation inhibition, which has to be faced when the two streams are mixed.

Finally, the technical and economical viability of recycling the anaerobically–aerobically treated PTA wastewater in the production process after further polishing has been demonstrated at full scale in a large PTA plant (Wong 2000). Such an example should become the rule and not remain an exception if the PTA industry wants to preserve water resources and achieve a sustainable development.

REFERENCES

Alzamora-Rumazo, C., Pryor, A., Ocampo-Mendoza, F., Campos-Villareal, J., Robledo, J.M. and Rodríguez-Mercado, E. (2000) Cleaner production in the chemical industry. *Wat. Sci. Technol.* **42**(5–6), 1–7.

Baron, C., Equihua, L.O. and Mestre, J.P. (2000) B.O.O. case: water management project for the use of reclaimed wastewater and desalted seawater for the "Antonio Dovali Jaime" refinery, Salina Cruz, Oaxaca, Mexico. *Wat. Sci. Technol.* **42**(5–6), 29–36.

Berné, F. and Cordonnier, J. (1995) *Industrial Water Treatment: Refining, Petrochemicals and Gas Processing Techniques*, Gulf Publishing, Houston, USA.

Blum, D.J.W. and Speece, R.E. (1991) A data base of chemical toxicity to environmental bacteria and its use in interspecies comparisons and correlations. *Res. J. Wat. Pollut. Cont. Fed.* **63**, 198–207.

Brugnaro, S. and Polo, B. (1985) Effect of some wastes on proliferation of filamentous bacteria in activated sludge. *Inquinamento* **27**(4), 65–68 (in Italian).

Bushway, M.H. and Gilman, G.L. (1986) Case history: development of a dedicated sludge application site by Amoco chemicals Co. and the associated permit to operate by South Carolina department of health and environmental control. *Proceedings of the 41st Purdue University Industrial Waste Conference*, Lewis Publishers Inc., Chelsea, MI, USA, pp. 726–735.

Carmichael, J.B. and Strzepek, K.M. (1987) *Industrial Water Use and Treatment Practices*. Cassell Tycooly, London, UK.

Chen, C.-L., Macarie, H., Ramirez, I., Olmos, A., Ong, S.L., Monroy, O. and Liu, W.-T. (2004) Microbial community structure in a thermophilic anaerobic hybrid reactor degrading terephthalate. *Microbiol.* **150**, 3429–3440.

Cheng, S.-S., Ho, C.-Y. and Wu, J.H. (1997) Pilot study of UASB process treating PTA manufacturing wastewater. *Wat. Sci. Technol.* **36**(6–7), 73–82.

Cheng, S.-S., Chen, Y.-N., Wu, K.-L., Chuang, H.-P. and Chen, S.-D. (2004) Study of a three-stage fluidized bed process treating acrylic-fiber manufacturing wastewater containing high-strength nitrogenous compounds. *Wat. Sci. Technol.* **49**(5–6), 113–120.

Chou, W.L., Speece, R.E. and Siddiqi, R.H. (1978a) Acclimation and degradation of petrochemical wastewater components by methane fermentation. *Biotechnol. Bioeng. Symp.* **8**, 391–414.

Chou, W.L., Speece, R.E., Siddiqi, R.H. and McKeon, K. (1978b) The effect of petrochemical structure on methane fermentation toxicity. *Prog. Wat. Tech.* **10**(5/6), 545–558.

Dean, J.A. (1992) *Lange's Handbook of Chemistry*, 14th edn. McGraw-Hill, New York, USA.

Ding, Z., Zhirong, S. and Wanli, H. (1996) A study on wastewater minimization in indirect electrochemical synthesis of benzaldehyde. *Wat. Sci. Technol.* **34**(10), 113–120.

Donlon, B.A., Razo-Flores, E., Field, J.A. and Lettinga, G. (1995) Toxicity of *N*-substituted aromatics to acetoclastic methanogenic activity in granular sludge. *Appl. Environ. Microbiol.* **61**, 3889–3893.

Driessen, W., Yspeert, P. and Engelaar, A. (2004) Anaerobic treatment of chemical industry effluents. *Proceedings of the 16th International Congress Chemical and Process Engineering, CHISA 2004*, Prague, Czech Republic, August 22–26, 10 pp.

Eckenfelder, W.W. and Englande, A.J. (1996) Chemical/petrochemical wastewater management – past, present and future. *Wat. Sci. Technol.* **34**(10), 1–7.

van Eeckert, M.H.A. and Schraa, G. (2001) The potential of anaerobic bacteria to degrade chlorinated compounds. *Wat. Sci. Technol.* **44**(8), 49–56.

Ely, S.R. and Olsen, G.P. (1989) Process for treatment of wastewater. *US Patent* 4,826,600 (assigned to Amoco Co).

Enger, W.A., van de Bergh, D.R., de Blank, P.B. and Olthof, I.A.B. (1989) Recalcitrante verbindingen bij de anaerobe zuivering van afvalwater van Shell Nederland Chemie, Moerdijk. *NVA Workshop Anaërobe afbreekbaarheid van recalcitrante verbindingen.* April 20–21, The Netherlands (in Dutch).

Evans, W.C. and Fuchs, G. (1988) Anaerobic degradation of aromatic compounds. *Ann. Rev. Microbiol.* **42**, 289–317.

Fajardo, C., Guyot, J.-P., Macarie, H. and Monroy, O. (1997) Inhibition of anaerobic digestion by terephthalic acid and its aromatic byproducts. *Wat. Sci. Technol.* **36**(6–7), 83–90.

Fewson, C.A. (1988) Biodegradation of xenobiotic and other persistent compounds: the causes of recalcitrance. *Trend. Biotechnol.* **6**, 148–153.

Field, J.A. (2002) Limits of anaerobic biodegradation. *Wat. Sci. Tech.* **45**(10), 9–18.

Field, J.A., Stams, A.J.M., Kato, M. and Schraa, G. (1995) Enhanced biodegradation of aromatic pollutants in cocultures of anaerobic and aerobic bacterial consortia. *Ant. van Leeuwen.* **67**, 47–77.

Fiorenza, S. and Rifai, H. (2003) Review of MTBE biodegradation and bioremediation. *Bioremed. J.* **7**, 1–35.

Fligg, J.E. (1996) Polyester growth and its impact on PTA supply. *Int. Fiber J.* **11**(5), 16–22.

Fortin, N., Morales, M., Nakagawa, Y., Focht, D. and Deshusses, M. (2001) Methyl tertbutyl ether (MTBE) degradation by a microbial consortium. *Environ. Microbiol.* **3**, 407–416.

Frankin, R.J., van Gils, W.M.A. and Wermeling, R.J.F. (1994) Full-scale anaerobic treatment of Shell wastewater containing benzoate with the Biothane® UASB process. *Poster Papers Proceedings of the 7th International Symposium Anaerobic Digestion*, Cape Town, South Africa, January, 23–27, pp. 248–251.

Fuchs, G., Mohamed, M.E.S., Altenschmidt, U., Koch, J., Lack, A., Brackmann, R., Lochmeyer, C. and Oswald, B. (1994) Biochemistry of anaerobic biodegradation of aromatic compounds. In: *Biochemistry of Microbial Degradation* (ed. C. Ratledge). Kluwer, Dordrecht, The Netherlands, pp. 513–553.

Gerike, P. and Fischer, W.K. (1979) A correlation study of biodegradability determinations with various chemicals in various tests. *Ecotoxicol. Environ. Safe.* **3**, 159–173.

Ghisalba, O. (1983) Microbial degradation of chemical waste, an alternative to physical methods of waste disposal. *Experientia* **39**, 1247–1257.

Gijzen, H.J., Bernal, E. and Ferrer, H. (2000) Cyanide toxicity and cyanide degradation in anaerobic wastewater treatment. *Wat. Res.* **34**, 2447–2454.

Gulyas, H. (1997) Processes for the removal of recalcitrant organics from industrial wastewaters. *Wat. Sci. Technol.* **36**(2–3), 9–16.

Han, N., Zhu, L., Wang, L. and Fu, R. (1999) Aqueous solubility of *m*-phthalic acid, *o*-phthalic acid and *p*-phthalic acid from 298 to 483 K. *Sep. Purif. Technol.* **16**, 175–180.

Harvey, L.M. and Rubiano, J.C. (1983) The anaerobic approach: a better answer. *Ind. Waste*. **29**, 22–25.

Horowitz, A., Shelton, D.R., Cornell, C.P. and Tiedje, J.M. (1982) Anaerobic degradation of aromatic compounds in sediments and digested sludge. *Dev. Ind. Microbiol.* **23**, 435–444.

Howerton, D.E. and Young, J.C. (1987) Two-stage cyclic operation of anaerobic filters. *J. Wat. Pollut. Cont. Fed.* **59**, 788–794.

Huber, L. (1967) Disposal of effluents from petroleum refineries and petrochemical plants. *Proceedings of the 22nd Purdue Industrial Waste Conference*, The Heckman Bindery Inc., North Manchester, IN, USA, pp. 1009–1024.

Jørgensen, S.E. (1979) *Industrial Waste Water Management.* Elsevier, Amsterdam, The Netherlands.

Kelly, M.D. (1999) Recovery process for oxidation catalyst in the manufacture of aromatic carboxylic acids. *US Patent* 5,955,394 (assigned to Mobile Process Technology Co.).

Khachane, P.K., Heesink, A.B.M., Versteeg, G.F. and Pangarkar, V.G. (2003) Adsorptive separation and recovery of organic compounds from purified terephthalic acid plant effluent. *Sep. Sci. Technol.* **38**, 93–111.

Kim, Y.-B., Lee, K. and Chung, J.-H. (2002) Optimum cleaning-in-place conditions for stainless steel microfiltration membrane fouled by terephthalic acid solids. *J. Memb. Sci.* **209**, 233–240.

Kleerebezem, R. (1999) *Anaerobic Treatment of Phthalates: Microbiological and Technological Aspects*. PhD Thesis, Wageningen University, Wageningen, The Netherlands.

Kleerebezem, R. and Macarie, H. (2003) Treating industrial wastewater: anaerobic digestion comes of age. *Chem. Eng.* **110**(4), 56–64.

Kleerebezem, R., Mortier, J., Hulshoff Pol, L.W. and Lettinga, G. (1997) Anaerobic pretreatment of petrochemical effluents: terephthalic acid wastewater. *Wat. Sci. Technol.* **36**(2–3), 237–248.

Kugelman, I.J. and Chin, K.K. (1971) Toxicity, synergism and antagonism in anaerobic waste treatment processes. *Adv. Chem. Ser.* **105**, 55–90.

Lau, C.M. (1978) Staging aeration for high efficiency treatment of aromatic acids plant wastewater. *Proceedings of the 32nd Purdue University Industrial Waste Conference*, Ann Arbor Science, Ann Arbor, MI, USA, pp. 63–74.

Levec, J. (1997) Wet oxidation processes for treating industrial wastewaters. *Chem. Biochem. Eng. Q.* **11**, 45–58.

Li, Y.-Y., Fang, H.H.P., Chen, T. and Chui, H.K. (1995) UASB treatment of wastewater containing concentrated benzoate. *J. Environ. Eng.* **121**, 748–751.

Liangming, X., Yuxiang, C. and Xiangdong, Z. (1991) The anaerobic biological treatment of high strength petrochemical wastewater by hybrid reactor. *Proc. Int. Conf. Petrol. Ref. Petrochem. Process.* (ed. H. Xianglin), Beijing, September 11–15, *Int. Acad. Pub.*, Beijing, China, pp.120–126.

Ludzack, F.J. and Ettinger, M.B. (1960) Chemical structures resistant to aerobic biochemical stabilization. *J. Wat. Pollut. Cont. Fed.* **31**, 1137–1200.

Lund, F.A. and Rodriguez, D.S. (1984) Acclimation of activated sludge to mono-substituted derivatives of phenol and benzoic acid. *J. Gen. Appl. Microbiol.* **30**, 53–61.

Macarie, H. (1992) *Anaerobic Treatment of the Wastewater from a Petrochemical Plant Producing Terephthalic Acid (1,4-Benzenedicarboxylic Acid)*. PhD Thesis, Université de Provence, Marseille, France.

Macarie, H. (2000) Overview of the application of anaerobic treatment to chemical and petrochemical wastewaters. *Wat. Sci. Technol.* **42**(5–6), 201–213.

Matsui, S., Murakami, T., Sasaki, T., Hirose, Y. and Igurria, Y. (1975) Activated sludge degradability of organic substances in the wastewater of the Kashima petroleum and petrochemical industrial complex in Japan. *Prog. Wat. Technol.* **7**(3/4), 645–659.

Moreno-Andrade, I. and Buitrón, G. (2003) Influence of the initial substrate to microorganisms concentration ratio on the methanogenic inhibition test. *Wat. Sci. Technol.* **48**(6), 17–22.

Noyola, A., Macarie, H., Varela, F., Landrieu, S., Marcelo, R. and Rojas, M.A. (2000) Upgrade of a petrochemical wastewater treatment plant by an upflow anaerobic pond. *Wat. Sci. Technol.* **42**(5–6), 269–276.

Olguín-Lora, P., Puig-Grajales, L. and Razo-Flores, E. (2003) Inhibition of the acetoclastic methanogenic activity by phenol and alkylphenols. *Environ. Technol.* **24**, 999–1006.

Olmos, A., Olguín, P., Fajardo, C., Razo-Flores, E. and Monroy, O. (2004) Physicochemical characterization of spent caustic from the OXIMER process and sour waters from Mexican oil refineries. *Ener. Fuel.* **18**, 302–304.

Page, I.C., Wilson, D.R., Cocci, A.A. and Landine, R.C. (1998) Anaerobic hybrid treatment of terephthalic acid wastewater. *Proceedings of the 71st Annual Water Environment Federation Conference*, Orlando, FL, USA, October 3–7, vol. 3. pp. 575–586.

Page, I.C., Cocci, A.A., Grant, S.R., Wilson, D.R. and Landine, R.C. (1999) Single-stage anaerobic hybrid treatment of a polyester-intermediate production wastewater. *Proceedings of the IAWQ International Conference Waste Minimisation and End of Pipe Treatment in Chemical and Petrochemical Industries* (eds. G. Buitrón and H. Macarie), Merida, Yucatan, Mexico, November 14–18, pp. 529–532.

Park, C.-M. and Sheehan, R.J. (1996) Phthalic acids and other benzenepolycarboxylic acids. In: *Kirk-Othmer Encyclopedia of Chemical Technology, 4th edn* (eds. R.E. Kirk,

D.F. Othmer, J. Kroschwitz and M. Howe-Grant), vol. 18. John Wiley and Sons, New York, USA, pp. 991–1043.

Pereboom, J.H.F., de Man, G. and Su, I.T. (1994) Start-up of full scale UASB-reactor for the treatment of terephthalic acid wastewater. *Poster Papers Proceedings of the 7th International Symposium Anaerobic Digestion*, Cape Town, South Africa, January 23–27, pp. 307–312.

Pitter, P. (1976) Determination of biological degradability of organic substances. *Wat. Res.* **10**, 231–235.

Primelles, J.A., Flores, V.E. and Díaz, L.M. (1999). Biolak extended aeration for a petrochemical wastewater effluent. *Proceedings of the IAWQ International Conference Waste Minimisation and End of Pipe Treatment in Chemical and Petrochemical Industries* (eds. G. Buitrón and H. Macarie), Merida, Yucatan, Mexico, November 14–18, pp. 581–584.

Providenti, M.A., Lee, H. and Trevors, J.T. (1993) Selected factors limiting the microbial degradation of recalcitrant compounds. *J. Ind. Microbiol.* **12**, 379–395.

Razo-Flores, E., Luijten, M., Donlon, B., Lettinga, G. and Field, J. (1997) Complete biodegradation of the azo dye azodisalicylate under anaerobic conditions. *Environ. Sci. Technol.* **31**, 2098–2103.

Razo-Flores, E., Lettinga, G. and Field, J.A. (1999) Biotransformation and biodegradation of selected nitroaromatics under anaerobic conditions. *Biotechnol. Prog.* **15**, 358–365.

Razo-Flores, E., Olguín-Lora, P., Alcántara, S. and Morales-Ibarría, M. (2004) Biotreatment of water pollutants from the petroleum industry. In: *Petroleum Biotechnology: Development and Perspectives* (eds. R. Vazquez-Duhalt and R. Quintero-Ramírez). *Studies in Surface Science and Catalysis*, vol. 151. Elsevier, Amsterdam, The Netherlands, pp. 513–536.

Rinzema, A. and Lettinga, G. (1988) Anaerobic treatment of sulfate containing wastewater. In: *Biotreatment Systems* (ed. D.L. Wise), vol. III. CRC Press, Boca Raton, FL, USA, pp. 65–109.

Roffia, P., Calini, P. and Motta, L. (1984) Byproduct identification in the terephthalic acid production process and possible mechanisms of their formation. *Ind. Eng. Chem. Prod. Res. Dev.* **23**, 629–634.

Schink, B. (1997) Energetics of syntrophic cooperation in methanogenic degradation. *Microbiol. Mol. Biol. Rev.* **61**, 262–280.

Seidell, A. (1941) *Solubilities of Organic Compounds: A Compilation of Quantitative Solubility Data from the Periodical Literature*, 3rd edn, vol. II. D. van Nostrand Co. Inc., NY, USA.

Shelley, S. (1991) And the winner is … Amoco Chemical. The company's innovative biodegradation technique – using anaerobic bacteria to degrade chemical waste – wins the 1991 Kirkpatrick chemical engineering achievement award. *Chem. Eng.* **98**(12), 90–93.

Sierra-Alvarez, R. and Lettinga, G. (1991) The effect of the aromatic structure on the inhibition of acetoclastic methanogenesis in granular sludge. *Appl. Microbiol. Biotechnol.* **34**, 544–550.

Storck, W.J., Tullo, A.H., Short, P.L. and Tremblay, J.F. (2005) World chemical outlook. *Chem. Eng. News.* **83**(2), 15–29.

Strotmann, U.J. and Weisbrodt, W. (1994) Wastewater treatment and integrated environmental protection at the BASF AG in Ludwigshafen, Germany. *Wat. Sci. Technol.* **29**(8), 185–192.

Subrahmanyam, P.V.R., Khadakkar, S.N., Chakrabarti, T. and Sundaresan, B.B. (1982) Wastewater treatment of a phthalate plasticizer, ethanolamine and morpholine manufacturing plant: a case study. *Proceedings of the 37th Purdue University Industrial Waste Conference*, Ann Arbor Science, Ann Arbor, MI, USA, pp. 13–20.

Trobisch, K.H. (1992) Recent development in the treatment of chemical waste water in Europe. *Wat. Sci. Technol.* **26**(1–2), 319–332.

Tur, M.-Y. and Huang, J.-C. (1997) Treatment of phthalic waste by anaerobic hybrid reactor. *J. Environ. Eng.* **123**, 1093–1099.

Vanduffel, J.M. (1993) Anaerobic treatment of organic acids. In *National Conference on Anaerobic Treatment of Complex Wastewaters*. Breda, The Netherlands (in Dutch).

Varela-Fuentes, F., Macarie, H., Marcelo-Valencia, R., Ramirez-Aguilar, G., González-Aleu, F.J. and Pérez-Vidal, A. (1998) Acidity, biodegradability and solids limitations in an anaerobic reactor fed with aromatic acids as COD source. *Proceedings of the V Latin-American Workshop-Seminar Wastewater Anaerobic Treatment*, Viña del Mar, Chile, October 27–30, 13 pp.

Wang, L.-C., Lee, W.-J., Tsai, P.-J. and Chen, S.-J. (2002) Potential method for reducing emissions of polycyclic aromatic hydrocarbons from the incineration of biological sludge for the terephthalic acid manufacturing industry. *Environ. Sci. Technol.* **36**, 3420–3425.

Wong, J.M. (2000) Testing and implementation of an advanced wastewater reclamation and recycling system in a major petrochemical plant. *Wat. Sci. Technol.* **42**(5–6), 23–27.

Wu, J.-H., Liu, W.-T., Tseng, I.-C. and Cheng, S.-S. (2001) Characterization of a 4-methylbenzoate-degrading methanogenic consortium as determined by small-subunit rDNA sequence analysis. *J. Biosci. Bioeng.* **91**, 449–455.

Xu, K., Xu, L., Dong, H., Sun, J. and Zhang, Q. (1988) Treatment of alkaline terephathalic acid waste water by flocculation sedimentation. *Wat. Treat.* **3**, 481–490.

Young, J.C. (1997) Anaerobic treatability of phthalic acid production wastewaters. *Proceedings of the 8th International Conference Anaerobic Digestion*, Sendai, Japan, May 25–29, vol. 1, pp. 380–387.

Young, J.C., Kim, I.S., Page, I.C., Wilson, D.R., Brown, G.J. and Cocci, A.A. (2000) Two-stage anaerobic treatment of purified terephthalic acid production wastewaters. *Wat. Sci. Technol.* **42**(5–6), 277–282.

Zitomer, D.H. and Speece, R.E. (1993) Sequential environments for enhanced biotransformations of aqueous contaminants. *Environ. Sci. Technol.* **27**, 227–244.

Zoutberg, G.R. and de Been, P. (1997) The Bioded® EGSB (expanded granular sludge bed) system covers shortcomings of the upflow anaerobic sludge blanket reactor in the chemical industry. *Wat. Sci. Technol.* **35**(10), 183–188.

9

Closing process water cycles and product recovery in textile industry: perspective for biological treatment

A.B. Dos Santos, I.A.E. Bisschops and F.J. Cervantes

9.1 INTRODUCTION

The textile industry represents an important economical sector around the world. In 2000, the European industry was composed of around 114 thousands companies (principally based in Italy, Germany, the UK, France and Spain) with an average turn-over of €198 billion a year, making it the world's leading exporter of textiles and the third largest exporter of clothing (IPPC 2003). When looking at the numbers for 2003, Asia was the largest textiles importer and exporter, and the leading exporter of clothing. The USA and Canada together are the largest clothing importers (WTO 2004). The worldwide distribution of imports and exports of textiles and clothing, in 2003, is shown in Figure 9.1. In Table 9.1, the values of textiles and clothing exports and imports, in 1995 and 2003, are given for the European Union (EU), China and the USA, showing the differences in how the textile market has developed.

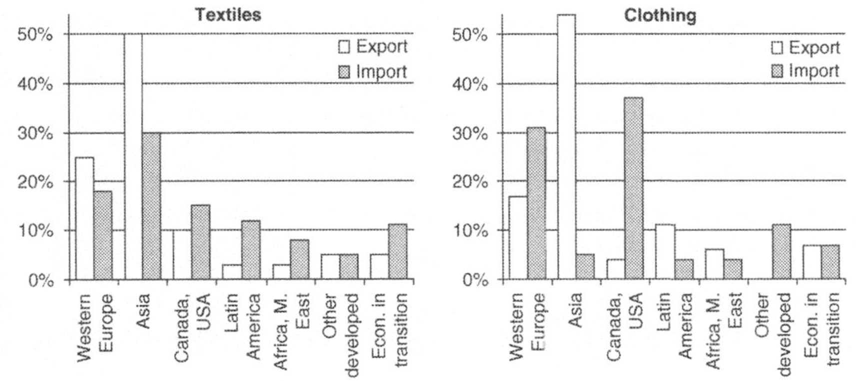

Figure 9.1 Distribution of imports and exports of textiles and clothing in the world, divided by region, in 2003 (WTO 2004).

Table 9.1 Exports and imports of textiles and clothing, in billion dollars, rounded off at 0.5 billion. EU (15) =Austria, Belgium, Denmark, Finland, France, Germany, Greece, Ireland, Italy, Luxembourg, The Netherlands, Portugal, Spain, Sweden and the UK (WTO 2004).

		Textiles		Clothing	
		Export	Import	Export	Import
World	1995	112	–	125	–
	2003	137	–	185	–
EU (15)	1995	22	17	15	40.5
	2003	26	20	19	60.5
China	1995	14	11	24	1
	2003	27	14	52	1.5
USA	1995	7.5	10.5	6.5	41.5
	2003	11	18.5	5.5	71.5

Textile processes, such as bleaching, dyeing, printing and finishing, give fabrics their visual, physical and aesthetic properties. From the different types of fibres that are processed, cotton represents the largest fraction (45%), followed by polyester (14%), viscose (12%), wool (8%) and acrylic (4%).

The increased demand for textile products over the last decades caused a proportional increase in the production of wastewater. A huge number of chemicals that have a negative effect on the environment and public health are released through textile industry wastewaters. Chemicals such as *alkyl phenol ethoxylates* (present in detergents, wetting and levelling agents), reported to disturb the reproduction of aquatic species; along with *sequestering agents* like *ethylenediaminetetracetic acid* (*EDTA*) and *diethylenetriaminepentacetic acid* (*DTPA*), capable of forming very stable complexes with metals, thus affecting their bioavailability; and the *dyestuffs*, which are recalcitrant by design and not readily degraded by common treatment methods, are examples of hazardous compounds present in textile wastewaters (DEPA 1997). Particularly the release of coloured compounds into water bodies is undesirable, not only because they reduce the transparency of water, which may drastically affect photosynthesis of aquatic plants, but also because many dyes and their degradation products are carcinogenic (Hao *et al.* 2000). Without adequate treatment these dyes may persist in the environment for an extended period of time. For instance, the half-life of hydrolysed Reactive Blue 19 is about 46 years at pH 7 and 25°C (Hao *et al.* 2000).

9.2 CHARACTERISATION OF TEXTILE WASTEWATER

Textile processing involves many different steps, in which wastewater is generated. The amount and composition of these wastewaters depend on many different factors, such as the type of fabric, the type of process, and the used chemicals. As previously mentioned, a large amount of hazardous compounds is emitted from the textile industry. Knowledge of what (and where) exactly is being emitted, is needed for different reasons: design of the wastewater treatment, assessment of the textile industry impact on the environment, ensuring good operation of the wastewater treatment facilities, enforcement of discharge standards, etc. Therefore, the adequate characterisation of textile effluents is important.

In the EU, the "Integrated Pollution Prevention and Control" directive is being implemented, as well as the best available techniques (BAT) have been defined with the objective to eliminate or reduce emissions. The textile BAT Reference Document (BREF) is finished (IPPC 2003), and probably contains the most recent and complete data concerning textile processes and their related emissions. It is suggested in the BREF that for several textile processes, where possible, closed-loop systems for water reuse and chemicals savings are BAT. The successful implementation of water treatment and reuse options will depend on the effective characterisation and monitoring of wastewater streams.

9.2.1 Textile processes and their wastewater

Figure 9.2 shows the sequence of textile processes as they generally take place in a factory. Some processes hardly generate wastewater, such as yarn manufacture, weaving (some machines use water), and singeing (just some lightly polluted cooling water). The amount of wastewater produced in a process like sizing is small, but very concentrated. On the other hand, processes like scouring, bleaching and dyeing generate large amounts of wastewater, varying much in composition. The processes that are very common and produce considerable amounts of wastewater will be briefly discussed.

9.2.1.1 Sizing and desizing

Cotton and some man made yarns are "sized" before spinning or weaving, to gain strength and minimise breaking of the fibres. Size liquid is pressed into the fibre and then the threads are dried (BTTG 1999). Starch or starch derivatives are applied in

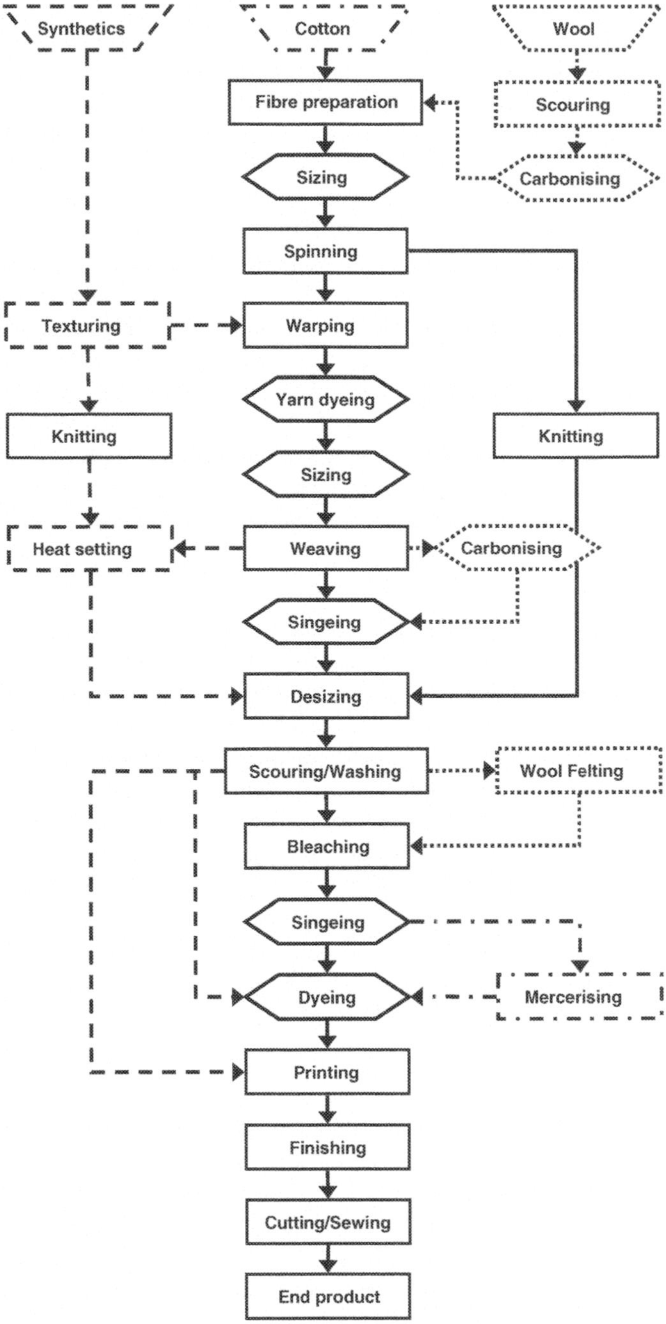

Figure 9.2 General flowchart for processes taking place in textile manufacturing. Square boxes are processes that always take place in that order, and "diamond" shaped boxes are processes that can occur at different places in the chain. The line style indicates if the process is only used for a certain fabric type: (—) for all fibres, (- - -) for synthetic fibres, (- -) for cotton and (…) for wool (Bisschops and Spanjers 2003).

75% of the sizing operations (Opwis *et al.* 1999), but also other substances are used, such as the synthetic sizing agents polyvinyl alcohol (PVA), polyacrylates and carboxymethyl cellulose (Correia *et al.* 1994; Opwis *et al.* 1999). Leftover sizing liquid is wasted, usually in small volumes but with very high levels of chemical oxygen demand (COD), biochemical oxygen demand (BOD) and suspended solids. The use of synthetic sizes diminishes the BOD load, and reclamation of recyclable sizes like PVA can lead to a 90% reduction of the total organic load (Delée *et al.* 1998). After weaving, the size is removed from the fabric in the desizing process, because it covers the yarn and might act as a barrier to dyes and other chemicals (Rucker and Smith 1988). Washing with detergents can be sufficient to remove some kinds of sizes, whereas starch is usually removed by using enzymes. In some special cases, the sizes washing may require the use of acids or alkalis. At the end of the process, fabrics are rinsed thoroughly. Desizing wastewater contains the used sizes and the agents applied for desizing. Its contribution to the overall BOD and total suspended solids (TSS) load of a mixed textile wastewater is usually very high. Up to 50% of the total BOD in woven fabric processing can originate from the desizing step when starch-based size is applied (Correia *et al.* 1994; BTTG 1999).

9.2.1.2 Scouring

Fibres contain different impurities that can interfere with dyeing and finishing, such as oils, fats, waxes, minerals and plant matter (natural fibres) spin finishes, and knitting oils (synthetic fibres). Also during processing contamination may occur, for instance with the grease used on equipments lubrication, factory dirt and temporary fabric markings (Tomasino 1992). These impurities are removed in the scouring process, either with water-containing scouring agents or with solvents. Water scouring is usually preferred over solvent scouring, as water is non-flammable, non-toxic and cheaper. Scouring agents include detergents, soaps and various assisting agents, such as alkalis and wetting agents (DEPA 1997). After scouring, the goods are thoroughly rinsed (or washed) to remove excess agents. Due to the used chemicals and the compounds released from the material during scouring, the generated wastewater is chemically aggressive and may be toxic. Characteristics are a high COD and solids content of the wastewater, and also the BOD level can be high.

9.2.1.3 Bleaching

Bleaching is commonly used to remove natural colouring from cotton, blend fabrics or yarn, and is sometimes required on wool and some synthetic fibres. Used chemicals include sodium hypochlorite, hydrogen peroxide, optical brighteners and some auxiliary compounds (DEPA 1997). BOD levels are low, but the solids content of the wastewater can be high (Correia *et al.* 1994). Denim processing results in an extremely high suspended solids content in the bleaching step due to the use of pumice stone (Orhon *et al.* 2001).

9.2.1.4 Dyeing

In the dyeing process large amounts of water are used, not only in the dyeing process but mainly also in the various rinsing steps, aiming to remove unfixed dyestuffs from the dyed fabric. In the dyebath, large amounts of dyes are used, because a considerable fraction (up to 40% for some types) remains unused instead of being fixed to the fabric. These unfixed dyes are mainly present in the effluents from the dyebath and first rinsing step, but they can also be found in the effluents of all following rinsing steps. Along with the unfixed dyes, the different auxiliary chemicals applied also end up in the wastewater (DEPA 1997). Examples of these auxiliaries are organic acids, fixing agents, defoamers, oxidising/reducing agents and diluents. The exact amount and kind of pollutants mainly depend on the applied dyes and the type of dyeing process (EPA 1997). Dyeing contributes to most of the metals and almost all of the salts and colour present in the overall textile effluent (EPA 1996). In some cases, up to 75% of the salts used may end up in the wastewater (DEPA 1997).

9.2.1.5 Printing

In this step, the dyes and auxiliaries are similar to those used in dyeing, with the difference that the colour is only applied to specific parts of the fabric. An important component in textile printing is the print paste, which consists of ingredients, such as water, thickeners, dyes, urea and various other chemicals (EPA 1997). Large amounts of residual pastes come from paste preparation and equipments cleaning (Bonomo *et al.* 1997). Other sources of pollutants are the leakage of dyestuffs and washing of textile products (Rigoni Stern *et al.* 1996). A common disposal method is the dilution of residual pastes and discharge with the other wastewater streams, where they considerably increase COD, nitrogen and dye loads (Malpei *et al.* 1998). The printing method determines the wastewater characteristics. Printing wastewaters are small in volume, but difficult to treat. They resemble textile-dyeing wastewaters but are more concentrated, and contain solvents and solids coming from print pastes (Bonomo *et al.* 1997; Kabdasli *et al.* 2000).

9.2.1.6 Finishing

In the finishing step, certain properties affecting the care, comfort, durability or human safety associated with the fabric, are altered (DEPA 1997). Finishes can consist of chemical or mechanical techniques (EPA 1996). As far as the mechanical finishes are concerned, no effluent is produced. In the chemical finishing steps only very low volumes of wastewater are produced, and there is hardly any information reported about the pollutants it contains.

9.2.1.7 Rinsing

Although rinsing is in fact not a separate textile process, it is interesting to regard the generated wastewater separately when looking at reuse possibilities. The wastewater from rinsing steps is much less polluted than the wastewater from the preceding baths of the same process. Almost all processes involve a number of hot or cold rinses. Rinsing is always needed to remove substances from the fabric that can influence the following processes, and after dyeing or printing to remove unfixed dyestuffs from the fabric. Depending on the used processes, rinsing can make up for 50% of the total water usage (DEPA 1997). For some dyeing processes, the rinsing water amounts to 70–80% of the total water consumption. Especially the last portions of rinsing water – amounting to 20–30% of total water consumption – have a quality indicating a possible direct reuse as first rinsing water (Wenzel *et al.* 1999). A relatively easy way of taking advantage of the water quality (and in some cases temperature) from the different rinsing steps is applying counter-flow rinsing. This strategy can be used in many cases, which will depend on the process and the type of machinery (Rucker and Smith 1988; EPA 1997).

9.2.2 Summary of effluent characteristics

Although effluent characteristics differ greatly even within the same process, some general values can be given. However, the individual wastewater characteristics of all textile steps, are not easily found. The ones generating smaller amounts of wastewater or less polluted effluents are hardly investigated. Table 9.2 gives an overview of the large fluctuation of important parameters from textile wastewaters. Mixed textile wastewater generally contains high levels of COD and colour, and usually has a high pH. As its composition depends on the types of processes and fabrics handled in the factory, it is difficult to present general values. For example, COD concentrations varying from approximately 150 mg/L to over 10,000 mg/L and pH values from 4.5 to 13, are reported (LeAF 2002).

9.2.3 Characterisation methods and data comparison

When looking at different publications regarding textile wastewater, it becomes apparent that the most widely used characterisation methods are for measuring COD, pH and colour. This is not surprising, as COD and pH are usually measured in all kinds of wastewater, and colour is a special characteristic of textile effluents.

Table 9.2 Average and peak values in mg/L for selected parameters of most reported textile processes.

Parameter	Desizing	Scouring	Bleaching	Dyeing
COD	3580–5900	3200–40,000	250–6000	550–8000
COD peak	11,000	90,000	13,500	40,000
BOD	200–5200	300–8000	80–400	11–2000
BOD peak		60,000	2800	27,000
TS	7600–42,000	1100–30,000	900–14,000	200–2000
TS peak		65,000		14,000
SS	400–800	200–20,000	35–900	25–200
SS peak	6000	40,000	25,000	
TDS	1600–6900		40–5000	
TDS peak			20,000[a]	
Lipids	190–750	100–9000		
Lipids peak		10,8000		
pH	6–9	7–14	6–13	3.5–12

TS: total solids; TDS: total dissolved solids; SS: suspended solids.
[a]Denim stonewash involving pumice stone.
Adapted from LeAF (2002).

Other commonly used parameters in characterisation are suspended solids and BOD (LeAF 2002). Most authors state that they used standard methods for all their analyses, but comparing the characteristics of an effluent with wastewaters reported in different studies is still often a difficult task. For the determination of most parameters different "standard methods" exist, and especially for colour this can cause problems because the resulting values of the different methods are practically incomparable. Also for other parameters this can be the case. Providing information on which method was exactly used is, therefore, very important to make data comparison possible.

Another difficulty is the lack of specification of effluent types: many authors do not clearly specify what kind of wastewater they used in their studies. Logically, the efficiency reported for a certain treatment technique applied to a dyebath containing disperse dyes, will not perform equally on a mixed wastewater containing effluents from bleaching, desizing and reactive dyeing processes. Therefore, it is important to make sure which effluent was studied, and which methods were used to characterise it, before using reported data for the selection of a specific wastewater treatment.

9.3 COD REMOVAL

To indicate the level of wastewater pollution as well as the efficiency of a wastewater treatment system, the parameter COD is probably the most widely used. As COD originates from almost all textile processes and due to the different characteristics of the latter, the options for COD removal should ideally be discussed for each separate wastewater stream. However, to keep the information presented in this chapter concise, the methods for COD removal will be described for textile wastewater in general.

Before deciding on a certain type of treatment, it is important to know the nature of the COD load with respect to the different fractions that contribute to the total COD, such as soluble, colloidal and suspended, and if they are biodegradable. For mixed textile effluents, generally all COD fractions are present. Therefore, these characteristics will help to determine which technique is suitable for the wastewater treatment. However, currently there is no single and economically attractive treatment that can remove COD from textile wastewaters. Therefore, the treatment plant will consist of different processes, which are following described.

9.3.1 Biological treatment

9.3.1.1 Aerobic COD removal

In aerobic wastewater treatments, biodegradable organic matter is oxidised by bacteria that use oxygen as a terminal electron acceptor. Aerobic systems such as activated sludge plants are widely used for the treatment of both domestic and industrial wastewaters (Metcalf and Eddy 2003). Also for textile wastewaters, aerobic systems are being used, but generally the treatment is not adequate. The wastewaters are usually highly concentrated, highly coloured, and too complex to reach complete removal of pollutants in a conventional aerobic treatment. As a result, the effluent of those systems treating textile wastewater, will contain high concentrations of residual COD and colour (Vandevivere *et al.* 1998). Many compounds present in textile effluents (e.g. dyes and certain surfactants) are resistant to aerobic biodegradation. Part of the colour removal in activated sludge systems is mainly due to settling of insoluble dyes in the primary settler, and adsorption of some types of soluble dyes onto the sludge (Delée *et al.* 1998). Starch and starch derivatives, being traditional sizing agents used in the sizing process, are usually readily biodegradable in aerobic systems. However, if desizing effluent forms a large fraction of the total wastewater stream treated in an activated sludge plant, this often causes a problem with bulking sludge (Delée *et al.* 1998). Surfactants of different types are extensively used in many textile processes, of which the class of alkyl phenol ethoxylates is the most important one. For example, nonylphenol ethoxylates (NPE) and its intermediate nonyl phenol, are more than 90% removed in activated sludge systems, but only one-third of that removal can be attributed to biological degradation: 60–65% are removed with the excess sludge (Vandevivere *et al.* 1998).

9.3.1.2 Anaerobic COD removal

Anaerobic treatment systems are capable of coping with high organic loads, and many dyes are susceptible to reductive transformation under anaerobic conditions (Delée *et al.* 1998). Additionally, in comparison with aerobic treatment, anaerobic treatment provides a better removal of AOX and heavy metals (Vandevivere *et al.* 1998). These characteristics make anaerobic systems more suitable for treatment of textile wastewater than aerobic systems (e.g. activated sludge systems), as textile wastewater usually has a high organic load and high concentrations of different dyes. Other advantages of this technology compared to aerobic treatment are the lower excess sludge production and the energy recovery in the form of biogas. However, the effluent from anaerobic bioreactors may still contain a considerable amount of COD (efficiency of about 60%), even if most of the originally present organic matter has been removed and a good colour removal is achieved (O'Neill *et al.* 2000). Therefore, the anaerobic treatment can be considered as a pre-treatment step, and a post-treatment is always needed to remove the remaining COD. A possible biological treatment strategy might be an aerobic post-treatment, in which the COD is ultimately converted into CO_2 and H_2O (Delée *et al.* 1998; O'Neill *et al.* 2000).

Under anaerobic conditions, many dye molecules can be reduced, for example, azo bond cleavage (azo dyes) or quinone-group reduction (anthraquinone dyes), but this reduction leads only to decolourisation. The reduction of anthraquinone dyes can be reversible, but azo dyes are irreversibly reduced, leading to the formation of degradation products such as aromatic amines. The dye reduction products, which still have a considerable amount of COD, are usually not degraded anaerobically. Furthermore, reaction products such as aromatic amines resulting from cleavage of the azo bond, are known to be carcinogenic (Vandevivere *et al.* 1998). Sizes such as starch and starch derivatives are readily biodegradable under anaerobic conditions, but synthetic sizes such as PVA are often not sufficiently removed (Delée *et al.* 1998). Another important constituent of textile wastewater are surfactants, which are known to strongly inhibit anaerobic treatment. Anionic surfactants such as alcohol sulphates are biodegradable, but they may become inhibitory to biological processes at high concentrations (Feitkenhauer 2003).

9.3.2 Non-biological treatment

9.3.2.1 Physical–chemical methods

In physical–chemical methods coagulant agents like ferric salts or aluminium polychloride are used to form flocs with the compounds present in the wastewater. These flocs are then separated by filtration or sedimentation. Polyelectrolyte can also be dosed during the flocculation phase to improve the flocs settleability (Vandevivere *et al.* 1998). The coagulation–flocculation method is one of the most widely used processes in textile wastewater treatment plants (WWTP) in many countries such as Germany and France. It can be used either as a pre-treatment, post-treatment, or even as a main treatment system (Gähr *et al.* 1994). On top of the problem of low colour removal efficiency with some dyes, physical–chemical methods demand large chemicals inputs, and produce high volumes of polluted sludge, which then must be treated (Robinson *et al.* 2001).

9.3.2.2 Chemical methods

Chemical oxidation typically involves the use of an oxidising agent such as ozone (O_3), hydrogen peroxide (H_2O_2), permanganate (MnO_4), etc. to change the chemical composition of a compound or a group of compounds. The usual low efficiency of both colour and COD removal of conventional chemical oxidation techniques have been overcome by the development of the so-called advanced oxidation processes (AOP). In this process, oxidising agents such as O_3 and H_2O_2 are used with catalysts (Fe, Mn and TiO_2), either in the presence or absence of an irradiation source (Alaton *et al.* 2002). Consequently, an improvement in the generation and use of the free hydroxyl radical ($HO^{\bullet}$) is obtained, representing a rate increase of one to several orders of magnitude, compared with normal oxidants in the absence of a catalyst (Hao *et al.* 2000).

The first example is the so-called *Fenton* reaction, in which hydrogen peroxide is added to an acid solution (pH 2–3) containing Fe^{+2} ions:

$$Fe^{+2} + H_2O_2 \rightarrow Fe^{3+} + HO^{\bullet} + HO^{-} \tag{9.1}$$

In comparison to ozonation, this method is relatively cheap and also presents high COD removal and decolourisation efficiencies. The main process drawbacks are the high sludge generation due to the flocculation of reagents and pollutants (Robinson *et al.* 2001), as well as the need for decreasing the bulk pH to acidic conditions. A pre-ozonation of coloured wastewaters prior to Fenton reaction not only considerably accelerated the overall colour removal rates, but also decreased the sludge generation (Hao *et al.* 2000).

In *H_2O_2/ultraviolet (UV) process* $HO^{\bullet}$ radicals are formed when water-containing H_2O_2 is exposed to UV light, normally in the range of 200–280 nm

(Metcalf and Eddy 2003). The H_2O_2 photolysis follows the reaction:

$$H_2O_2 + UV\ (\text{or hv}, \lambda \approx 200\text{–}280\,\text{nm}) \rightarrow HO^{\bullet} + HO^{\bullet} \quad (9.2)$$

This process is the most widely used AOP technology for the treatment of hazardous and refractory pollutants present in wastewaters, mainly because no sludge is formed and a high COD removal in a short retention time is achieved. In some cases, however, the H_2O_2/UV process presents low COD and colour removal efficiency drawbacks due to inefficient use of UV light (mainly for highly coloured wastewaters) (Moraes *et al.* 2000), or because of the low molar extinction coefficient of H_2O_2 (specific oxidation capacity), requiring high dosages of the latter.

The *UV-based methods* in the presence of a catalyst, for example, a semiconductive material such as TiO_2 also generate $HO^{\bullet}$ radicals that react with the wastewater pollutants, but are economically less attractive. However, when solar radiation could be used instead of artificial UV light, the operational costs of this method would drop considerably (Alaton *et al.* 2002).

A drawback of AOP is their sensitivity towards dyebath constituents like sodium salts, detergents and sequestering agents, as has been shown in studies performed with dyebath effluents rather than with dyestuff solutions (Alaton *et al.* 2002).

9.3.2.3 Physical methods

Filtration methods such as ultrafiltration, nanofiltration and reverse osmosis have been used for water reuse and chemical recovery. In the textile industry these filtration methods can be used for both filtering and recycling not only pigment-rich streams, but also mercerising and bleaching wastewaters. The specific temperature and chemical composition of the wastewater determine the type and porosity of the filter to be applied (Porter 1997). The main drawbacks of membrane technology are its high investment costs, the potential membrane fouling and the production of a concentrated stream which needs to be treated (Robinson *et al.* 2001). The recovery of process ingredients from membrane concentrates can attenuate the treatment costs, for instance the recovery of sodium hydroxide from mercerising effluents, or the recovery of sizing agents such as PVA (Porter 1997).

Adsorption methods, for instance using activated carbon (AC), have been mainly investigated for the removal of colour from textile effluents and will be further discussed in Section 9.4.2.3.

9.4 COLOUR REMOVAL

The effluents from the dyeing and rinsing steps mainly represent the coloured fraction of textile wastewaters, which may contain large amounts of dyes. The different dye classes are described in Table 9.3.

A dye molecule may have one or more chromophores, that is, structures that are responsible for its colour, as well as electron withdrawing or donating substituents that cause or intensify the colour of the chromophores, called auxochromes (Christie 2001). The most important chromophores are azo (—N=N—), carbonyl (—C=O), methine (—CH=), nitro (=NO_2) and quinoid groups. Azo dyes represent about

Table 9.3 Dye classes and their respective annual production in the western part of the EU (W-EU) and in the rest of the world.

Dye class	Description	W-EU (10^6kg)	World (10^6kg)
Acid (and mordant)	Anionic compounds that are mainly used in the nitrogen-containing fabrics such as wool, polyamide and silk. They represent the largest class of dyes.	24	100
Azoic	They allow colours with outstanding fastness, but their use has decreased because of the application costs and process complexity.	2	48
Basic	Cationic compounds that are used for dyeing acid-group containing fibres, usually synthetic fibres like modified polyacryl.	8	44
Direct	Widely applied for cellulose fibres usually at high salt concentrations. Around 75% of the total consumption is for cotton and viscose substrates.	9	64
Disperse	Usually non-soluble dyes that penetrate in the synthetic fibres such as cellulose and polyesters by using high temperatures or chemical softeners.	22	157
Reactive	About 35% of the dyes used in the cellulose fibres are reactive dyes. They usually require a high pH, high salt levels, and high temperatures for fixation.	13	114
Sulphur	Normally insoluble in water, unless an alkaline medium and reducing agents are used. Afterwards, they come back to the insoluble form remaining fixed into the fibre by applying an oxidation step. Used for cellulose and cellulose–polyester fibres.	3	101
Vat	Similar application as sulphur dyes. Their biggest use is with cellulose fibres.	4	40
	Σ	85	668

Adapted from Van der Zee (2002); and IPPC (2003).

70% of the dyes consumed worldwide by weight, followed by the anthraquinone and phthalocyanine dyes. The most important auxochromes are amine (—NH_3), carboxyl (—COOH), sulphonate (—SO_3H) and hydroxyl (—OH) groups. The different decolourisation processes are following described.

9.4.1 Biological treatment

9.4.1.1 Colour removal by aerobic microorganisms

Bacteria Aromatic compounds are susceptible to biological degradation under both aerobic and anaerobic conditions. Under aerobic conditions, the enzymes mono- and dioxygenase catalyse the incorporation of oxygen from O_2 into the aromatic ring of organic compounds prior to ring fission (Field *et al.* 1995). Although azo dyes are aromatic compounds, their substituents containing mainly nitro and sulphonic groups, are quite recalcitrant to aerobic bacterial degradation (Claus *et al.* 2002). This fact is probably related either to the electron-withdrawing nature of the azo bond and its resistance to oxygenases attack, or to the fact that oxygen is a more effective electron acceptor, therefore, having more preference for reducing equivalents than the azo dye (Knackmuss 1996). However, in the presence of specific oxygen-catalysed enzymes called azo reductases, some aerobic bacteria are able to reductively cleave not only the carboxylated growth substrates of the bacteria but also the sulphonated structural analogues (Stolz 2001).

Fungi The capacity of fungi to decolourise dyes is related to the formation of exoenzymes such as peroxidases and phenoloxidases, and subsequent oxidation of the dyes. Lignin and manganese peroxidases have a similar reaction mechanism, which starts with the enzyme oxidation by H_2O_2 to an oxidised state during their catalytic cycle. Afterwards, in a mechanism involving two successive electron transfers, substrates such as azo dyes can reduce the enzyme to the original form (Stolz 2001). Eighteen fungal strains able to degrade lignocellulosic material or lignin derivatives were tested with the azo dyes Reactive Orange 96, Reactive Violet 5 and Reactive Black 5. Only three strains, viz. *Bjerkandera adusta, Trametes versicolor* and *Phanerochaete chrysosporium*, were able to decolourise all azo dyes (Heinfling *et al.* 1997). Phenoloxidases, which can be divided into tyrosinases and laccases, are oxidoreductases that can catalyse the oxidation of phenolic and other aromatic compounds without the use of cofactors (Duran *et al.* 2002). Laccases are Cu-containing enzymes that have very broad substrate specificity with respect to electron donors, for example, dyes. However, despite that laccases from *Trametes Versicolor, Polyporus pinisitus* and *Myceliophthora thermophila* were found to decolourise anthraquinone and indigoid-based dyes at high rates, the azo dye Direct Red 29 (Congo Red) was a very poor substrate for laccases (Claus *et al.* 2002). It has been reported that the azo dye must be electron-rich to be susceptible to oxidation by laccase of *Pyricularia oryzae*. This situation is suitable for the generation of a phenoxy radical, with consequent azo bond cleavage and the release of molecular nitrogen (Chivukula and Renganathan 1995).

9.4.1.2 Colour removal by strictly anaerobic or facultative microorganisms incubated under anaerobic conditions

Under anaerobic conditions a low redox potential ($<-50\,mV$) can be achieved, which is necessary for the effective decolourisation of dyes. Colour removal under anaerobic conditions is also referred as dye reduction, of which literature mostly covers the biochemistry of azo dye reduction. The azo bond (—N=N—) cleavage involves a transfer of four electrons (reducing equivalents), which proceeds through two stages, to the azo linkage. In each stage two electrons are transferred to the azo dye, which acts as a final electron acceptor. The reductive decolourisation of azo dyes under anaerobic conditions is a combination of both biological and chemical mechanisms. The *biological* contribution can be divided in specialised enzymes called azo reductases, which are present in bacteria that are able to grow using only azo dye as a carbon and energy source; or non-specific enzymes that catalyse the reduction of a wide range of electron-withdrawing contaminants, including azo dyes (Stolz 2001). However, up to date there is no clear evidence of anaerobic azo reductase; or non-specific enzymes that catalyse the reduction of a wide range of electron-withdrawing contaminants, including azo dyes. Thus, a co-metabolic reaction is probably the main mechanism of dye reduction, in which the reducing equivalents or reduced cofactors like NADH, NAD(P)H, $FMNH_2$ and $FADH_2$ acting as secondary electron donor, channel electrons to cleave the azo bond (Stolz 2001). The *chemical* contribution to the reductive decolourisation of azo dyes under anaerobic conditions may involve biogenic reductants like sulphide, cysteine, ascorbate or Fe^{2+} (Van der Zee *et al.* 2001b; Yoo 2002).

Even though anaerobic azo dye reduction could be readily achieved with different microorganisms, there is no strain reported so far that is able to decolourise a broad range of azo dyes. Therefore, the use of a specific strain or enzymes on azo dye reduction does not make much sense in treating textile wastewater, which is usually composed of many kinds of dyes. The use of mixed cultures, such as anaerobic granular sludge, which is composed of stable microbial pellets with a high activity, is probably a more logical alternative. Different reactor configurations, such as the widely used upflow anaerobic sludge bed (UASB) system and expanded granular sludge bed (EGSB) system, are used to immobilise high concentrations of biomass. Indeed, the different microbial consortia present in anaerobic granular sludge can carry out tasks that no individual pure culture can undertake successfully (Dos Santos 2005). The choice of the primary electron donor to be used in the reductive decolourisation of azo dyes is extremely important. For instance, acetate and other volatile fatty acids are normally poor electron donors, whereas ethanol, glucose, H_2/CO_2 and formate are more effective electron donors for colour removal.

It was found that reductive decolourisations were improved by using *redox mediators*, which might increase decolourisation rates from one up to several orders of magnitude (Cervantes *et al.* 2001; Van der Zee *et al.* 2001a). Many flavin-based vitamins and quinones containing humus have been reported to accelerate colour removal. The use of redox mediators as an electron shuttle showed to be extremely effective in enhancing colour removal under mesophilic conditions, but their catalytic effect on decolourisation rates is distinctly decreased under thermophilic conditions (Dos Santos *et al.* 2004). Moreover, a significant enhancement of the electron transfer capacity and subsequent increase in colour removal of azo dyes by simply applying high temperature (55°C) was demonstrated in continuous flow experiments at different hydraulic retention time's (HRT's) (Dos Santos *et al.* 2005). This phenomenon is of considerable interest if one realizes that textile wastewaters are generally discharged at high temperatures. Therefore, a very compact thermophilic reactor could be an option as a pre-treatment unit for textile wastewaters.

9.4.2 Non-biological treatment

The basic mechanisms of non-biological treatment methods have been described in Section 9.3.2 and will not be repeated here.

9.4.2.1 Physical–chemical methods

In physical–chemical methods coagulant agents like ferric salts or aluminium polychloride are used to form flocs with the dyes, which are then separated by filtration or sedimentation. Coagulation–flocculation methods are successfully applied for colour removal of sulphur and disperse dyes, whereas acid, direct, reactive and vat dyes present very low coagulation–flocculation capacity (Hao *et al.* 2000; Mattioli *et al.* 2002).

9.4.2.2 Chemical methods

Among all the oxidants explained in Section 9.3.2.2, ozone is the most widely used because of its high reactivity with many dyes, usually providing good colour removal efficiencies (Alaton *et al.* 2002). However, disperse dyes and those insoluble in water represent a drawback for the process, as well as the high cost of ozone (Hao *et al.* 2000). AOP and combinations of them have been investigated for colour removal, all of which are capable of producing the free hydroxyl radical ($HO^{\bullet}$). The Fenton process, for example, is relatively cheap in comparison to ozonation, and presents high decolourisation efficiencies (Robinson *et al.* 2001). The H_2O_2/UV process has been successfully applied for colour removal. For instance, more than 95% decolourisation was achieved in treating reactive, basic, acid and direct dyes at pH 5, whereas disperse and vat dyes were only partially decolourised (Hao *et al.* 2000). A comparative study between ozone and H_2O_2/UV was carried out in treating a concentrated reactive dyebath from a textile factory. The H_2O_2/UV system presented decolourisation rates close to those rates obtained with ozone but with a lower cost (Alaton *et al.* 2002).

The *UV-based methods* in the presence of a catalyst such as TiO_2, have also been shown to distinctly enhance colour removal (Hao *et al.* 2000; So *et al.* 2002). Thus, different combinations such as Ozone/TiO_2, Ozone/TiO_2/H_2O_2 and TiO_2/H_2O_2, have been investigated, but are enormously influenced by the type of dye, dye concentration and pH (Galindo *et al.* 2000). Recently, the utilisation of solar technologies instead of UV-based methods has been attracting attention.

9.4.2.3 Physical methods

Filtration methods such as ultrafiltration, nanofiltration and reverse osmosis can be used for elimination of colour from textile wastewaters. Water reuse from dyebath effluents has been successfully achieved by using reverse osmosis. However, a coagulation and micro-filtration pre-treatment was necessary to avoid membrane fouling (Vandevivere *et al.* 1998).

Adsorption methods for colour removal are based on the high affinity of many dyes for adsorbent materials. Decolourisation by adsorption is influenced by some physical–chemical factors such as dye–adsorbent interactions, adsorbent surface area, particle size, temperature, pH and contact time (Mattioli *et al.* 2002). The main criteria for the selection of an adsorbent should be based on characteristics such as high affinity and capacity for target compounds and the possibility of adsorbent regeneration (Karcher *et al.* 2001). AC is the most common adsorbent and can be very effective with many dyes. However, its efficiency is directly dependent on the type of carbon material used and the wastewater characteristics, that is, types of dyes present in the stream. Additionally, AC is relatively expensive and has to be regenerated off-site with losses of about 10% in the thermal regeneration process. In order to decrease the adsorbent losses during regeneration, new adsorbent materials have been tested for their ability for on-site regeneration. Karcher *et al.* (2001) studied alternative materials, such as zeolites, polymeric resins, ion exchangers and granulated ferric hydroxide. It was found that zeolites and microporous resins were unsuitable due to their low sorption capacity. Although the ion exchangers provided good sorption capacity, regeneration was sometimes difficult. A number of low-cost adsorbent materials such as peat, bentonite clay, fly ash, etc. have been investigated on colour removal (Robinson *et al.* 2001). However, the efficiency of these materials varied with the dye class. For instance, fly ash presented high sorption affinity for acid dyes, whereas peat and bentonite presented high affinity for basic dyes.

9.5 INDUSTRIAL FULL-SCALE APPLICATIONS FOR TREATMENT OF TEXTILE WASTEWATERS

9.5.1 Case study 1

9.5.1.1 Description

The textile factory is placed in Brazil, and produces blue and black indigo dyed fabrics. Its production is destined for both Brazilian and international markets.

9.5.1.2 WWTP set-up

The WWTP is composed of a preliminary treatment, equalisation tank (with aerator and pH adjustment), and a physical–chemical treatment composed of fine screening, rapid mixing, slow mixing, primary settler, sand filtration and AC column. In the physical–chemical treatment aluminium sulphate (or aluminium polychloride) and a cationic polymer are used as coagulants. The average influent flow is 336 m^3/day, which was only composed of the industrial fraction. The domestic sewage generated in the factory was treated separately. The main characteristics of the wastewater were its dark colour (blue or black), caused by the different dyes and pigments used in the dyeing process; and high concentration of organic matter, mainly resulting of the starch used in the sizing step (Table 9.4). Even though about 50% of the total sodium hydroxide amount used for mercerising was recovered in the factory, the pH of the effluent was alkaline. Detergents, soaps, surfactants and wetting agents were also present in the wastewater.

9.5.1.3 Possible improvements with the aim of products recovery and water reuse

Some approaches not only aiming the products recovery, but also the improvement of the wastewater treatment are following described. As can be implied, the main

Table 9.4 Wastewater characterisation for a textile factory producing blue and black indigo dyed fabrics.

Parameter	Unit	Range of values
COD	(mg/L)	887–14,200
BOD	(mg/L)	400–4000
Temperature	(°C)	45–60
Conductivity	(μS/cm)	3670–6060
Hardness	(mg $CaCO_3$/L)	155–264
TS	(mg/L)	5367–6708
TSS	(mg/L)	588–781
TDS	(mg/L)	4624–5926
FSS	(mg/L)	73–182
VSS	(mg/L)	406–709

TS: total solids; FSS: fixed suspended solids; VSS: volatile suspended solids, TDS: total dissolved solids. The other parameters are described in Table 9.2.

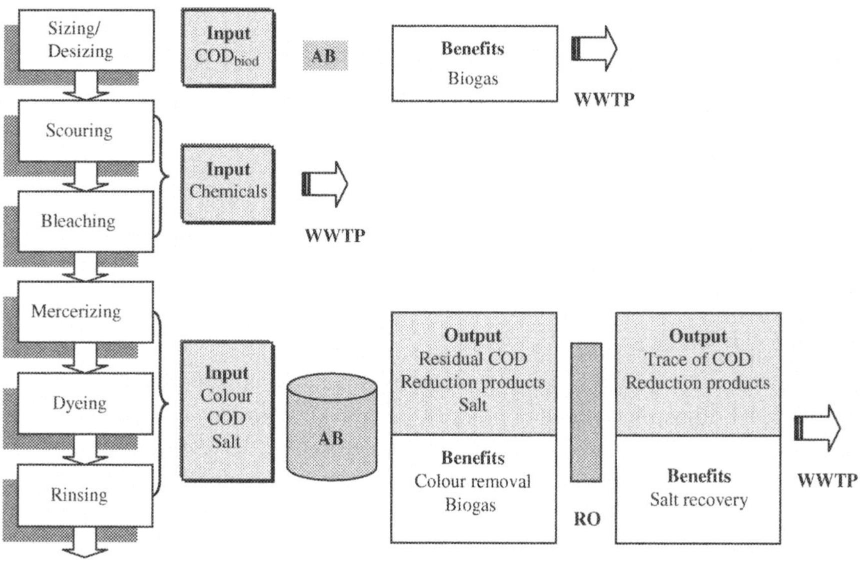

Figure 9.3 Schematic of the possible improvements in the wastewater treatment located in Brazil with the goal of product recovery. AB: anaerobic bioreactor; RO: reverse osmosis.

concern of the case study is the high COD level and electrical conductivity of the wastewater. The first strategy to be considered is to increase the recovery efficiency in the mercerising step, in order to decrease the discharge of salt (NaOH) in the wastewater. Treating separately the effluent from the mercerising, dyeing and rinsing steps, would improved the removal of salts and colour (Figure 9.3). The colour removal could be accomplished in an anaerobic bioreactor, and the salt that ends up in the wastewater removed by using a physical treatment such as reverse osmosis. However, the removal of colour leads to the production of reductive products such as aromatic amines, which have to be removed afterwards in the main WWTP, the latter working as a polishing step. In some textile factories, the dyeing process is done in semi-continuous systems, which discharge a heavily coloured and salty wastewater in a very short time. For those cases, the inclusion of an equalisation tank is convenient, from which the wastewater will be pumped constantly besides providing the possibility for making pH adjustments. In situations where the industrial flow is not so high, another approach is mixing the industrial wastewater with the domestic wastewater in order to dilute the salt content. However, this strategy is not suitable in the current case study because the volume of industrial wastewater is by far higher than the volume of domestic sewage.

For the other steps, for instance the wastewater generated in sizing/desizing processes, could be treated separately by using an anaerobic bioreactor, which is capable of coping with high concentrations of organics (Figure 9.3). The main benefits from the anaerobic treatment are the biogas, which has a calorific value, as well as a low sludge production and a stabilized sludge. Afterwards, the effluent would be sent to the main WWTP. An aerobic process such as an activated sludge system could be an interesting option followed by methods such as ozone and physical–chemical treatment. For the wastewater generated in the scouring/bleaching steps, they could be sent directly to the main WWTP (Figure 9.3). However, the chemicals used in these steps will play an important role, as they can be extremely toxic to microorganisms. Thus, the selection of less aggressive, and when possible biodegradable agents, is one of the most important recommendation here.

9.5.2 Case study 2

9.5.2.1 Description

The textile factory described in this case study is located in The Netherlands and produces fabrics and technical textiles for clothing. A wide range of processes take place in the factory, including weaving, desizing and scouring, bleaching, mercerising, dyeing and finishing.

9.5.2.2 WWTP set-up

Part of the factory's wastewater, consisting of a mixture of effluents from dyeing, scouring, bleaching and desizing processes, is treated in an anaerobic–aerobic treatment plant to reduce its COD load and colour. In the plant, approximately 360 m^3 of combined textile wastewater (with a COD concentration average of 4500 mg/L) is treated per day. The dyes used in the factory are mainly disperse, vat and reactive dyes, of which 70% are insoluble. Additionally, the wastewater contains concentrations of sulphate varying from 200 to 1000 mg/L. The first part of the treatment plant is an equalisation/pre-acidification tank, from which wastewater is pumped into a mixing tank for pH and temperature control (Figure 9.4).

After the necessary adjustments the wastewater is transferred to a mesophilic anaerobic Internal Circulation (IC®) reactor (see Figure 9.5), resulting in a COD removal efficiency of 35–55% and a decolourisation of 80–95%. COD removal was affected by the presence of sulphate; at high sulphate concentrations less COD was removed. Post-treatment of the anaerobic effluent takes place in an aerobic

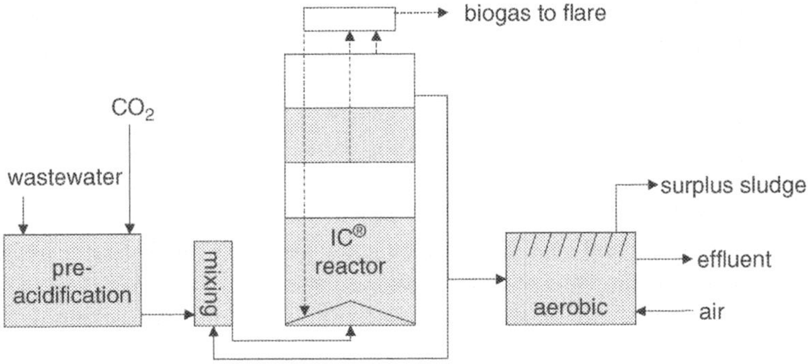

Figure 9.4 Scheme of the wastewater treatment plant located in The Netherlands.

Figure 9.5 Photograph of the anaerobic reactor at the textile factory located in The Netherlands.

basin equipped with plate settlers, after which a total COD reduction of 80–90% is achieved. During the aerobic treatment step a large part of the anaerobically non-degradable dyes and dye degradation products are either degraded or adsorbed to the sludge. Finally, the treated effluent is discharged into the sewer, which takes the wastewater to the municipal WWTP. The remaining factory wastewater that is not treated in the previously described treatment plant is aerated and neutralised in a separate basin.

Many dyes and their degradation products are toxic, as well as other chemicals used in the textile industry, and therefore the toxicity of the wastewater before and after biological treatment was investigated. Using the Microtox® method, it was shown that both the pre-acidified wastewater and the anaerobic effluent were toxic, but that after aerobic treatment the toxicity had disappeared. When the anaerobic reactor was bypassed, the aerobic effluent was found to be toxic, indicating a main contribution of anaerobic treatment in detoxifying the textile effluent.

9.5.2.3 Possible improvements with the aim of water reuse

The wastewater can be treated up to a level that the water itself can be recovered as process water. However, this requires a lot of effort, as the standards for process water quality are very high. Regarding the COD and colour removal reached by the treatment plant of the factory studied, additional polishing would still be necessary to reach a good enough water quality for reuse. Possible post-treatment options could be by using filtration, adsorption, or a physical–chemical method. The option of using a post-treatment consisting of a sand filter and a Membrane Assisted Affinity Separation (MAAS) system was investigated on pilot scale in the factory discussed here. The Dutch research institute TNO has developed the MAAS technology for the removal of organic and inorganic matter, which combines adsorption with membrane separation. The conducted pilot tests showed a COD reduction of over 90%. A reusability assessment showed that the recovered water would still need additional treatment to lower its hardness, before it could serve as a substitute for the process water used in the factory. The hardness of process water is very important in the textile industry, as changes in hardness influence the reproducibility of the resulting colour of dyed products. Process recipes are based on a certain constant process water quality, with the objective of always obtaining the same product quality. Therefore, changing to different process water would involve changes in all recipes of the factories as well. In theory this is possible, but only when the applied treatment can guarantee that the reclaimed water always has exactly the same quality.

9.5.2.4 Possible improvements with the aim of products recovery

In the textile industry, products recovery can mainly take place directly after the effluents have been generated, before mixing of the different process streams. We do not know the level of products recovery within the factory at this moment, so that we cannot suggest real improvements. However, assuming that no products recovery is taking place, some possible measures can be indicated for the processes carried out in this factory. Depending on the used size, 80–85% of it can be recovered from the desizing effluent via ultrafiltration and reused in the sizing process, leading to a large decrease in wastewater COD (IPPC 2003). From the mercerisation effluent, alkali can be recovered by evaporation and lye purification, reducing the alkaline load of the wastewater and subsequently the use of acids for pH adjustments in the treatment plant (IPPC 2003). Due to the complex nature of textile factory effluents, products recovery from mixed wastewater is practically impossible.

9.5.2.5 Acknowledgements

Information for the preparation of Sections 9.5.2.1–9.5.2.3 of this case study was kindly provided by Mrs Carla Frijters of Paques B.V.

REFERENCES

Alaton, I.A., Balcioglu, I.A. and Bahnemann, D.W. (2002) Advanced oxidation of a reactive dyebath effluent: comparison of O_3, H_2O_2/UV-C and TiO_2/UV-A processes. *Wat. Res.* **36**, 1143–1154.

Bisschops, I.A.E. and Spanjers, H. (2003) Literature review on textile wastewater characterisation. *Environ. Technol.* **24**, 1399–1411.

Bonomo, L., Rozzi, A. and Malpei, F. (1997) Feasibility analysis of print pastes co-disposal in anaerobic sludge digesters. *Wat. Sci. Technol.* **36**, 181–188.

BTTG (1999) *Report 3: Textile Processing Techniques.* British Textile Technology Group.

Cervantes, F.J., Van der Zee, F.P., Lettinga, G. and Field, J.A. (2001) Enhanced decolourisation of Acid Orange 7 in a continuous UASB reactor with quinones as redox mediators. *Wat. Sci. Technol.* **44**, 123–128.

Chivukula, M. and Renganathan, V. (1995) Phenolic azo dye oxidation by laccase from *Pyricularia oryzae. Appl. Environ. Microbiol.* **61**, 4374–4377.

Christie, R. (2001) *Colour Chemistry.* The Royal Society of Chemistry, Cambridge, United Kingdom.

Claus, H., Faber, G. and Koenig, H. (2002) Redox-mediated decolorization of synthetic dyes by fungal laccases. *Appl. Microbiol. Biotechnol.* **59**, 672–678.

Correia, V.M., Stephenson, T. and Judd, S.J. (1994) Characterisation of textile wastewaters: a review. *Environ. Technol.* **15**, 917–929.

Delée, W., O'Neill, C., Hawkes, F.R. and Pinheiro, H.M. (1998) Anaerobic treatment of textile effluents: a review. *J. Chem. Technol. Biotechnol.* **73**, 323–335.

DEPA (1997) *Environmental Assessment of Textiles.* Danish Environmental Protection Agency, 220 pp.

Dos Santos, A.B. (2005) *Reductive Decolourisation of Dyes by Thermophilic Anaerobic Granular Sludge.* Ph.D. Thesis, Wageningen University, Wageningen, The Netherlands.

Dos Santos, A.B., Bisschops, I.A.E., Cervantes, F.J. and Van Lier, J.B. (2004) Effect of different redox mediators during thermophilic azo dye reduction by anaerobic granular sludge and comparative study between mesophilic (30°C) and thermophilic (55°C) treatments for decolourisation of textile wastewaters. *Chemosphere* **55**, 1149–1157.

Dos Santos, A.B., Traverse, J., Cervantes, F.J. and Van Lier, J.B. (2005) Enhancing the electron transfer capacity and subsequent colour removal in bioreactors by applying thermophilic anaerobic treatment and redox mediators. *Biotechnol. Bioeng.* **89**, 42–52.

Duran, N., Rosa, M.A., D, A.A. and Gianfreda, L. (2002) Applications of laccases and tyrosinases (phenoloxidases) immobilized on different supports: a review. *Enz. Microbiol. Technol.* **31**, 907–931.

EPA (1996) *Manual Best Management Practices for Pollution Prevention in the Textile Industry.* Cincinnati: National Risk Management Research Laboratory, Center for Environmental Research Information.

EPA (1997). *Profile of the Textile Industry.* Washington, USA, Environmental Protection Agency.

Feitkenhauer, H. (2003) Anaerobic digestion of desizing wastewater: influence of pretreatment and anionic surfactant on degradation and intermediate accumulation. *Enz. Microbiol. Technol.* **33**, 250–258.

Field, J.A., Stams, A.J.M., Kato, M. and Schraa, G. (1995) Enhanced biodegradation of aromatic pollutants in cocultures of anaerobic and aerobic bacterial consortia. *Antonie van Leeuwenhoek* **67**, 47–77.

Gähr, F., Hermanutz, F. and Oppermann, W. (1994) Ozonation – an important technique to comply with new German laws for textile wastewater treatment. *Wat. Sci. Technol.* **30**, 255–263.

Galindo, C., Jacques, P. and Kalt, A. (2000) Photodegradation of the aminoazobenzene acid orange 52 by three advanced oxidation processes: UV/H_2O_2, UV/TiO_2 and VIS/TiO_2; Comparative mechanistic and kinetic investigations. *J. Photochem. Photobiol. A: Chem.* **130**, 35–47.

Hao, O.J., Kim, H. and Chang, P.C. (2000) Decolorization of wastewater. *Crit. Rev. Env. Sci. Tec.* **30**, 449–505.

Heinfling, A., Bergbauer, M. and Szewzyk, U. (1997) Biodegradation of azo and phthalocyanine dyes by *Trametes versicolor* and *Bjerkandera adusta. Appl. Microbiol. Biotechnol.* **48**, 261–266.

IPPC (2003) Integrated Pollution Prevention and Control in Textile Industry.

Kabdasli, I., Gürel, M. and Tünay, O. (2000) Characterisation and treatment of textile printing wastewaters. *Environ. Technol.* **21**, 1147–1155.

Karcher, S., Kornmuller, A. and Jekel, M. (2001) Cucurbituril for water treatment. Part I: Solubility of cucurbituril and sorption of reactive dyes. *Wat. Res.* **35**, 3309–3316.

Knackmuss, H.J. (1996) Basic knowledge and perspectives of bioelimination of xenobiotic compounds. *J. Biotechnol.* **51**, 287–295.

LeAF (2002) Literature review on textile wastewater characterisation, EU Project Towef. Wageningen: Lettinga Associates Foundation.

Malpei, F., Andreoni, V., Daffonchio, D. and Rozzi, A. (1998) Anaerobic digestion of print pastes: a preliminary screening of inhibition by dyes and biodegradability of thickeners. *Biores. Technol.* **63**, 49–56.

Mattioli, D., Malpei, F., Bortone, G. and Rozzi, A. 2002. Water minimization and reuse in textile industry. *Water Recycling and Resource Recovery in Industry: Analysis, Technologies and Implementation.* IWA Publishing, Cornwall, UK p. 677.

Metcalf and Eddy (2003) *Wastewater Engineering: Treatment and Reuse,* 4th edn. McGraw-Hill, New York, USA.

Moraes, S.G., Freire, R.S. and Duran, N. (2000) Degradation and toxicity reduction of textile effluent by combined photocatalytic and ozonation processes. *Chemosphere* **40**, 369–373.

O'Neill, C., Hawkes, F.R., Hawkes, D.W., Esteves, S. and Wilcox, S.J. (2000) Anaerobic–aerobic biotreatment of simulated textile effluent containing varied ratios of starch and azo dye. *Wat. Res.* **34**, 2355–2361.

Opwis, K., Knittel, D., Kele, A. and Schollmeyer, E. (1999) Enzymatic recycling of starch-containing desizing liquors. *Starch/Stärke* **51**, 348–353.

Orhon, D., Babuna, F.G. and Insel, G. (2001) Characterization and modelling of denim-processing wastewaters for activated sludge. *J. Chem. Technol. Biotechnol.* **76**, 919–931.

Porter, J.J. 1997. *Filtration and Recovery of Dyes from Textile Wastewater, Treatment of Wastewaters from Textile Processing.* Schriftenreihe Biologische Abwasserreinigung, Berlin, Germany.

Rigoni Stern, S., Szpyrkowicz, L. and Zilio Grandi, F. (1996) Treatment of silk and lycra printing wastewaters with the objective of water reuse. *Wat. Sci. Technol.* **33**, 95–104.

Robinson, T., McMullan, G., Marchant, R. and Nigam, P. (2001) Remediation of dyes in textile effluent: a critical review on current treatment technologies with a proposed alternative. *Biores. Technol.* **77**, 247–255.

Rucker, J.W. and Smith, C.B. (1988). *Troubleshooting in Preparation – A systematic approach.* Raleigh: Department of Textile Engineering, Chemistry and Science College of Textiles, North Carolina State University.

So, C.M., Cheng, M.Y., Yu, J.C. and Wong, P.K. (2002) Degradation of azo dye Procion Red MX-5B by photocatalytic oxidation. *Chemosphere* **46**, 905–912.

Stolz, A. (2001) Basic and applied aspects in the microbial degradation of azo dyes. *Appl. Microbiol. Biotechnol.* **56**, 69–80.

Tomasino, C. (1992). *Chemistry and Technology of Fabric Preparation and Finishing.* Raleigh: Department of Textile Engineering, Chemistry and Science College of Textiles, North Carolina State University.

Van der Zee, F. (2002) *Anaerobic Azo Dye Reduction.* PhD Thesis, Wageningen University, Wageningen, The Netherlands.

Van der Zee, F.P., Bouwman, R.H.M., Strik, D.P.B.T.B., Lettinga, G. and Field, J.A. (2001a) Application of redox mediators to accelerate the transformation of reactive azo dyes in anaerobic bioreactors. *Biotechnol. Bioeng.* **75**, 691–701.

Van der Zee, F.P., Lettinga, G. and Field, J.A. (2001b) Azo dye decolourisation by anaerobic granular sludge. *Chemosphere* **44**, 1169–1176.

Vandevivere, P.C., Bianchi, R. and Verstraete, W. (1998) Treatment and reuse of wastewater from the textile wet-processing industry: review of emerging technologies. *J. Chem. Technol. Biotechnol.* **72**, 289–302.

Wenzel, H., Sojka-Ledakowicz, J., Knudsen, H.-H., Machnowski, V., Koprowska, J., Grzywacz, K., Hansen, J., Birch, H., Pedersen, B.M. and Jozwik, A. (1999) *Cleaner Technology Transfer to the Polish Textile Industry.* Danish Environmental Protection Agency.

WTO (2004) *Background Statistical Information with Respect to Trade in Textiles and Clothing.* World Trade Organization. 107 pp.

Yoo, E.S. (2002) Kinetics of chemical decolorization of the azo dye C.I. Reactive Orange 96 by sulfide. *Chemosphere* **47**, 925–931.

10

Heavy metal removal with biogenic sulphide: advancing to full-scale

J. Weijma, A. Veeken, H. Dijkman, J. Huisman and P. Lens

10.1 INTRODUCTION

The capacity of sulphate-reducing bacteria (SRB) to remove or recover dissolved heavy metals from wastewaters has been long recognised. Heavy metals, such as copper, zinc, cadmium, lead, nickel and iron, precipitate with sulphide, the product of sulphate reduction, to form insoluble metal sulphides, thereby concentrating the metals into a separable and sometimes valuable form. Metal removal and recovery based on biological sulphate reduction has matured into a full-scale technology, demonstrating that it is a "working biotechnology". In the past decade it was also shown that SRB can reduce specific heavy metals, metalloids and radionuclides to insoluble forms, a potential that has not been exploited in practice yet.

In their free form, heavy metals at ppm concentrations are toxic for nearly all forms of life. Accordingly, discharge of heavy metals into the environment can have devastating effects on aquatic and terrestrial ecosystems (Johnson 2000). Therefore,

 Advanced Biological Treatment Processes for Industrial Wastewaters edited by F.J. Cervantes, S.G. Pavlostathis and A.C. van Haandel. ISBN: 1843391147. Published by IWA Publishing, London, UK.

discharge limits for dissolved heavy metals are generally well below 1 ppm in many industrialised countries. Well-known examples of heavy metal-contaminated wastewater are acid mine drainage and wastewaters generated by the metallurgical, electronics and pharmaceutical industries (Brooks 1991). Also industrial activities in the past left a legacy of metal-contaminated ground and surface waters.

Methods for metals removal include precipitation, coagulation–flocculation, ion exchange, solvent extraction, complexation, adsorption, filtration and membrane processes (Brooks 1991). Precipitation with hydroxide is the most widely method used, mainly for its simplicity of process control and low cost of chemicals (lime, limestone, caustic soda). However, hydroxide precipitation is still costly due to high investments necessary for dewatering and disposal of the produced bulky sludge (Peters *et al.* 1985).

As legislation for metal discharge becomes more stringent and metal resources become increasingly scarce, heavy metal removal technologies are needed with a better performance than those currently adopted. Technology based on metal precipitation with sulphide is a strong candidate for this because it offers some fundamental advantages over hydroxide precipitation (Peters *et al.* 1985):

(1) effluent concentrations are orders of magnitude lower: μg/L vs. mg/L;
(2) the interference of chelating agents in the wastewater is less problematic;
(3) selective metal removal gives better opportunities for metal reuse;
(4) metal sulphide sludges settle, thicken and dewater better than hydroxide sludges;
(5) existing smelters can process sulphide precipitates, enabling metal recovery.

Earlier objections against the use of sulphide due to its toxic, malodorous and corrosive nature have been overcome by adequate safety measures and the use of modern corrosion-resistant construction materials (i.e. plastic). Sulphide chemicals, such as Na_2S, NaHS, CaS, FeS and H_2S, are available but are relatively expensive. Moreover, the hazards that accompany transport, handling and storage of the chemical sulphides in bulk volumes lead to additional costs for safety measures. These drawbacks are overcome by on-site production of biogenic sulphide. Biogenic sulphide can be generated from various sulphur sources such as sulphate, sulphite, thiosulphate and elemental sulphur. With sulphate, sulphide generation followed by metal sulphide precipitation proceeds according to:

$$8e^- + SO_4^{2-} + 9H^+ \Rightarrow HS^- + 4H_2O \tag{10.1}$$

$$HS^- + Me^{2+} \Rightarrow MeS + H^+ \tag{10.2}$$

$$4H_2 + SO_4^{2-} + Me^{2+} \Rightarrow MeS + 4H_2O \tag{10.3}$$

The metal sulphides formed are highly insoluble at neutral pH, resulting in dissolved metal levels of 1 ppb or below.

10.2 BACKGROUND

10.2.1 Microbiology

Although many naturally occurring bacteria generate sulphide from various sulphur sources, only two bacterial groups do so at a sufficient rate for use in high-rate processes: (i) the SRB, which conserve energy by electron transfer from hydrogen and organic substrates to sulphur oxyanions like sulphate, sulphite, thiosulphate and (ii) the sulphur-reducing bacteria which transfer electrons to elemental sulphur (Widdel and Hansen 1992). Both bacterial groups form a heterogeneous group, with a wide range of physiological characteristics, metabolising numerous organic compounds, growing at −5°C to 80°C, and at salinities up to 18% NaCl.

The use of SRB in environmental biotechnology for sulphate removal is already well established (Oude Elferink *et al.* 1994; Hulshoff Pol *et al.* 1998). Severe inhibition of the bacteria already occurs at free metal concentrations of only a few ppm (Bharathi *et al.* 1990; Jalali and Baldwin 2000; Sani *et al.* 2001a, b), which is a critical consideration for process design of heavy metal removal with SRB.

10.2.2 Precipitation kinetics

Success of the precipitation process not only depends on removal of metal ions from the soluble phase but also on the separation of the solid phase (i.e. metal sulphide precipitate) from the liquid phase. Therefore, solid–liquid separation processes, such as sedimentation or filtration, are of key importance. The settling and dewatering characteristics of metal precipitates are directly related to the morphology, density and particle size distribution of the precipitate. Little is known about factors that affect the morphology of metal sulphide precipitates, whereas density mostly has a characteristic value. The particle size distribution is determined by the kinetics of precipitation, that is, the competition between nucleation, crystal growth and agglomeration (Zauner and Jones 2000). Nucleation produces only very small particles that are difficult to separate from the liquid phase, whereas crystal growth and agglomeration of crystals result in larger particles. For precipitates with a very low solubility such as heavy metal sulphides, very small particles normally prevail as (i) the saturation level cannot be controlled at low levels (Mersmann 1999) and (ii) local supersaturation at the feed points due to micromixing limitation cannot be prevented (Zauner and Jones 2000). Precipitation kinetics are poorly studied for metal sulphides, and accordingly, particle sizes are not predictable. However, by controlling the sulphide level and pH in the precipitation vessel, particle sizes can be manipulated (see Box 10.1).

10.3 ENGINEERING METAL PRECIPITATION WITH BIOGENIC SULPHIDE

The feasibility of sulphate reduction technologies for heavy metal removal has already been demonstrated in numerous bench-scale and some pilot-scale studies (Table 10.1), which can be categorised in low-rate and high-rate processes. In

Box 10.1 Control and selectivity in metal sulphide precipitation

Veeken *et al.* (2003) developed a new process for heavy metal precipitation in which the sulphide addition is controlled by the combination of a sulphide-selective electrode (which measures the sulphide activity S^{2-}) and a pH electrode. The control strategy combines the simplicity and flexibility of hydroxide precipitation and low effluent concentration of sulphide precipitation. Cd, Cu, Ni, Pb and Zn are removed to levels <0.05 mg/L at pH 6.0 while maintaining the total sulphide concentration <0.02 mg/L avoiding the corrosive and odorous character of sulphides. The sulphide level at which heavy metals precipitate is unique for each heavy metal and is directly related to the solubility product of the corresponding heavy metal sulphide, presented by $K_{SP} = (M^{2+})(S^{2-})$. Thus, the (S^{2-}) level represents a control parameter to selectively precipitate heavy metals. This is shown in the figure below for a mixture of Cu and Zn (nitrate) in a batch and completely mixed stirred tank reactor (CSTR) precipitation process and pH control at 6.0. In analogy to pH measurements, the term pS is introduced where $pS = -\log(S^{2-})$. For the batch titration, the metals are precipitated consecutively, first the metal with the lower solubility (Cu) and then Zn. Both metals showed two constant pS levels during precipitation, showing that the levels of pS are unique. The selectivity of precipitation is 100% for Cu and Zn. Precipitation in CSTR also showed 100% selective precipitation of Cu and Zn in the S^{2-} range 10^{-37}–10^{-30}.

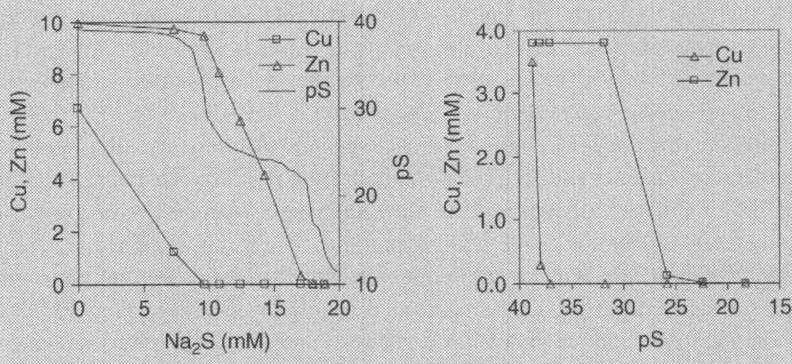

Cu and Zn precipitation in batch

high-rate processes the metal removal and recovery takes place in engineered bioreactor systems. Studies on low-rate metal removal in low-tech systems mostly deal with metal decontamination of acid mine or rock drainage polluted with a “cocktail” of heavy metals in for instance wetlands (Gazea *et al.* 1996). The low purity and quantities, and the poor accessibility make recovery of metals in a reusable form hardly possible in these processes.

This chapter focuses on engineered high-rate metal removal systems, for which investment and operational costs generally are higher than low-rate systems.

Table 10.1 Selected bench-scale and pilot studies on sulphate reduction technology for metal removal.

Treatment system	Target metal	Electron donor	Reference
Low rate			
Extractive membrane reactor	Zn	Ethanol	Chuichulcherm *et al.* (2001)
Solid substrate reactor	Al, Cd, Fe, Mn, Ni, Zn	Compost	Dvorak *et al.* (1992)
	As, Cu, Cd, Fe, Mn, Zn	Cow manure, sawdust, cheese whey	Drury (1999)
	Cu, Fe, Mn, Zn	Wood chips, compost, sludge, soil	Chang *et al.* (2000)
	Fe	Methanol	Tsukamoto and Miller (1999)
Permeable reactive barrier	Fe, Ni	Compost, leaf mulch, wood chips	Benner *et al.* (1999)
High rate			
Bioreactor	Zn	Hydrogen	Copini *et al.* (2000)
	Al, Fe, Ni, Zn	Methanol	Glombitza (2001)
Bioreactor and precipitator	Cu, Zn	Ethanol	Hammack and Dijkman (1999)
	Al, Co, Cu, Fe, Mn, Ni, Zn	Molasses	
	Al, Cu, Fe, Mn, Zn	Lactate	Hammack *et al.* (1994)
Upflow sludge bed reactor	Ni	Sewage	De Lima *et al.* (2001)

Therefore, application of high-rate technology is particularly competitive when legislation for metal discharge is stringent or when the metal-containing precipitate has considerable economic value. In some cases, the revenues of the recovered metals may offset the investment and operational costs of the plant. The main aspects of the design of high-rate processes are discussed below.

10.3.1 Selection of electron donor

Reduction of sulphur compounds consumes electron donors (Equation (10.1)). Metal-contaminated wastewaters typically contain very few electron donors suitable for SRB and so electron donors need to be added. The selection of electron donor is crucial as it greatly affects the operational costs and process performance. The cheapest option seemingly represents organic waste materials such as those from the food processing industry (molasses, whey, etc.), activated sludge, compost, manure, leaves, wood chips, plant material, algal biomass, etc. (Dvorak *et al.* 1992; Bechard *et al.* 1994; Christensen *et al.* 1996; Waybrant *et al.* 1998; Prasad *et al.* 1999; Chang *et al.* 2000; Ludwig *et al.* 2000; De Lima *et al.* 2001; Russell *et al.* 2001).

However, organic waste materials usually are available in only relatively small quantities, limiting their use to processes with relatively low sulphate loads. Moreover, organic waste materials usually are of highly complex and variable composition, thereby hampering process predictability and stability.

For high-rate metal precipitation, selected electron donors must be relatively pure, readily biodegradable and cheap. Chemicals that meet these conditions are methanol, ethanol and hydrogen. Hydrogen is produced (if not available on-site) by cracking methanol or by means of a natural gas reformer. In both cases a mixture of H_2, CO_2 and some (1–5%) of CO is produced. For low sulphate loads (<100 kg S/h) at mesophilic temperatures (20–40°C), ethanol is most economical. Although the chemical costs are somewhat higher compared with hydrogen, the investment cost is lower since no natural gas reformer is needed and less safety measures are required. For higher sulphate loads at mesophilic temperatures (25–40°C), hydrogen gas is the best option. Under thermophilic conditions (55–65°C) the use of hydrogen is less efficient due to the formation of the by-products methane and acetate (Van Houten *et al.* 1997). At these high temperatures, methanol is the most efficient electron donor because methane formation is negligible (Weijma *et al.* 2000).

10.3.2 Selection of sulphur source

The amount of electron donor required for reduction to sulphide varies among the sulphur species: from eight electrons in sulphate to only two in elemental sulphur.

Figure 10.1 Full-scale (500 m^3) sulphate-reducing bioreactor for zinc and sulphate removal at Pasminco Budel Zink (zinc refinery) in the Netherlands.

Therefore, use of elemental sulphur minimises the amount of electron donor per produced sulphide. The solid form of elemental sulphur also lowers transport costs. However, sulphate is already present as sulphur source in many metal-containing waste streams, especially in the metallurgical and mining industries. It may be more favourable to use sulphate as a sulphur source, especially when sulphate removal is also required. In other situations, there might be sulphur-polluted waste streams (such as SO_2) available on-site that require treatment anyway. In the process flow sheets for biological metal removal presented here some excess sulphide may be produced, which can be oxidised by chemical or biological methods to elemental sulphur (Janssen *et al.* 1997). The produced elemental sulphur may be reintroduced into the process as a sulphur source.

10.3.3 Process flow sheets

Three flow sheets (Figure 10.2) exist to combine biological sulphide production with heavy metal precipitation in a continuous process, which are further discussed below.

10.3.3.1 Single-stage sulphide generation and precipitation

In the single-stage flow sheet, biological sulphate reduction and metal sulphide precipitation proceed simultaneously in one reactor (Figure 10.2A). Environmental conditions should be kept optimal for biological activity, that is, a circumneutral pH and a temperature of 20–40°C (optimally 35°C) should be maintained. Furthermore, in order to prevent accumulation of inhibitory dissolved metal, and to guarantee the proper redox conditions for the SRB, an excess sulphide concentration of 200–400 mg S/L must be maintained in the reactor.

Unfortunately, the metal sulphide crystallisation may not proceed optimally under these conditions. Most of the common heavy metals, such as iron, cobalt, nickel, copper, zinc and lead, will precipitate below discharge standards at such conditions, but the relatively high sulphide level results in formation of very small (colloidal) precipitates, resulting in poor settling and dewatering characteristics of the metal sulphide sludge (Veeken *et al.* 2003). Also organic compounds excreted by the bacteria or bacteria themselves might interfere with crystallisation, phenomena that still are poorly studied for most metal sulphides.

The advantage of the single-stage process is that it represents a relatively simple flow sheet. However, it is less suitable when metal concentrations are low ($<\sim$50 mg/L) because the process is rapidly hydraulically limited, resulting in low metal removal rates per reactor volume. The single-stage flow sheet has already been applied for 10 years on full-scale for sulphate and zinc removal from polluted groundwater at the Pasminco Budel Zink refinery in the Netherlands (Scheeren *et al.* 1993). The process is also applicable to treat wastewater with a high sulphuric acid content (e.g. wash tower acid at metal ore refineries), provided that a basic heavy metal precipitate (e.g. metal oxide/hydroxide) is available on-site, which is often the case at metal refineries. The sulphuric acid is neutralised with the basic heavy metals and the resulting metal sulphate solution is treated in a biological sulphate-reducing reactor, yielding a pure metal sulphide that can be returned to the refinery smelter.

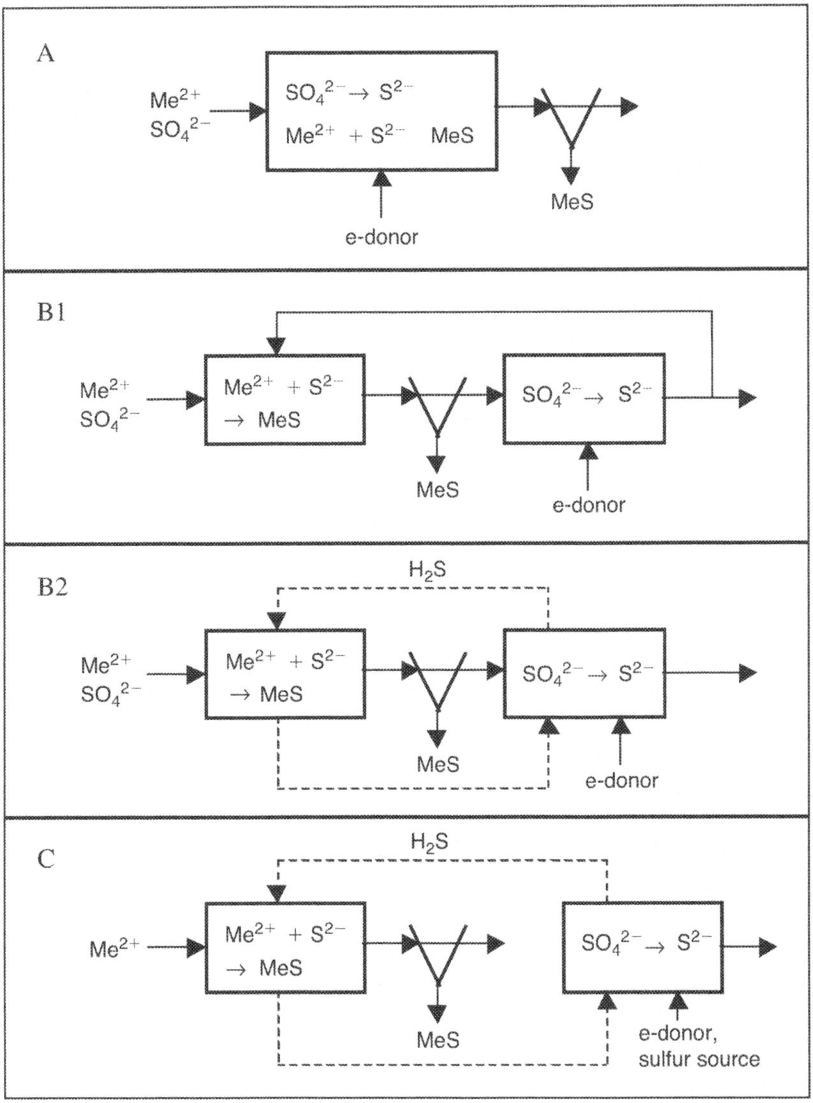

Figure 10.2 Flow sheets for metal precipitation with biogenic sulphide. (A) Single-stage sulphide generation and precipitation. (B) Sulphide generation and precipitation in series, with sulphide transfer via a liquid (B1) or gas (B2) recycle. (C) Separated sulphide generation and metal sulphide precipitation. Continuous lines represent liquid flows, dashed lines represent gas flows.

Such a process is presently in operation on full-scale (500 m^3) at the Pasminco Budel Zink refinery in the Netherlands (Copini 2000; Figure 10.1). The single-stage process is also applicable for converting insoluble heavy metal salts to the corresponding metal sulphides. One example is the conversion of lead waste from car batteries, consisting mainly of $PbSO_4$, to PbS (Olper 2000; Weijma *et al.* 2002).

10.3.3.2 Sulphide generation and precipitation in series

In two-stage processes, biological sulphate reduction and metal sulphide precipitation proceed separately. Transfer of sulphide from the bioreactor to the precipitator can be accomplished with a liquid or gas recycle (Figure 10.2B), or with both. The metal sulphide sludge produced might be retained in the precipitation vessel (e.g. by a filter) or it might be separated from the effluent, for example, by means of sedimentation in a clarifier, as shown in Figure 10.2B. In the bioreactor the sulphur source is then reduced to sulphide, which is recycled to the precipitator.

This method allows independent control of environmental conditions in the precipitator (especially pH and sulphide level) and the sulphate-reducing bioreactor. Metals are precipitated selectively by installing precipitators and clarifiers in series, controlling the pH or sulphide level independently in each precipitator. Feeding only a fraction of the flow from the precipitator to the bioreactor limits the hydraulic loading of the bioreactor, making treatment of dilute waste streams more economical. The two-stage process has been studied in bench-scale (Hammack *et al.* 1994a, b) and pilot-scale (Boonstra *et al.* 1999; Glombitza 2001), mostly aiming to selectively precipitate metals from "cocktails" but it appears that this technology has not yet reached full-scale applications.

10.3.3.3 Separated sulphide generation and metal sulphide precipitation

In this flow sheet, the liquid circuits for precipitation and sulphate reduction are entirely separated (Figure 10.2C). H_2S is stripped from the biological solution with an oxygen-free carrier gas and fed to the precipitator. Possible undesired interferences of compounds present in the biological solution on metal sulphide precipitation, such as co-precipitation of metal phosphates and carbonates, are avoided. In addition, there is no direct contact between the bacteria and possible toxic or inhibiting compounds present in the metal-bearing wastewater. Selective metal precipitation can be established by using precipitators in series, controlling the conditions in each precipitator for optimal precipitation of a specific metal. Clarifiers between the precipitators can be used to obtain relatively pure metal sulphides. This process is currently in operation at Kovohute Pribram (Czech Republic) to treat wastewater containing sulphate, lead, zinc, tin, and high concentrations of arsenic and antimony from a slag dump leachate (Dijkman *et al.* 1999).

10.4 DISSIMILATORY METAL REDUCTION

The technologies presented thus far relate to metals that can be removed as insoluble metal sulphides (e.g. zinc, copper, led and cadmium). Several heavy metals, metalloids and even radionuclides, which do not react with sulphide, can still be removed from waste streams by SRB-based technologies via a direct reduction (Table 10.2). Although this capacity is also found in other groups of bacteria (Lovley 1995), SRB are among the more rapid metal-reducing species (Lloyd *et al.* 1999).

Given that most of these compounds are present in only very low concentrations (up to 1 mg/L), special reactor designs are needed to cope with the hydraulic

Table 10.2 Microbial metal reduction.

Reaction	Microorganism	References
$Tc^{7+} \rightarrow Tc^{4+}$	*Desulfovibrio desulfuricans*	Lloyd *et al.* (1999)
$U^{6+} \rightarrow U^{4+}$	*Geobacter metallireducens*	Tucker *et al.* (1998)
	Shewanella putrefaciens	Abdelouas *et al.* (1998)
	Desulfovibrio spp.	Lovley and Phillips (1994)
$Se^{6+} \rightarrow Se^{0}$	*Clostridium* spp.	Tomei *et al.* (1995)
	Flavobacterium spp.	
	Thauera selenatis	
	Desulfovibrio desulfuricans	
$Cr^{6+} \rightarrow Cr^{3+}$	Many bacteria	Smith and Gadd (2000)
	Enterobacter cloacae HO1	Tucker *et al.* (1998)
	Desulfovibrio spp.	Fude *et al.* (1994)
		Lovley and Phillips (1994)
$Mo^{6+} \rightarrow Mo^{5+}$	*Desulfovibrio desulfuricans*	Tucker *et al.* (1997)
	Desulfovibrio vulgaris	
	Pseudomonas sp.	
	Micrococcus sp.	
$As^{5+} \rightarrow As^{3+}$	*Geospirillum arsenophilus*	Macy *et al.* (2000)
	Chrysiogenes arsenatis	Newman *et al.* (1997)
	Desulfotomaculum auripigmentum	
$V^{5+} \rightarrow V^{3+}$	Mixed sulphate-reducing culture	Uhrie *et al.* (1996)
$Hg^{2+} \rightarrow Hg^{0}$	Many bacteria	Von Canstein *et al.* (1999)
	Pseudomonas putida	
$Pd^{2+} \rightarrow Pd^{0}$	*Desulfovibrio desulfuricans*	Yong *et al.* (2002)
$Au^{3+} \rightarrow Au^{0}$	*Geobacter metallireducens*	Lovley (1995)
	B. subtilis	
	Desulfovibrio sp.	
$Ag^{+} \rightarrow Ag^{0}$	*Geobacter metallireducens*	Lovley (1995)

limitations of bioreactors treating these dilute wastewaters. For instance, Tucker *et al.* (1998) immobilised SRB cells in acrylamide gel beads to guarantee cell retention. Another option is to immobilise the bacteria on sand grains in a continuous sand filter. This concept has been successfully tested on pilot-scale for removal of dissolved uranium (Weijma and Schouten 2005). In this study, dissolved uranium was biologically reduced under sulphate-reducing conditions to insoluble uraninite (UO_2). The uranium precipitate was simultaneously separated from the water in the sand filter. When such pilot-studies transit into full-scale works, this will place SRB as one of the most important microbial species in metal bioremediation.

10.5 OUTLOOK

The perception that metal-bearing waste "always" is a low-value material for which it is not economical to invest in highly engineered technology is outdated now. Despite the fact that several sulphide-based heavy metal removal/recovery technologies can be considered as proven technology, its penetration into the market still proceeds slowly. One factor that might have a function in this still poor acceptance is the notorious, "bad-smell" reputation of sulphide. Also the fact that industries are not familiar with biological technologies adds to the reluctance to implement them. However, the present acceptance of anaerobic digestion systems for the treatment of organic waste in the chemical industry proves that eventually, such reluctance will be overcome.

REFERENCES

Abdelouas, A., Lu, Y., Lutze, W. and Nuttall, E. (1998) Reduction of U(VI) to U(IV) by indigenous bacteria in contaminated ground water. *J. Contam. Hydrol.* **35**, 217–233.

Bechard, G., Yamazaki, H., Gould, W.D. and Bedard, P. (1994) Use of cellulosic substrates for the microbial treatment of acid mine drainage. *J. Environ. Qual.* **23**, 111–116.

Benner, S.G., Blowes, D.W., Gould, W.D., Herbert Jr, R.B. and Ptacek, C.J. (1999) Geochemistry of a permeable reactive barrier for metals and acid mine drainage. *Environ. Sci. Technol.* **33**, 2793–2799.

Bharathi, P.A.L., Sathe, V. and Chandramohan, D. (1990) Effect of lead, mercury and cadmium on a sulfate-reducing bacterium. *Environ. Pollut.* **67**, 361–374.

Boonstra, J., Van Lier, R., Janssen, G., Dijkman, H. and Buisman, C.J.N. (1999) Biological treatment of acid mine drainage. *Process Metall.* **9B**, 559–567.

Brooks, C.L. (1991) *Metal Recovery from Industrial Waste.* Lewis Publishers, Chelsea Michigan, USA.

Chang, I.S., Shin, P.K. and Kim, B.H. (2000) Biological treatment of acid mine drainage under sulphate-reducing conditions with solid waste materials as substrate. *Water Res.* **34**, 1269–1277.

Christensen, B., Laake, M. and Lien, T. (1996) Treatment of acid mine water by sulfate-reducing bacteria; results from a bench scale experiment. *Water Res.* **30**, 1617–1624.

Chuichulcherm, S., Nagpal, S., Peeva, L. and Livingston, A. (2001) Treatment of metal-containing wastewaters with a novel extractive membrane reactor using sulfate-reducing bacteria. *J. Chem. Technol. Biotechnol* **76**, 61–68.

Copini, C.F.M., Janssen, G.H.R., Buisman, C.J.N. and Vellinga, S. (2000) Recovery of sulfides from sulfate containing bleed streams using a biological process. In: *Proceedings of the Lead-Zinc Symposium October 22–25, 2000* (eds. J.E. Dutrizac, J.A. Gonzalez, D.M. Henke, S.E. James and A.H.-J. Siegmund). TMS, Pittsburgh, USA, pp. 891–901.

De Lima, A.C.F., Goncalves, M.M.M., Granato, M. and Leite, S.G.F. (2001) Anaerobic sulphate-reducing microbial process using UASB reactor for heavy metals decontamination. *Environ. Technol.* **22**, 261–270.

Dijkman, H., Buisman, C.J.N. and Bayer, H.G. (1999) Biotechnology in the mining and metallurgical industries: cost savings through selective precipitation of metal sulfides, Proceedings of the COPPER 99-COBRE 99 International Conference, 4th April, 113–126.

Drury, W.J. (1999) Treatment of acid mine drainage with anaerobic solid-substrate reactors. *Water Environ. Res.* **71**, 1244–1250.

Dvorak, D.H., Hedin, R.S., Edenborn, H.M. and McIntire, P.E. (1992) Treatment of metal-contaminated water using bacterial sulfate reduction: results from pilot-scale reactors. *Biotechnol. Bioeng.* **40**, 609–616.

Fude, L., Harris, B., Urrutia, M.M. and Beveridge, T.J. (1994) Reduction of Cr(VI) by a consortium of sulfate-reducing bacteria (SRB III). *Appl. Environ. Microbiol.* **60**, 1525–1531.

Gazea, B., Adam, K. and Kontopoulos, A. (1996) A review of passive systems for the treatment of acid mine drainage. *Miner. Eng.* **9**, 23–42.

Glombitza, F. (2001) Treatment of acid lignite mine flooding water by means of microbial sulfate reduction. *Waste Manage.* **21**, 197–203.

Hammack, R.W. and Dijkman, H. (1999) The application of bacterial sulfate reduction treatment to severely contaminated mine waters: results of three years of pilot plant testing. *Proceedings of the COPPER 99-COBRE 99 International Conference, 4th April*, 97–111.

Hammack, R.W., Dvorak, D.H. and Edenborn, H.M. (1994a) Selective metal recovery using biogenic hydrogen sulfide: Rio Tinto Mine, Nevada. Extr. Process. Treat. Minimization Wastes 1994, *Proceedings of International Symposium*, 747–760.

Hammack, R.W., Edenborn, H.M. and Dvorak, D.H. (1994b) Treatment of water from an open-pit copper mine using biogenic sulfide and limestone: a feasibility study. *Water Res.* **28**, 2321–2329.

Hulshoff Pol, L.W., Lens, P.N.L., Stams, A.J.M. and Lettinga, G. (1998) Anaerobic treatment of sulphate-rich wastewaters. *Biodegradation* **9**, 213–224.

Jalali, K. and Baldwin, S.A. (2000) The role of sulphate reducing bacteria in copper removal from aqueous sulphate solutions. *Water Res.* **34**, 797–806.

Janssen, A.J.H., Ma, S.C., Lens, P. and Lettinga, G. (1997) Performance of a sulphide oxidizing expanded bed reactor supplied with dissolved oxygen. *Biotechnol. Bioeng.* **53**, 32–40.

Johnson, D.B. (2000) Biological removal of sulphurous compounds from inorganic wastewaters. In: *Environmental Technologies to Treat Sulfur Pollution* (eds. P.N.L. Lens and L.W. Hulshoff Pol.) IWA, London, pp. 175–205.

Lloyd, J.R, Thomas, G.H., Finlay, J.A., Cole, J.A. and Macaskie, L.E. (1999a) Microbial reduction of technetium by *Escherichia coli* and *Desulfovibrio desulfuricans* enhancement via the use of high-activity strains and effect of process parameters. *Biotechnol. Bioeng.* **66**, 122–130.

Lloyd, J.R., Ridley, J., Khizniak, T., Lyalikova, N.N. and Macaskie, L.E. (1999b) Reduction of technetium by *Desulfovibrio desulfuricans*: biocatalyst characterization and use in a flowthrough bioreactor. *Appl. Environ. Microbiol.* **65**, 2691–2696.

Lovley, D.R. (1995) Bioremediation of organic and metal contaminants with dissimilatory metal reduction. *J. Ind. Microb.* **14**, 85–93.

Lovley, D.R. and Phillips, E.J.P. (1994) Reduction of chromate by *Desulfovibrio vulgaris* and its c3 cytochrome. *Appl. Environ. Microbiol.* **60**, 726–728.

Ludwig, R., Mountjoy, K., McGregor, R. and Blowers, D. (2000) Treatment of heavy metals using an organic sulfate reducing PRB. Chem. Oxid. React. Barriers: Rem. Chlorinated Recalcitrant Compd., 2nd International Conference 361–368.

Macy, J.M., Santini, J.M., Pauling, B.V., O' Neill, A.H. and Sly, L.I. (2000) Two new arsenate/sulfate-reducing bacteria: mechanisms of arsenate reduction. *Arch. Microbiol.* **173**, 49–57.

Mersmann, A. (1999) Crystallization and precipitation. *Chem. Eng. Process* **38**, 345–353.

Newman, D.K., Kennedy, E.K., Coates, J.D., Ahmann, D., Ellis, D.J., Lovley, D.R. and Morel, F.M. (1997) Dissimilatory arsenate and sulfate reduction in *Desulfotomaculum auripigmentum* sp. nov. *Arch. Microbiol.* **168**, 380–388.

Olper, M., Maccagni, M., Buisman, C.J.N. and Schultz, C.E. (2000) Electrowinning of lead battery paste with the production of lead and elemental sulphur using bioprocess technologies. In: *Lead-Zinc 2000, Proceedings of the Symposium* (eds. J.E. Dutrizac, J.A. Gonzalez, D.M. Henke, S.E. James and A.H.-J. Siegmund). TMS, Pittsburgh, USA, pp. 803–813.

Oude Elferink, S.J.W.H., Visser, A., Hulshoff Pol, L.W. and Stams, A.J.M. (1994) Sulfate reduction in methanogenic bioreactors. *FEMS Microbiol. Rev.* **15,** 119–136.

Peters, R.W., Ku, Y. and Bhattacharyya, D. (1985) Evaluation of recent treatment techniques for removal of heavy metals from industrial wastewaters. *AIChE Symp. Ser.* **81**, 165–203.

Prasad, D., Wai, M., Berube, P. and Henry, J.G. (1999) Evaluating substrates in the biological treatment of acid mine drainage. *Environ. Technol.* **20**, 449–458.

Russell, R.A., Holden, P.J., Wilde, K.L. and Neilan, B.A. (2001) Demonstration of the use of algal biomass to drive bacterial sulfate reduction using wetland microflora. *Process Metall.* **11B,** 307–315.

Sani, R.K., Peyton, B.M. and Brown, L.T. (2001a) Copper-induced inhibition of growth of *Desulfovibrio desulfuricans* G20: assessment of its toxicity and correlation with those of zinc and lead. *Appl. Environ. Microbiol.* **67**, 4765–4772.

Sani, R.K., Geesey, G. and Peyton, B.M. (2001b) Assessment of lead toxicity to *Desulfovibrio desulfuricans* G20: influence of components of lactate C medium. *Adv. Environ. Res.* **5**, 269–276.

Scheeren, P.J.H., Koch, R.O. and Buisman, C.J.N. (1993) Geohydrological containment system and microbial water treatment plant for metal-contaminated groundwater at Budelco. In: *World Zinc '93*. Hobart, Tasmania, pp. 373–384.

Smith, W.L. and Gadd, G.M. (2000) Reduction and precipitation of chromate by mixed culture sulphate-reducing bacterial biofilms. *J. Appl. Microbiol.* **88**, 983–991.

Tomei, F.A., Barton, L.L., Lemanski, C.L., Zocco, T.G., Fink, N.H. and Sillerud, L.O. (1995) Transformation of selenate and selenite to elemental selenium by *Desulfovibrio desulfuricans*. *J. Ind. Microbiol.* **14**, 329–336.

Tsukamoto, T.K. and Miller, G.C. (1999) Methanol as a carbon source for microbiological treatment of acid mine drainage. *Water Res.* **33**, 1365–1370.

Tucker, M.D., Barton, L.L. and Thomson, B.M. (1997) Reduction and immobilization of molybdenum by *Desulfovibrio desulfuricans*. *J. Environ. Qual.* **26**, 1146–1152.

Tucker, M.D., Barton, L.L. and Thomson, B.M. (1998) Reduction of Cr, Mo, Se and U by *Desulfovibrio desulfuricans* immobilized in polyacrylamide gels. *J. Ind. Microbiol. Biotechnol.* **20**, 13–19.

Uhrie, J.L., Drever, J.I., Colberg, P.J.S. and Nesbitt, C.C. (1996) *In situ* immobilization of heavy metals associated with uranium leach mines by bacterial sulfate reduction. *Hydrometallurgy* **43**, 231–239.

Van Houten, R.T., Yun, S.Y. and Lettinga, G. (1997) Thermophilic sulphate and sulphite reduction in lab-scale gas-lift reactors using H_2 and CO_2 as energy and carbon source. *Biotechnol. Bioeng.* **55**, 807–814.

Veeken, A.H.M., de Vries, S., van der Mark , A. and Rulkens. (2003). Selective precipitation of heavy metals as controlled by a sulphide-selective electrode. *Sep. Sci. Technol.* **38**, 1–19.

Von Canstein, H., Li, Y., Timmis, K.N., Deckwer, W.D. and Wagner-Dobler, I. (1999) Removal of mercury from chloralkali electrolysis wastewater by a mercury-resistant *Pseudomonas putida* strain. *Appl. Environ. Microbiol.* **65**, 5279–5284.

Waybrant, K.R., Blowes, D.W. and Ptacek, C.J. (1998) Selection of reactive mixtures for use in permeable reactive walls for treatment of mine drainage. *Environ. Sci. Technol.* **32**, 1972–1979.

Weijma, J. and Schouten, G.P. (2005). Pilot Study for Biological Uranium Removal with ThioMet Technology. Paper presented at *Uranium Mining and Hydrogeology IV, International Conference*, Freiberg, 12–16 September 2005.

Weijma, J., Stams, A.J.M., Hulshoff Pol, L.W. and Lettinga, G. (2000) Thermophilic sulfate reduction and methanogenesis with methanol in a high rate anaerobic reactor. *Biotechnol. Bioeng.* **67**, 354–363.

Weijma, J., de Hoop, K., Bosma, W. and Dijkman, H. (2002) Biological conversion of anglesite (PbSO4) and lead waste from spent car batteries to galena (PbS). *Biotechnol. Prog.* **18**, 770–775.

Widdel, F. and Hansen, T.A. (1992) The dissimilatory sulfate- and sulfur-reducing bacteria. In: *The prokaryotes*, 2nd edn (eds. A. Balows, H.G. Trüper, M. Dworkin, W. Harder and K.-H. Schleifer). Springer-Verlag, New York, pp. 583–624.

Yong, P., Farr, J.P.G., Harris, I.R. and Macaskie, L.E. (2002) Palladium recovery by immobilized cells of *Desulfovibrio desulfuricans* using hydrogen as the electron donor in a novel electrobioreactor. *Biotechnol. Lett.* **24**, 205–212.

Zauner, R. and Jones, A.G. (2000) Determination of nucleation, growth, agglomeration and disruption kinetics from experimental precipitation data: the calcium oxalate system. *Chem. Eng. Sci.* **55**, 4219–4232.

Index

References to tables are bold and to figures are italic

IWA Publishing's authorised EU representative for General Product Safety Regulations is Diane D'Arras, 15 rue Duret, 75116 Paris, France, e-mail: safety@iwap.co.uk.

Printed and bound by CPI Group (UK) Ltd, Croydon, CR0 4YY
12/05/2026
02108867-0004